This book gives an up-to-date, systematic account of the microscopic theory of Bose-condensed fluids developed since the late 1950's. In contrast to the usual phenomenological discussions of superfluid ^{4}He, the present treatment is built on the pivotal role of the Bose broken symmetry and a Bose condensate.

The many-body formalism is developed with emphasis on the one- and two-particle Green's functions and their relation to the density-response function. These are all coupled together by the Bose broken symmetry, which provides the basis for understanding the elementary excitations and response functions in the hydrodynamic and collisionless regions. It also explains the difference between excitations in the superfluid and normal phases. The distinction between single-particle excitations and collective zero sound modes, and their hybridization by the Bose condensate, is a central unifying theme. Throughout, an attempt is made to bring out the essential physics behind the formal field-theoretic analysis. A detailed discussion is given of recent high-resolution neutron and Raman scattering data over a wide range of temperatures. Chapter 4 gives the first critical assessment of the experimental evidence for a Bose condensate in liquid ^{4}He, based on high-momentum neutron-scattering data.

This volume will act as a stimulus and guide to work on excitations in superfluid ^{4}He and ^{3}He–^{4}He mixtures over the next decade. It will also lead to a greater understanding of the dynamical implications of Bose condensation in other condensed matter systems (high-temperature superconductors, exciton gases, spin-polarized hydrogen, etc.).

Excitations in a Bose-condensed liquid

CAMBRIDGE STUDIES IN LOW TEMPERATURE PHYSICS

Cambridge Studies in Low Temperature Physics is an international series which contains books at the graduate text level and above on all aspects of low temperature physics. This includes the study of condensed state helium and hydrogen, condensed matter physics studied at low temperatures, superconductivity and superconducting materials and their applications.

1 D. S. Betts *An Introduction to Millikelvin Technology*
2 A. B. Pippard *Magnetoresistance in Metals*
3 R. J. Donnelly *Quantized Vortices in Helium II*
4 A. Griffin *Excitations in a Bose-Condensed Liquid*

Excitations in a Bose-condensed liquid

ALLAN GRIFFIN
University of Toronto

CAMBRIDGE UNIVERSITY PRESS
Cambridge, New York, Melbourne, Madrid, Cape Town, Singapore, São Paulo

Cambridge University Press
The Edinburgh Building, Cambridge CB2 2RU, UK

Published in the United States of America by Cambridge University Press, New York

www.cambridge.org
Information on this title: www.cambridge.org/9780521432719

First published 1993
This digitally printed first paperback version 2005

A catalogue record for this publication is available from the British Library

Library of Congress Cataloguing in Publication data

Griffin, Allan.
Excitations in a Bose-condensed liquid / Allan Griffin.
p. cm. – (Cambridge studies in low temperature physics : 4)
Includes bibliographical references and index.
ISBN 0 521 43271 5
1. Bose–Einstein condensation. 2. Helium. 3. Superfluidity.
I. Title. II. Series.
QC175.47.B65G75 1993
530.42–dc20 92-33320
CIP

ISBN-13 978-0-521-43271-9 hardback
ISBN-10 0-521-43271-5 hardback

ISBN-13 978-0-521-01998-9 paperback
ISBN-10 0-521-01998-2 paperback

Contents

Preface

The well known Landau theory of the low-temperature properties of superfluid ^{4}He starts from a weakly interacting gas of phonons and rotons. This theory is very successful but it is essentially phenomenological since it makes no reference to the Bose condensate. The core of this book is a discussion of the modern microscopic theory of Bose-condensed systems based on finite-temperature Green's function techniques (the dielectric formalism). My emphasis is on developing the language and concepts of this formalism in a way that brings out the essential physics. This book is the first general account of the progress made in the last two decades toward understanding the excitations in superfluid ^{4}He specifically within the framework of a Bose-condensed liquid. I hope it will be a guide and stimulus to a new generation of experimentalists and theorists studying superfluid ^{4}He. The book should also be of interest to a much wider audience, since the phenomenon of Bose condensation, with its associated macroscopic quantum effects, plays a central role in modern condensed matter physics (Anderson, 1984).

The goal of this book is two-fold: (a) to summarize the field-theoretic analysis of a Bose-condensed fluid and (b) to use this formalism to understand the nature of the excitations in superfluid ^{4}He. I emphasize how a Bose broken symmetry inevitably leads to certain characteristic features in the structure of various correlation functions, the most spectacular being the phenomenon of superfluidity. A major theme is the way in which a Bose condensate mixes the elementary excitation and density fluctuation spectra. Related to this is an attempt to understand the relation between the excitations in liquid ^{4}He below and above the superfluid transition. Most previous theoretical accounts have been limited to low temperatures ($T \lesssim 1$ K).

Apart from Section 9.1, I do not discuss the results based on variational many-body wavefunctions. This latter approach has been very successful, but it is effectively limited to zero temperature as far as dynamical properties are concerned and, moreover, the role of the Bose broken symmetry is somewhat implicit. It is expected that this correlated-basis-function technique will be the subject of a separate monograph in the Cambridge Studies in Low Temperature Physics.

Comparing the present book with the recent exhaustive account of superfluid ^{3}He by Vollhardt and Wölfle (1990), one can only be struck by how completely the various excitations in superfluid ^{3}He are now understood, and what an enormous amount of work is still needed before we reach a similar quantitative stage in dealing with superfluid ^{4}He as a Bose-condensed phase.

At one level, the field-theoretic analysis of Bose fluids is a well developed area of many-body theory (beginning with the work of Beliaev, 1958a,b). But we are not yet at a stage where we can make quantitative numerical predictions about superfluid ^{4}He at finite temperatures using this approach, mainly because of the difficulties of dealing with a liquid, whether Bose-condensed or not. Consequently, we depend crucially on experimental data for guidance as to which of several theoretically possible scenarios is actually realized in superfluid ^{4}He. This explains why, in what is essentially a book about theory, I devote considerable space to neutron-scattering experiments, the most powerful probe we have of the excitations in superfluid ^{4}He. To help experimentalists understand the essential concepts of the microscopic theory, I often try to summarize the theoretical analysis in physical terms.

When I need to recall some result of many-body theory or finite-temperature Green's function techniques, I usually give a specific reference to Mahan (*Many-Particle Physics*, Second Edition, 1990). One deficiency of this otherwise excellent book is that it does not include any discussion of the Beliaev Green's function formulation of Bose-condensed fluids. Classic accounts of the Beliaev–Dyson equations for interacting Bose fluids are given in the well known texts by Abrikosov, Gor'kov and Dzyaloshinskii (1963), Fetter and Walecka (1971) and Lifshitz and Pitaevskii (1980). I also recommend the brief account in Inkson (1984). The reader who wishes to follow the theoretical analysis in detail will need to be familiar with one of these accounts and have some background in many-body theory (such as given in the early sections of Chapters 2 and 3 of Mahan, 1990).

The first three chapters of the book can be viewed as introductory, and include considerable background material for the non-expert. Chapter 4 gives a status report of the direct experimental evidence for a Bose condensate in superfluid ^{4}He. I feel this is necessary in a book devoted to the dynamical consequences of such a condensate. An extended outline of the chapters is given in Section 1.3.

Within the general approach based on the role of the Bose condensate, several sections of this book contain material which has not been published before in the literature. I call specific attention to:

(a) The detailed comparison of the nature of phonons in the two-fluid and collisionless regions, at both $T = 0$ and finite temperatures (Section 6.3).
(b) The parameterization of the dielectric formalism developed as a concrete illustration of the recent Glyde–Griffin scenario (Section 7.2).
(c) The f-sum rules specific to Bose-condensed fluids (Sections 8.1 and 8.3).
(d) The unified discussion of how the single-particle, particle–hole (density) and two-particle (pair) excitation branches are hybridized via the Bose order parameter (Chapter 10).
(e) The comparison between the microscopic theories of excitations in superfluid and solid ^{4}He (Chapter 11).

Acknowledgements

The present book grew out of a project initiated by Henry Glyde for an introductory book on the theory of neutron scattering from quantum fluids and solids (^{4}He and ^{3}He). We eventually decided that a modern account of Bose-condensed liquids really deserved a separate monograph. Several chapters have benefited as a result of extensive discussions with Henry. Our collaboration also led to a new interpretation of the phonon–maxon–roton dispersion curve (Glyde and Griffin, 1990) which is developed in the present account. I would also like to thank Emile Talbot for his useful contributions at the early stages in the writing of this book.

Over the last twenty years, I have been strongly influenced by the manuscript preprint of Nozières and Pines, written in 1964 but only published in 1990. The present book is, in many ways, a development of their approach. I would like to acknowledge several stimulating discussions with Philippe Nozières.

Experimental data taken by the neutron-scattering groups at the Chalk River Laboratories of AECL in Canada and the Institut Laue–Langevin in Grenoble, France have been a continual guide and stimulus to my theoretical work. I would like to single out Eric Svensson for countless discussions about the data. I would also like to thank Dave Woods, Varley Sears, Bill Buyers and Peter Martel at Chalk River. At Grenoble, I thank Bill Stirling (now at Keele, U.K.) and Ken Andersen for their assistance. In particular, I am very grateful to them and their colleagues at ILL for making available some of their unpublished high-resolution time-of-flight neutron data for use in this book. I also specifically acknowledge Björn Fåk (now at CENG, Grenoble) for his careful reading of the manuscript and many useful suggestions for its improvement.

My research work on Bose fluids has been generously supported over many years by grants from the Natural Sciences and Engineering Research Council (NSERC) of Canada.

I would like to express my appreciation to Karl Schroeder and Jennifer Tam, who expertly keyboarded the manuscript in TEX and endless revisions of it. Karl had the extra strain of dealing with the many changes which arose as the theoretical interpretations evolved dynamically during the writing of the book.

It is with particular pleasure that I thank my wife, Christine McClymont, for her very valuable editorial polishing of the final manuscript as well as for her advice and interest over the three years it took to complete this book.

Allan Griffin
Toronto, Ontario

1

Theory of excitations in superfluid ^{4}He: an introduction

The major goal of the present book is to outline the field-theoretic analysis of the dynamical behaviour of a Bose-condensed fluid that has developed since the late 1950's. While we often use the weakly interacting dilute Bose gas (WIDBG) for illustrative purposes, the emphasis is on the dynamical properties of a specific Bose-condensed liquid, superfluid ^{4}He. We attempt to develop a coherent picture of the excitations in liquid ^{4}He which is consistent with, and rooted in, an underlying Bose broken symmetry. Recent high-resolution neutron-scattering studies in conjunction with new theoretical studies have led to considerable progress and it seems appropriate to summarize the current situation. The only other systematic account of superfluid ^{4}He as a Bose-condensed liquid is the classic monograph by Nozières and Pines (1964, 1990).

The phenomenon of Bose condensation plays a central role in many different areas of modern condensed matter physics (Anderson, 1984). Historically it was first studied in an attempt to understand the unusual properties of superfluid ^{4}He (London, 1938a). In a generalized sense, however, it also underlies much of the physics involved in superconductivity in metals and the superfluidity of liquid ^{3}He, in which Cooper pairs play the role of the Bosons (see, for example, Leggett, 1975 and Nozières, 1983). In recent years, there has been increased research on the possibility of creating a Bose-condensed gas, involving such exotic composite Bosons as excitons in optically excited semiconductors, spin-polarized atomic Hydrogen, and positronium atoms. This research, in turn, has re-stimulated theoretical interest in Bose condensation in general and its implications in both liquids and gases.

Although ^{4}He was first liquefied by Kammerlingh Onnes in 1908, it was not until 1928 that Wolfke and Keesom found it could exist in *two* phases, now called Helium-I and Helium-II, separated by the transition

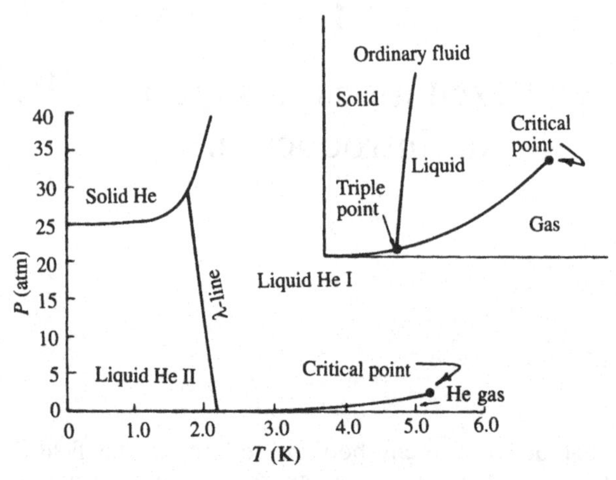

Fig. 1.1. The P–T diagram for the condensed phases of ^{4}He contrasted to that for a normal liquid.

temperature $T_\lambda = 2.17$ K at SVP. Kapitza (1938) and, independently, Allen and Misener (1938), found that the fluid flowed without any apparent viscosity below T_λ. At higher temperatures ($T > T_\lambda$), the liquid exists as a "normal" liquid and is called liquid Helium-I. Below T_λ, liquid ^{4}He is a "superfluid" (a term introduced by Kapitza for a fluid having, in a certain sense, zero viscosity) and is called Helium-II. Most of our discussion will deal with this low-temperature superfluid phase, but we will also be interested in how it differs from Helium-I (above T_λ). The phase diagram is shown in Fig. 1.1.

The study of superfluid ^{4}He has played a central role in developing our understanding of many-body systems. The modern idea of quasiparticles started with Landau's original theory of superfluid ^{4}He (1941). Bogoliubov's (1947) derivation of the phonon spectrum in a weakly interacting dilute Bose gas was one of the first treatments of a broken symmetry. In this chapter, we give a brief review of theories about the excitations of superfluid ^{4}He and how they developed since the pioneering work of London, Tisza, Landau, Bogoliubov and Feynman. In this mini-history, we also introduce the kind of questions with which the rest of this book will be concerned. We assume the reader is already familiar with an elementary account of the properties of superfluid ^{4}He, such as the first three chapters of Nozières and Pines (1964, 1990). In Section 1.1, we sketch the well known Landau–Feynman quasiparticle picture (which makes no reference to an underlying Bose condensate). In Section 1.2, we review the development of the theoretical approach in which the condensate plays a crucial role. It is with this latter approach that this book is concerned.

Many experimental probes have given valuable information about the excitations in superfluid ^{4}He, including Brillouin and Raman light scattering, sound propagation, as well as thermodynamic and transport studies. However, the most powerful technique has been inelastic neutron scattering. Through the dynamic structure factor $S(\mathbf{Q}, \omega)$, this is a direct probe of the density fluctuations which turn out to be the elementary excitations of superfluid ^{4}He (but not, as we shall see, of *normal* liquid ^{4}He). Moreover, recent high-resolution neutron-scattering data over a wide range of temperatures and pressures have stimulated the theoretical developments which are a major theme of this book. For these reasons, we shall discuss neutron-scattering data at some length. As background, we summarize in Chapter 2 some general properties of the dynamic structure factor which hold for any system and also the main features that $S(\mathbf{Q}, \omega)$ exhibits in superfluid ^{4}He. References to neutron-scattering results in the overviews given in Sections 1.1 and 1.2 are more fully explained in Chapter 2. For a general account of experimental studies on excitations in superfluid ^{4}He, the classic review article by Woods and Cowley (1973) is highly recommended.

In Section 1.3, we conclude with an extended outline of the contents of the succeeding chapters of this book.

1.1 Landau–Feynman picture

The superfluid characteristics of Helium-II are well described by the two-fluid model, first suggested by Tisza (1938) and later brought to fruition by Landau (1941). In this model (for authoritative accounts, see Landau and Lifshitz, 1959; Khalatnikov, 1965), the liquid consists of two inter-penetrating fluids, a normal fluid and a superfluid, each having its own density and velocity fields. The normal fluid has a finite viscosity, contains all the entropy of the liquid, and has a mass density ρ_N and velocity $\mathbf{v}_N$. The superfluid component has no entropy and it has a mass density ρ_S and a velocity $\mathbf{v}_S$. If $\mathbf{v}_S$ is less than some appropriate critical velocity, the superfluid component flows with zero viscosity. The local mass density and the momentum current density are given by the equations

$$\left.\begin{aligned} \rho(\mathbf{r}) &= \rho_N(\mathbf{r}) + \rho_S(\mathbf{r}) \, , \\ \mathbf{J}(\mathbf{r}) &= \rho_N(\mathbf{r})\mathbf{v}_N(\mathbf{r}) + \rho_S(\mathbf{r})\mathbf{v}_S(\mathbf{r}) \, . \end{aligned}\right\} \tag{1.1}$$

In He-II, the normal fluid fraction ρ_N/ρ is a strong function of temperature. As the temperature decreases, the normal fluid fraction goes to zero. Above T_λ, the normal fluid fraction is unity, corresponding to the fact that there is no superfluid component. There are many ways of measuring ρ_S and ρ_N, but the most precise experimental results are from measurements of ρ_N from the moment of inertia of a rotating disk (Wilks, 1967).

The two-fluid model, consisting of (1.1), the continuity equation, and a set of three other equations describing the flow produced by gradients in the temperature, pressure, and chemical potential, has been extremely successful in describing the macroscopic transport properties and the low-frequency, low-wavelength hydrodynamic modes in Helium-II (Khalatnikov, 1965). One of the great successes of the field-theoretic analysis of Bose-condensed fluids (see Chapter 6) is to show how this two-fluid description can be a natural *consequence* of a Bose order parameter (Bose condensation).

London (1938a,b) proposed that superfluidity in liquid ^{4}He is a manifestation of Bose–Einstein condensation. Tisza (1938) also conjectured that the superfluid component in his phenomenological two-fluid model could be interpreted as the fraction of the Helium atoms that are Bose-condensed. As we discuss in more detail in Chapters 4 and 6, the condensate fraction and the superfluid fraction are *not* the same in a Bose-condensed liquid like superfluid ^{4}He. At zero temperature, the superfluid fraction is 100% (since no quasiparticles are thermally excited) while the condensate fraction is only about 9%. In his classic papers of 1941 and 1949, Landau strongly argued against any connection between a Bose condensate and the two-fluid model. The modern view vindicates *both* Landau and London in that the microscopic basis of Landau's quasiparticle picture and the two-fluid description lies in the existence of a Bose condensate.

The Landau (1941, 1947) theory of superfluidity is based on the low-lying excited states of a Bose *liquid*. He showed that the *low-temperature* thermodynamic and transport properties of superfluid ^{4}He could be understood in terms of a weakly interacting gas of Bose "quasiparticles" (phonons and rotons). This description is in many ways analogous to the phonon picture of an anharmonic crystal (see Chapter 11). This was developed at some length by Peierls, Frenkel and others in the early 1930's. In superfluid ^{4}He, Landau had no well defined scheme to calculate the dispersion relation of the quasiparticles; i.e., how to relate it to the microscopic forces between the ^{4}He atoms. Although certain qualitative features could be argued to be consequences of the Bose statistics of the ^{4}He atoms, the dispersion relation of the quasiparticles in superfluid ^{4}He was viewed by Landau (1947) as something to be determined from experiment. We recall his strong resistance (Landau, 1941, 1949) to trying to use a WIDBG as a model for superfluid ^{4}He, as London (1938a,b) and Tisza (1938) tried to do. A quasiparticle spectrum which was acoustic at low Q but has a roton minimum at large Q has precisely the features needed to explain the temperature dependence of two-fluid thermodynamic functions as well as the velocities of first and second sound. In the two-fluid picture, the normal fluid density describes the thermally excited quasiparticles, while the superfluid density can be thought of (in a rough sort of way) as the probability that the liquid is in its ground state.

At low temperatures ($T \sim 1$ K), the dominant feature of the dynamic structure factor $S(\mathbf{Q}, \omega)$ in the range 0.1 Å$^{-1} < Q < 2.4$ Å^{-1} is an extraordinarily sharp resonance. This is illustrated by the representative data in Fig. 1.2 as well as in many other figures in this book. (It is to be emphasized that plots of $S(\mathbf{Q}, \omega)$ data almost always still contain instrumental resolution broadening, with a width typically of order 1–2 K.) As first pointed out by Cohen and Feynman (1957), the peaks in $S(\mathbf{Q}, \omega)$ correspond to the creation of the elementary excitations which Landau (1941, 1947) originally postulated to describe the thermodynamic and transport properties of superfluid ^{4}He. The roton was first observed using neutron scattering by Palevsky *et al.* (1957). The complete quasiparticle dispersion relation ω_Q (see Fig. 1.3) was first determined by Yarnell, Arnold, Bendt and Kerr (1959) and Henshaw and Woods (1961) at $T = 1.1$ K. For Q values below about 0.6 Å^{-1}, the dispersion relation is phonon-like ($\omega_Q = cQ$), with a slope slightly larger than the thermodynamic speed of sound. It bends slightly upward (anomalous dispersion) before bending over to a maximum energy (referred to as the "maxon") of about 14 K at $Q_M = 1.13$ Å^{-1}. The minimum at $Q_R = 1.93$ Å^{-1} occurs at an energy $\Delta = 8.62$ K at SVP. The dispersion relation near the minimum is called the "roton" region. For Q within about 0.25 Å^{-1} of Q_R, ω_Q is found to be very well described by the simple Landau expression (Woods, Hilton, Scherm and Stirling, 1977)

$$\omega_Q = \Delta + \hbar(Q - Q_R)^2/2\mu_R \tag{1.2}$$

where the roton "mass" is $\mu_R = 0.13m$ (m is the bare mass of a ^{4}He atom).

An important early study by Bendt, Cowan and Yarnell (1959) showed that various thermodynamic functions of superfluid ^{4}He up to about

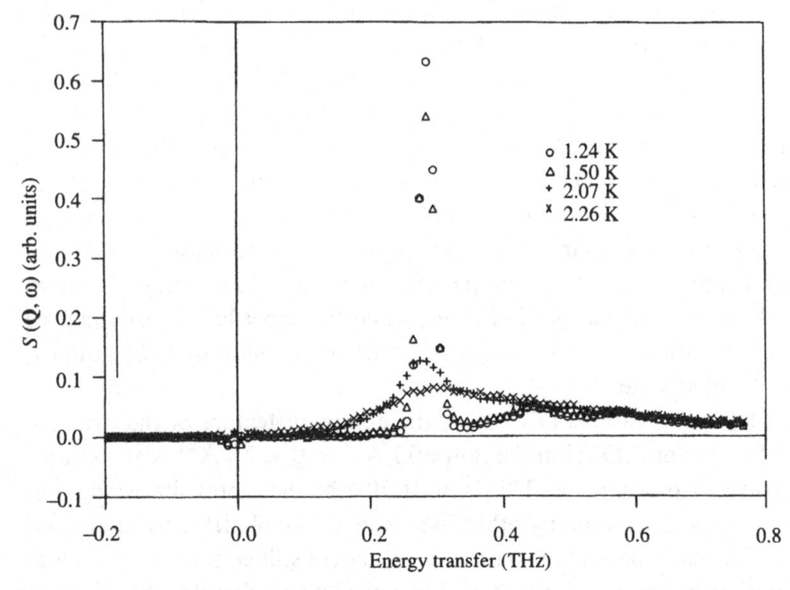

Fig. 1.2. A plot of the neutron-scattering intensity vs. the energy at a momentum transfer $Q = 1$ Å^{-1}, as a function of the temperature [Source: Stirling, 1991; Andersen, Stirling *et al.*, 1991].

1.8 K could be calculated (with an accuracy of a few per cent) by treating it as a non-interacting gas of Bose quasiparticles with the ω_Q dispersion relation determined by the peak position in $S(\mathbf{Q}, \omega)$ measured at 1.1 K. This work is an empirical proof that, at least up to 1.8 K, the resonances in $S(\mathbf{Q}, \omega)$ do indeed give the energies of the elementary excitations of superfluid ^{4}He. This is *not* the case in normal liquid ^{4}He or in liquid ^{3}He.

In his original paper, Landau (1941) introduced phonons and rotons as two quite distinct kinds of excitations. However, beginning with a later addendum (Landau, 1947), it has been traditional to assume that the quasiparticle spectrum in Fig. 1.3 describes a single excitation branch ω_Q. From this point of view, while different parts of the dispersion curve are described as phonons and rotons, they are not thought to be qualitatively different types of excitations. In spite of this, while the physics behind the phonon part has been viewed as "obvious" (i.e., a compressional sound wave), the physical nature of the roton has been the subject of much discussion over the years (see, for example, Feynman, 1954; Miller, Pines and Nozières, 1962; Chester, 1963, 1969, 1975).

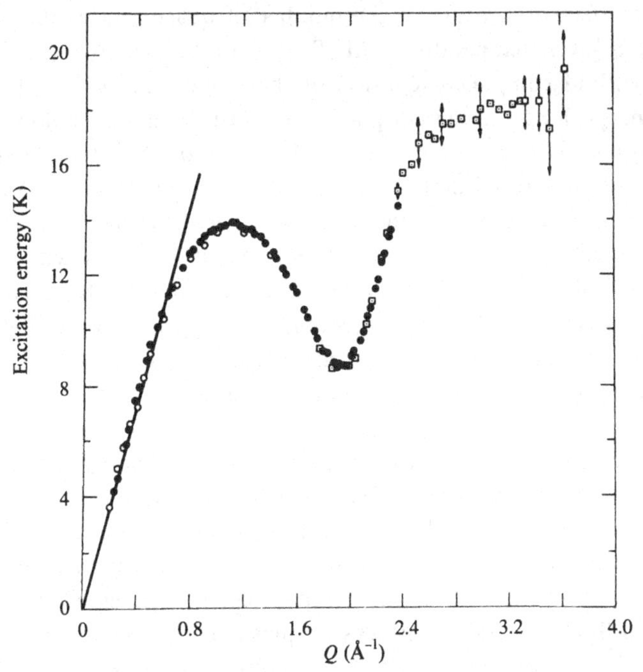

Fig. 1.3. The quasiparticle dispersion curve at $T = 1.1$ K and SVP. This is a compilation of data from rotating crystal (RCS) and triple axis (TACS) spectrometers [Source: Cowley and Woods, 1971].

The region around $Q \sim 1\text{Å}^{-1}$ is often referred to as a "maxon". This region, however, has such a high energy ($\sim$ 14 K) that maxons are never thermally excited and thus they play no direct role in the thermodynamic or transport properties of superfluid ^{4}He.

In a rotating vessel of Helium-II, only the gas of excitations rotates with the vessel; thus the angular momentum of the excitations determines the moment of inertia of the system. At temperatures less than about 1 K, when there are few excitations thermally excited, the moment of inertia is greatly reduced from the classical prediction that the entire liquid rotates with the vessel. Landau (1941) was able to express the normal fluid density ρ_N explicitly in terms of the Bose distribution function $N(\omega)$ of the thermally excited phonon-roton excitations,

$$\rho_N(T) = \int \frac{d^3p}{(2\pi)^3} \frac{p^2}{3} \left[-\frac{\partial N(\omega_p)}{\partial \omega_p} \right] . \tag{1.3}$$

The normal fluid fraction $\rho_N(T)$ turns out to be a useful quantity since it gives an effective measure of the number of quasiparticles thermally excited at a given temperature. At $T = 0$, of course, $\rho_N=0$ and the entire superfluid component consists of the liquid. Below $T \lesssim 0.6$ K, the low-energy, long-wavelength phonon excitations make the dominant contribution to ρ_N (we note that $\omega_Q \sim 1$ K at $Q \sim 0.1$ Å^{-1}). On the other hand, ρ_N/ρ is very small until the temperature reaches about 1.2 K, and then the thermally excited rotons make the overwhelming contribution. The roton's relatively high energy ($\Delta \sim 8$ K) is compensated by the high density of states at $Q \sim Q_R$. Finally, at about 1.7 K or so, the quasiparticle damping becomes appreciable and one expects that the simple *non-interacting* quasiparticle gas picture, and hence (1.3), will increasingly break down as we approach the lambda point $T_\lambda = 2.17$ K (SVP).

Landau (in collaboration with Khalatnikov) also developed the formalism of "quantum hydrodynamics" to calculate the effect of weak quasiparticle interactions (three-phonon and four-phonon processes, phonon–roton scattering, etc.) and to use these results to obtain the temperature dependence of the characteristic transport coefficients associated with the normal fluid (the interacting gas of quasiparticles). At temperatures $T \lesssim 1.7$ K, the Landau–Khalatnikov picture gives a very satisfactory account of the transport properties of superfluid ^{4}He, as described in detail in the well known monographs by Khalatnikov (1965) and Wilks (1967).

Landau was the first to make a clear distinction between collective modes (such as first and second sound) and elementary excitations (the phonon–roton quasiparticles). In particular, Landau emphasized that the phonon quasiparticle is conceptually quite different from the first sound hydrodynamic mode, even though the speeds are quite close in magnitude. This distinction between collective modes and quasiparticles is equally important in Fermi liquids like normal liquid ^{3}He (see Chapter 1 of Lifshitz and Pitaevskii, 1980) and was later incorporated into the description of the excitations of an anharmonic crystal (see Chapter 11).

The microscopic basis of Landau's picture was developed by Feynman in the period 1953–1957. Feynman dealt directly with the excited states of a Bose *liquid*, as opposed to Bogoliubov's work on the excitations of a Bose *gas*. Feynman (1953b, 1954) showed that if atoms obeyed Boltzmann statistics, he could construct wavefunctions describing the motion of a single atom with energy $Q^2/2m^*$, where m^* is some effective mass arising from the interactions with other atoms. On the other hand, he found that even at low Q, such single-particle states develop an energy gap (relative to the energy of the liquid in its ground state). This gap is a result of the required Bose symmetry of the wavefunction (a pairwise interchange of atoms does not alter the wavefunction) and the fact that in a *liquid*, motion of an ^{4}He atom requires that other nearby atoms move out of the way. Feynman estimated that this single-particle energy gap is of the order of the potential well depth any given ^{4}He atom moves in ($\sim$ 10 K). In contrast with these *high-energy* single-particle-like excited states, the excited states corresponding to collective long-wavelength density fluctuations (which involve a large number of atoms moving a small amount coherently) had a sound wave dispersion relation. These brilliant papers by Feynman (1953b, 1954) still deserve careful study. For an assessment of Feynman's work on liquid ^{4}He from a modern perspective, see Pines (1989). For a lucid summary, we refer to the discussion by Wilks (1967).

Feynman (1954) presented a variety of physical arguments to the effect that the excited-state wavefunctions are all described by the expression

$$|\Phi_F\rangle = \sum_{j=1}^{N} f(\mathbf{r}_j)\,|\Phi_0\rangle \,, \tag{1.4}$$

where $|\Phi_0\rangle$ is the ground-state many-particle wavefunction and the sum is over the position operators $\mathbf{r}_j$ of the ^{4}He atoms. Carrying out a $T = 0$ variational calculation with (1.4), it is found that $f(\mathbf{r}_j) = \exp(i\mathbf{Q}\cdot\mathbf{r}_j)$ minimizes the total energy. Since the Fourier transform of the density operator $\hat{\rho}(\mathbf{r})$ is (see Section 2.1)

$$\hat{\rho}^+(\mathbf{Q}) = \sum_{j=1}^{N} e^{i\mathbf{Q}\cdot\mathbf{r}_j} \,, \tag{1.5}$$

this means that the Feynman ansatz (1.4) corresponds to the creation of a single density fluctuation of wavevector $\mathbf{Q}$. The energy of this state (relative to the ground state $|\Phi_0\rangle$) is found to be given by the Feynman–Bijl relation

$$\omega_Q^F = \frac{\varepsilon_Q}{S(\mathbf{Q})} \,, \tag{1.6}$$

where $\varepsilon_Q = Q^2/2m$ and $S(\mathbf{Q})$ is the static structure factor (a quantity whose importance had only recently been emphasized by van Hove (1954) at the time of Feynman's work). Feynman made the following, logically distinct, remarks concerning $|\Phi_F\rangle = \hat{\rho}^+(\mathbf{Q})|\Phi_0\rangle$:

(a) Considering Q as a variational parameter, (1.6) has a local minimum at $Q \simeq Q_0 = 2\pi/r_0$ corresponding to the local maximum in $S(\mathbf{Q})$. Here r_0 is the mean distance between the ^{4}He atoms.

(b) In the limit of low Q, (1.6) can be shown to give the correct energy of a state with a single phonon present.

(c) $|\Phi_F\rangle = \hat{\rho}^+(\mathbf{Q})|\Phi_\circ\rangle$ is an exact eigenstate of the total momentum operator, which commutes with the total Hamiltonian. It follows that (1.6) gives an *upper* bound to the energy of *any* excited state with momentum Q.

These three arguments led Feynman to suggest that $|\Phi_F\rangle = \hat{\rho}^+(\mathbf{Q})|\Phi_0\rangle$ is a good approximation for an excited state of momentum Q, at *all* values of Q. Somewhat surprisingly, the variational ansatz of Feynman describes the low-energy, long-wavelength collective phonon modes and the high-energy, short-wavelength roton modes (which are more single-particle-like). As even Feynman (1954) noted, while the quite different atomic motions involved in phonons and rotons seem to be captured in his wavefunction, it does not appear to shed much light on the microscopic nature of a roton. The question "what is a roton?" has been a recurring one in superfluid ^{4}He research (see also Section 12.1).

If the exact excited states are approximated by Feynman's wavefunctions $|\Phi_F\rangle$, the $T = 0$ dynamic structure factor (see Section 2.1) is trivially given by

$$S(\mathbf{Q}, \omega) = S(\mathbf{Q})\, \delta(\omega - \omega_Q^F) \ . \qquad (1.7)$$

Since we are assuming that the *exact* eigenstates are density fluctuations, the resulting density fluctuation spectrum (1.7) is a delta function at ω_Q^F. By generalizing the variational form (1.4) to include the possibility of creating an admixture of two density fluctuations (Feynman and Cohen, 1956; Miller, Pines and Nozières, 1962), one is led to improved approximations to the excited states which include "backflow" effects. In sophisticated versions of this kind of calculation (see, for example, Manousakis and Pandharipande, 1984, 1986), the predicted quasiparticle spectrum is in good agreement with the experimental dispersion curve shown in Fig. 1.3. In addition, such studies give rise to the kind of high-energy multiparticle structure exhibited by $S(\mathbf{Q}, \omega)$ at low temperatures. This variational approach is referred to as the correlated-basis-function (CBF) method and is reviewed by Woo (1976) and Campbell (1978). We discuss such calculations further in Section 9.1.

Feynman's penetrating analysis of the low-lying excited states of superfluid ^{4}He makes crucial use of Bose statistics in constructing wavefunc-

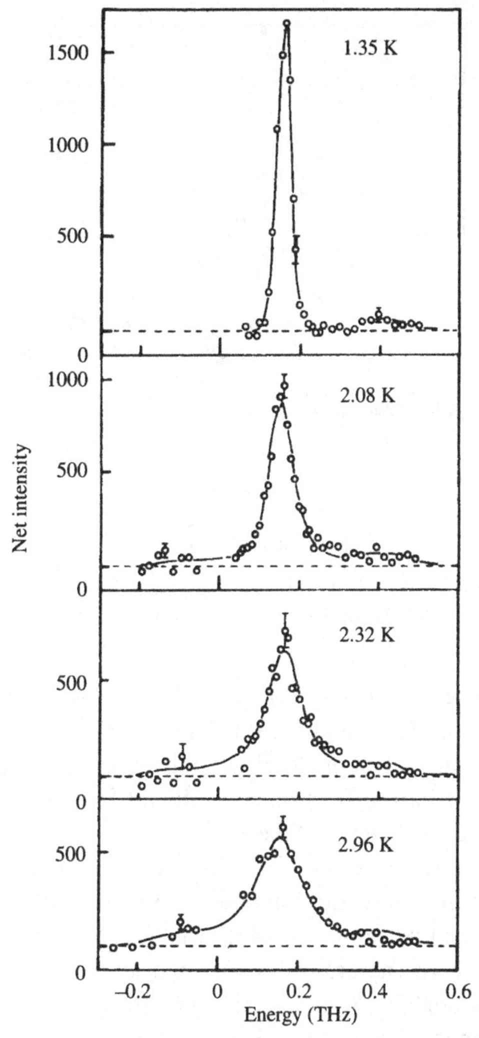

Fig. 1.4. $S(\mathbf{Q},\omega)$ vs. ω for $Q = 0.4$ Å^{-1} at SVP in both the superfluid and normal phases. These recent results confirm the pioneering work of Woods (1965b) [Source: Stirling and Glyde, 1990].

tions – but never mentions the role of a Bose broken symmetry (see also Section 12.1). Bose–Einstein condensation did play a crucial role in his discussion (Feynman, 1953a) of the nature of the superfluid phase transition at T_λ but no connection was ever made between this work and his later studies of the excitation spectrum at low temperatures (Feynman, 1953b, 1954). It is relevant to note that Feynman's work was completed

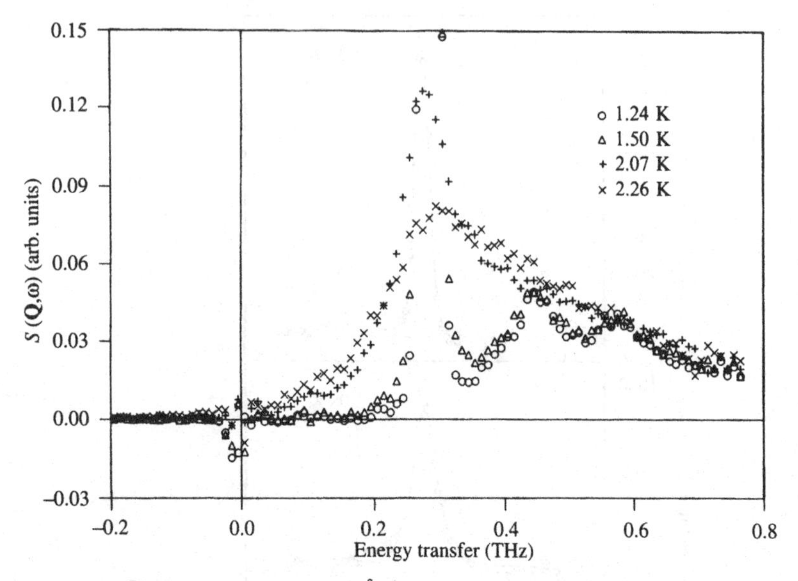

Fig. 1.5. $S(\mathbf{Q}, \omega)$ vs. ω for $Q = 1$ Å^{-1} at SVP showing the high-energy multi-particle structure. This is an expanded version of Fig. 1.2 [Source: Andersen, Stirling *et al.*, 1991].

before the appearance of the seminal paper by Penrose and Onsager (1956) in which the idea of a Bose broken symmetry was extended to a Bose *liquid.* Their numerical estimate of 8% for the condensate fraction was based on a simple hard-sphere model of liquid ^{4}He at $T = 0$. This initial estimate has stood the test of time, with the best current theoretical value being about 9% (for a review, see Whitlock and Panoff, 1987).

Successful as it is, however, the Landau–Feynman scenario sketched above offers little insight into how these excitations change as the temperature increases or how the nature of the quasiparticle spectrum must be altered as we pass through T_λ. It gives no insight into the result of Woods (1965b) that the phonon peak at $Q \simeq 0.38$ Å^{-1} in superfluid ^{4}He is almost unchanged in energy as we go from well below to well above T_λ, apart from a steady increase in the width. Does this mean that the phonons are quasiparticles above T_λ as well? For example, if we consider the density fluctuations at low Q, we note that experimental $S(\mathbf{Q}, \omega)$ line shapes such as in Fig. 1.4 look very similar at 2.1 K and 2.3 K. Why? What are the reasons for the differences *and* similarities between $S(\mathbf{Q}, \omega)$ in superfluid ^{4}He and that observed in normal liquid ^{3}He and in solid Helium?

In addition, even at T =1 K, the scattering intensity $S(\mathbf{Q}, \omega)$ contains more than just a sharp resonance at ω_Q. As first noted by Cowley and

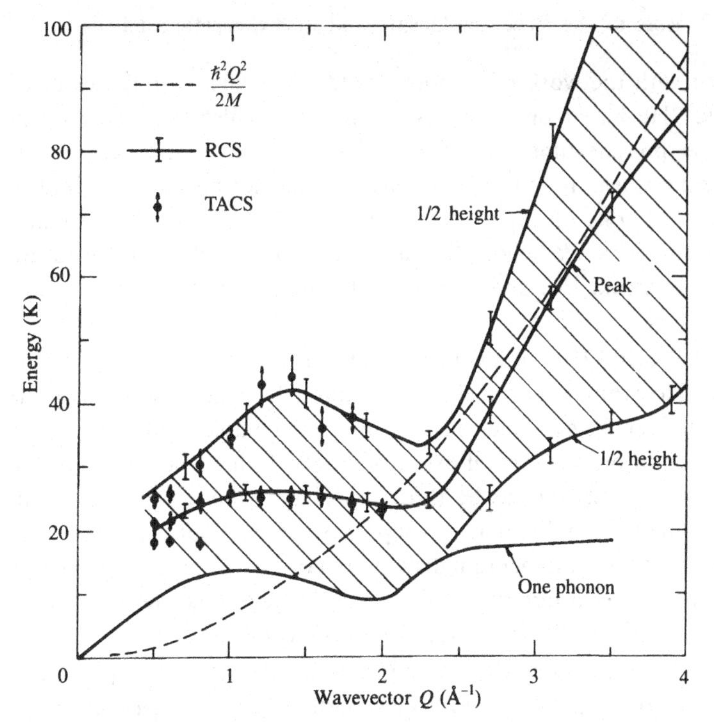

Fig. 1.6. The broad high-energy multiparticle (or multiphonon) structure in $S(\mathbf{Q}, \omega)$ at intermediate Q at 1.1 K and SVP. The "peak" position and its width are indicated [Source: Cowley and Woods, 1971].

Woods (1971), in the region up to about $Q \sim 2.5$ Å^{-1} there is always an additional broad "multiparticle" structure centred around ~20-25 K, with an intensity which increases with Q. This is obvious from results such as those shown in Fig. 1.5. This multiparticle (or multiphonon) spectrum shown in Fig. 1.6 makes it very difficult to clearly separate out the quasiparticle peak – especially as we go to higher temperatures ($T \gtrsim 1.3$ K) where it becomes increasingly broadened. To carry out this separation, we must have some theoretical guidance as to what the quasiparticle and the multiparticle structure involve. The same kind of problem occurs with extracting the phonon frequencies from the $S(\mathbf{Q}, \omega)$ data in solid Helium, as we discuss in Chapter 11.

These sorts of questions force us to search for a more fundamental understanding of the elementary excitations of superfluid ^{4}He and their relation to the structure exhibited by $S(\mathbf{Q}, \omega)$.

1.2 Role of the Bose condensate and field-theoretic analysis

Beginning with the work of London (1938a,b), a microscopic theory of superfluid ^{4}He based on a Bose condensate has also been developed. Up to the present time, however, it has not played a large role in experimental research and for this reason it has not been emphasized in standard textbooks on liquid ^{4}He, which concentrate on the Landau–Feynman scenario. We believe that it will be a dominant theme of future work on superfluid ^{4}He and consequently give it the central role in this book.

Since the early 1960's, it has become increasingly accepted that insofar as liquid ^{4}He exhibits macroscopic quantum effects (as described by the two-fluid model, for example), this was due to a complex order parameter $\langle\hat{\psi}(\mathbf{r})\rangle$. In addition, the critical exponents near T_λ have been found, to a high degree of precision, to be those associated with a two-component order parameter. Finally, many groups have carried out experiments to measure directly the momentum distribution of ^{4}He atoms by high-energy inelastic neutron scattering and from this to determine the fraction of atoms in the zero-momentum state (the Bose condensate).

However, a question hardly ever addressed is how this order parameter is related to the standard picture of superfluid ^{4}He as a gas of weakly interacting phonon–roton quasiparticles. To ignore the role of the Bose order parameter is equivalent to discussing spin waves in a Heisenberg ferromagnet without mentioning that they are associated with fluctuations of the spontaneous magnetic order $\langle S_z\rangle \neq 0$, or discussing phonons in a crystal without reference to the fact that they are vibrations of an underlying lattice. In any modern account of superfluid ^{4}He, the Bose order parameter must play a central, unifying role. In this book, we show how the field-theoretic analysis of Bose fluids can relate the Landau–Feynman quasiparticle picture to the underlying Bose condensate and how one can understand what happens to the "excitations" when liquid ^{4}He passes through the lambda transition.

What is the direct evidence for a Bose condensate in liquid ^{4}He? Miller, Pines and Nozières (1962) and Hohenberg and Platzman (1966) first suggested that the condensate density n_0 could be obtained from $S(\mathbf{Q},\omega)$ data in the high-momentum limit. This stimulated many experimental studies (for a recent review, see Sokol and Snow, 1991) at increasingly high momentum transfers (up to Q $\sim$ 23 Å^{-1}). To the extent that the impulse approximation is valid, the momentum distribution $n(\mathbf{p})$ of ^{4}He atoms can be obtained, as discussed in Section 2.3 and Chapter 4.

The magnitude and temperature dependence of the condensate fraction $n_0(T)/n$ can then be extracted from $n(\mathbf{p})$. In Chapter 4, we discuss the problems involved in this analysis in some detail. While there is still some uncertainty about the precise magnitudes, recent experimental results convincingly show that n_0 is zero above T_λ and builds up to a value (as T is lowered) which is consistent (within about 1%) with the best Monte Carlo computer simulation values at $T = 0$.

A major recent development concerning the Bose condensate is the Monte Carlo study at finite temperatures carried out by Ceperley and Pollock (1986) and Pollock and Ceperley (1987). This uses an imaginary-time path-integral formulation as initiated by Feynman (1953a). Their work is the culmination of a long history of research using Monte Carlo techniques to study liquid ^{4}He, starting with the estimate of n_0 at $T = 0$ by McMillan (1965). Ceperley and Pollock found:

(a) The specific heat shows the beginnings of the characteristic λ-singularity close to $T_\lambda = 2.17$ K (see Fig. 1.7).
(b) The calculated condensate fraction $n_0(T)/n$ is essentially zero above T_λ and then rapidly builds up below T_λ to a value consistent with 9% at $T = 0$ (see Fig. 1.8).
(c) The calculated normal fluid density $\rho_N(T)$ is in excellent agreement with experimental results (see Fig. 1.9).

Although the small sample size (64 atoms) leads to finite-size rounding near T_λ, these results are a *major watershed* in our understanding of superfluid ^{4}He. The successful *ab initio* calculation of the normal and superfluid densities is especially important. Pollock and Ceperley (1987) started with an exact relation which expresses $\rho_S(T)$ as the difference between the long-wavelength, zero-frequency momentum current-response functions to transverse and longitudinal perturbations. As we discuss in Section 6.1, this fundamental definition of $\rho_S(T) \equiv \rho - \rho_N(T)$ has been understood since the early sixties, as well as the fact that $\rho_S \neq 0$ is a direct consequence of a Bose broken symmetry $\langle\hat{\psi}(\mathbf{r})\rangle \neq 0$. This generalized definition of $\rho_S(T)$ and $\rho_N(T)$ is not dependent on a *weakly* interacting quasiparticle picture being valid (as it isn't near T_λ) and thus is much more general than Landau's formula in (1.3).

The results of Ceperley and Pollock give perhaps the most direct "evidence" that the superfluid transition in liquid ^{4}He is associated with a Bose broken symmetry. Their work, in conjunction with the experimental determination of the condensate from high-momentum neutron scattering (see Chapter 4), forces us to examine an aspect of superfluid ^{4}He

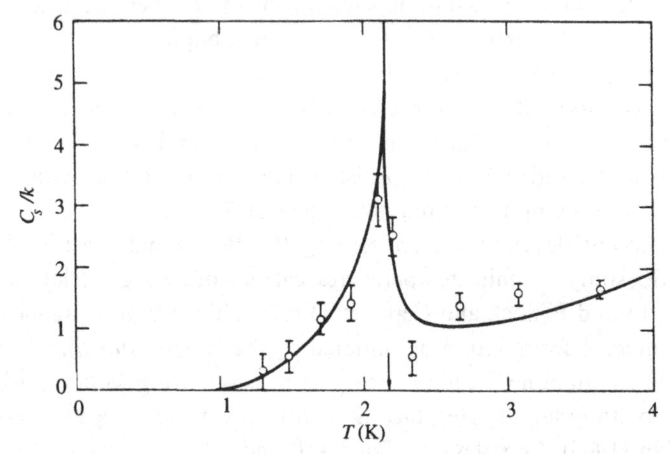

Fig. 1.7. PIMC values of the specific heat at SVP in the region near T_λ. The full line is the experimental data as given in Wilks (1967). The unlabelled arrow shows the superfluid transition temperature T_λ [Source: Ceperley and Pollock, 1986].

which has been largely untouched – namely, the changing properties as the temperature increases through T_λ as the *direct* consequence of the condensate decreasing in magnitude and vanishing at T_λ.

In concluding this section, we summarize some highlights of the development of the field-theoretic analysis of a Bose-condensed liquid. Fritz London first understood that the superfluidity of liquid ^{4}He must have its explanation as a macroscopic quantum effect (see the introduction of the monograph by London, 1950). Unfortunately, the ideal Bose gas (the remarkable properties of which were first delineated by London) does not exhibit the needed long-range phase coherence. London never appreciated that the next step in his programme was in fact given by the work of Bogoliubov (1947) and Penrose (1951), in which the order parameter $\langle\hat{\psi}(\mathbf{r})\rangle$ formed the microscopic foundation for understanding superfluid ^{4}He (see Section 3.1). More specifically, the $T = 0$ study by Bogoliubov (1947) of a weakly interacting dilute Bose gas (WIDBG) was pivotal in that it showed how a Bose condensate in an interacting gas could change the dispersion relation from particle-like to phonon-like at low momentum. As we discussed earlier, this is sufficient to lead to superfluidity (i.e., a finite critical velocity, by Landau's argument). Bogoliubov's work was immediately appreciated by Landau (1949) as an example of a microscopic calculation which exhibited the required

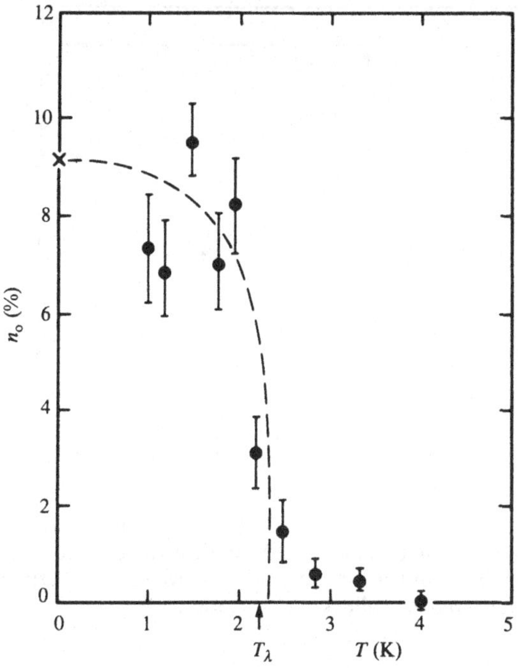

Fig. 1.8. Percentage of ^{4}He atoms with zero momentum at SVP, as given by PIMC calculations. The dashed line is a guide to the eye. The cross is the GFMC ground-state value at $T = 0$ as calculated by Kalos, Lee, Whitlock and Chester (1981) [Source: Ceperley and Pollock, 1986].

phonon spectrum at low Q. It was far ahead of its time, being one of the first studies of broken symmetry with a symmetry-restoring Goldstone mode (Anderson, 1984).

As an historical aside, however, it seems that Bogoliubov's work was largely unappreciated in the West until about 1956, when Lee, Yang and Huang rederived many of the results independently by a different approach (the latter work is nicely reviewed by Huang, 1964). There is not a single reference to Bogoliubov's work in London's well known monograph (London, 1954) or in any of Feynman's papers, and only a single passing reference in Lee, Huang and Yang (1957) in the well known series of papers associated with these authors.

Bogoliubov's discussion of a dilute Bose gas was generalized by Beliaev (1958a) using Green's function techniques so as to be applicable to Bose *liquids*. The next five years or so was a period of intense research developing this approach, the most important papers being those of

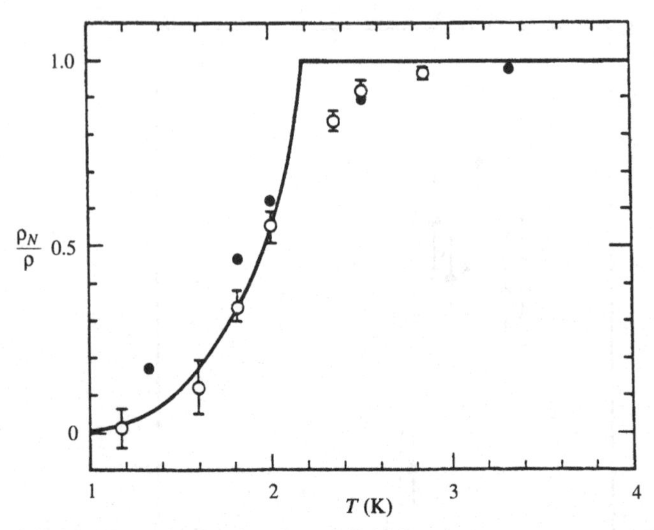

Fig. 1.9. PIMC results for the normal-fluid fraction vs. temperature. The full line is the experimentally measured value. The theoretical values were obtained from evaluating the "winding number" (open circles) as well as the current-response functions (solid dots) [Source: Pollock and Ceperley, 1987].

Hugenholtz and Pines (1959), Bogoliubov (1963, 1970), Gavoret and Nozières (1964) and Hohenberg and Martin (1965). Still useful are the excellent summaries of the implications and accomplishments of the field-theoretic approach up to 1965 given by Hohenberg and Martin (1965), Martin (1965), Pines (1965) and Nozières and Pines (1964, 1990). In particular, these studies gave us an understanding of:

(a) How the two-fluid equations are a natural consequence of a Bose broken symmetry, without the need for a well defined quasiparticle spectrum as in Landau's original theory.

(b) How the Bose broken symmetry leads to the poles of the single-particle Green's function $G(\mathbf{Q}, \omega)$ being the same as the poles of the density correlation function $S(\mathbf{Q}, \omega)$. In particular, this was exhibited by the *tour de force* calculation of Gavoret and Nozières (1964) who explicitly proved by diagrammatic methods that the phonon is the dominant excitation in *all* correlation functions in the long-wavelength, $T = 0$ limit. This explained, at least in one important limit, why one could measure the elementary excitation spectrum by inelastic neutron scattering, but *only* in the Bose-condensed phase (this point seems to have been first explicitly noted by Hohenberg and Martin, 1965). In normal liquids (including liquid ^{4}He and liquid ^{3}He), the density fluctuation modes are not the elementary excitations.

An important development was initiated by Ruvalds and Zawadowski (1970), who pointed out that any structure in the two-quasiparticle spectrum would also be coupled into the single-particle Green's function *through* the effect of the Bose broken symmetry. As a result, the single-quasiparticle spectrum is hybridized with the two-quasiparticle spectrum, with a strength related to the condensate density n_0. Their work was a natural continuation of the pioneering field-theoretic analysis of Pitaevskii (1959) on the stability of excitations at large Q. It led to a new interpretation of the results shown in Fig. 1.6, in terms of a single-particle excitation (such as given by (1.6)) crossing a relatively dispersionless pair-excitation branch describing two rotons. In this interpretation, the natural continuation of the unrenormalized roton dispersion curve is the free-particle excitation at large wavevectors as in (1.6), rather than as the broad resonance at 2Δ for $Q \geq 2.5$ Å^{-1} assumed in Fig. 1.3. Ruvalds and Zawadowski emphasized the crucial role that a Bose broken symmetry has on the coupling of the one- and two-excitation spectrum at *large* ω (see Chapter 10), complementing the work of Gavoret and Nozières (1964) and others at low ω.

The next major period of research progress after the "golden era" of the early 1960's was the early 1970's when the dielectric formalism (this name arose because the theory was first developed by Ma and Woo (1967) for a *charged* Bose gas) was used to analyse the structure of the one- and two-particle Green's functions and shows more clearly the role of the Bose order parameter. Key papers were by Ma, Gould and Wong (1971), Wong and Gould (1974, 1976), Griffin and Cheung (1973), and Szépfalusy and Kondor (1974). This work led to:

(a) Consistent finite-temperature approximations for both $G(\mathbf{Q},\omega)$ and $S(\mathbf{Q},\omega)$ in terms of irreducible, proper diagrammatic contributions, which allowed one to see how a Bose order parameter leads to mixing of the density fluctuations with the more fundamental single-particle spectrum.

(b) A realization of the qualitative differences between the spectrum exhibited by a Bose-condensed gas at $T = 0$, in which almost all the atoms are in the condensate, and at finite temperatures where the dynamics of the non-condensate atoms play a crucial role. This led to simple models which were more appropriate to compare with superfluid ^{4}He than the traditional weakly interacting dilute Bose gas at $T = 0$ with $n_0 \simeq n$.

(c) Understanding how the excitations of the superfluid *merge* with those of the normal phase as the liquid passes through the transition temperature.

In the traditional Landau–Feynman quasiparticle picture (Section 1.1), the quasiparticle dispersion relation in superfluid ^{4}He continues to exist right up to about $Q \sim 2.5$ Å^{-1} (see Fig. 1.3). That is, it is assumed that there is a smooth transition from the low-Q phonon to the high-Q maxon–roton part of the spectrum. In a scenario developed by Glyde and Griffin (1990) based on the dielectric formalism, the (zero sound) phonon branch at low Q broadens substantially for $Q \gtrsim 1$ Å^{-1} and effectively disappears. In contrast, the sharp component which is observed in $S(\mathbf{Q}, \omega)$ at $Q \gtrsim 1$Å^{-1} is viewed as a single-particle excitation branch which has been mixed into the density fluctuation spectrum via the Bose condensate. In this new interpretation of the excitations in superfluid ^{4}He, the existence of the Bose broken symmetry leads to a hybridization of the single-particle and collective density excitations, with the observed phonon–maxon–roton spectrum being the final result. This scenario is developed at length in Chapter 7. The condensate-induced hybridization of single-particle and density excitations makes $S(\mathbf{Q}, \omega)$ for superfluid ^{4}He especially interesting from the point of view of neutron scattering, since it explains why one can probe the elementary excitations through a study of $S(\mathbf{Q}, \omega)$. The microscopic theory needed to deal with this condensate-induced hybridization is reviewed in Chapters 5 and 6. At a more phenomenological level, of course, the excited states in superfluid ^{4}He at $T = 0$ are well described by variational many-body wavefunctions first introduced by Feynman (1954) and developed by many others since. However, as we discuss in Section 9.1, the underlying physics involved in such variational wavefunctions is not easily extracted, especially as concerns the role of the condensate. Moreover, it is very difficult to use this wavefunction approach at finite temperatures, when damping of excitations is important.

1.3 Outline of book

Few substances have been investigated as thoroughly as superfluid ^{4}He. It is a vast subject and any monograph must have focus. Our major theme will be the difference between single-particle modes (elementary excitations) and density fluctuations (collective modes) and how these are intertwined in superfluid ^{4}He due to the Bose broken symmetry.

As should be clear from Section 1.2, we put emphasis on the field-theoretic approach which has led to what we call the dielectric formalism (Ma and Woo, 1967). The recent scenario of Glyde and Griffin (1990) concerning the nature of the phonon–maxon–roton excitations is a natural development within this dielectric formulation, elements of which have been developing over the last 30 years. It starts with the premise that superfluid ^{4}He has a Bose condensate and that this must underlie the unique nature and role of the excitations in this liquid.

In this book, we will always work with a spatially uniform Bose order parameter. In particular, we ignore all effects related to the appearance of quantized vortex filaments or the associated vortex oscillations. For further discussion of this topic, see Nozières and Pines (1964, 1990), Lifshitz and Pitaevskii (1980), and the recent monograph by Donnelly (1991). The dielectric formalism can be applied to situations involving vorticity but this must be left to future studies.

The dielectric formalism is a well defined many-body formalism which allows detailed calculations in a consistent manner. It can be used at any temperature (above and below T_λ) and is valid in the hydrodynamic two-fluid region as well as the high-energy collisionless region probed by neutrons. Needless to say, controlled microscopic calculations are difficult once we go away from a weakly interacting dilute Bose gas. However such model calculations (at finite T) do indicate the general structure imposed by the Bose broken symmetry on the various response functions in any Bose-condensed fluid. This structure must be allowed for in variational and phenomenological approaches if one is to come to grips with the underlying Bose broken symmetry of superfluid ^{4}He.

Because it is this aspect which has been crucial in recent research in understanding the physical basis of the excitations in superfluid ^{4}He, we also emphasize the difference between the dynamic structure factor above *and* below the superfluid transition temperature. Related to this, we give a detailed analysis of the neutron-scattering data as a function of the temperature. This is in contrast to most of the theoretical literature on superfluid ^{4}He, which has concentrated on either the low-temperature region ($\lesssim$ 1 K) or the extreme critical region (within a few mK of the superfluid transition).

Using the renormalization group formalism and the concept of universality classes, the static and dynamic behaviour of superfluid ^{4}He near T_λ appears to be very well understood in terms of a fluid with a two-component order parameter. We will not be concerned with this subject in the present book, but note that it gives additional powerful evidence that the superfluid phase is indeed associated with a Bose broken symmetry.

Neither will we discuss the extensive work on the phonon dispersion relation and damping at very low T ($\lesssim$ 0.6 K), where one can work with phenomenological model Hamiltonians describing well defined weakly interacting quasiparticles. As discussed by Khalatnikov (1965), in this region one can calculate the thermodynamic and transport properties of superfluid ^{4}He with considerable success, as well as the stability of excitations due to spontaneous decay processes allowed by kinematics.

We now briefly outline the contents of each chapter. As we have mentioned, Chapter 2 gives a brief summary of the neutron-scattering data for $S(\mathbf{Q},\omega)$ in superfluid and normal liquid ^{4}He. We also introduce the impulse approximation for $S(\mathbf{Q},\omega)$ valid at large momentum transfers, which will be needed in Chapter 4.

In Chapter 3, we review the concept of Bose broken symmetry and how it modifies the nature of excitations. We also introduce a major theme of this book: namely, the difference between a Bose-condensed *gas* at low T and a Bose-condensed *liquid* like superfluid ^{4}He. In the latter, the condensate fraction is never more than about 10%, and thus it is crucial to include density fluctuations arising from the non-condensate atoms. The latter are neglected in the simple Bogoliubov approximation discussed in most textbooks.

We discuss the atomic momentum distribution $n(\mathbf{p})$ in liquid ^{4}He in Chapter 4, both theoretically and experimentally. High-momentum neutron-scattering studies in recent years (especially by Sokol and coworkers) have given very strong confirmation of the Monte Carlo estimates of the condensate fraction. We also give a detailed review of the condensate-induced changes in $n(\mathbf{p})$ at low but finite momentum.

Chapter 5 is the heart of this book. We introduce the diagrammatic analysis in terms of irreducible, proper contributions and show how the single-particle Beliaev Green's functions $G_{\alpha\beta}(\mathbf{Q},\omega)$ and the density-response function $\chi_{nn}(\mathbf{Q},\omega)$ are coupled via the Bose order parameter. This formalism is illustrated using simple Bose-gas model calculations and we show it can lead to a coupling of the single-particle (SP) pole of $G(\mathbf{Q},\omega)$ with the collective zero sound (ZS) pole of $\chi_{nn}(\mathbf{Q},\omega)$ when $n_0 \neq 0$. The related analysis (at $T = 0$) of Gavoret and Nozières (1964) is also summarized. We argue that the dielectric formalism is the method of choice for future microscopic studies of the dynamics of Bose-condensed fluids.

The difference between low-frequency hydrodynamic and high-frequency collisionless phonon excitations is the main subject of Chapter 6. It is important that the field-theoretic analysis we use to discuss the collisionless region (as probed by neutrons) also leads to hydrodynamic correlation functions which exhibit the characteristic behaviour associated with superfluidity (as probed by Brillouin light scattering and ultrasonic studies). We compare first and second sound in the low-ω region with phonon excitations in the large-ω region, both in a Bose gas and in Bose liquid. In Section 6.3, we also give a critical analysis of the Gavoret–Nozières results in the limit of $Q, \omega \to 0$. Somewhat surprisingly, the physical interpretation of the low-energy phonon region in a Bose-condensed liquid is more complicated than the high-energy roton region.

In Chapter 7, we review high-resolution neutron-scattering data for $S(\mathbf{Q}, \omega)$, with emphasis on how the line shape varies with temperature from 1 K to well above T_λ. Recent experiments at ILL (Talbot *et al.*, 1988; Stirling and Glyde, 1990; Andersen, Stirling *et al.*, 1991) have confirmed earlier work at Chalk River (Woods, 1965; Woods and Svensson, 1978) that the effect of temperature is quite different on the phonon and maxon–roton excitations as one goes from the superfluid to normal phase. The dielectric formalism naturally leads to a picture in which the elementary excitations below T_λ involve the hybridization of a low-Q zero sound (ZS) phonon with a high-Q high-energy single-particle (SP) maxon–roton. This scenario of Glyde and Griffin (1990) gives a plausible connection between the underlying Bose condensate and the phonon–maxon–roton dispersion curve.

The well known f-sum rule for $S(\mathbf{Q}, \omega)$ plays a very important role in analysing neutron-scattering data as well as constraining phenomenological theories. In Chapter 8, we discuss similar sum rules which are unique to Bose-condensed fluids (Wagner, 1966; Wong and Gould, 1974; Stringari, 1992). We also critically review the Woods–Svensson (1978) two-component formula for $S(\mathbf{Q}, \omega)$. Their ansatz is important historically as the first attempt to describe the fact that the maxon–roton has a weight in $S(\mathbf{Q}, \omega)$ which vanishes at T_λ.

In Chapter 9, we review three alternative theoretical approaches to describing the excitations in superfluid ^{4}He, mainly from the point of view of making some contact with the dielectric formalism. These are variational many-body wavefunctions (the correlated-basis-function method), the polarization potential approach of Pines and coworkers, and the memory function formalism. The role of the Bose order parameter is somewhat implicit in all these techniques (which so far have been mainly developed at $T \simeq 0$).

In Chapter 10, we discuss the high-energy multiparticle spectrum (including the two-roton bound state) which shows up in the neutron-scattering intensity but which can also be studied more directly by Raman light scattering. We review the condensate-induced hybridization between the two-particle and the single-particle spectrum, as first studied by Pitaevskii (1959) and later by Ruvalds and Zawadowski (1970).

In Chapter 11, we review the distinction between the poles of the displacement correlation function (phonons) and the density fluctuations described by $S(\mathbf{Q}, \omega)$ in highly anharmonic crystals (Ambegaokar, Conway and Baym, 1965). We call attention to certain similarities between the modern theory of excitations in quantum crystals like solid ^{4}He and the dielectric formalism used to describe superfluid ^{4}He, and suggest that this may be the basis for understanding the striking similarity of $S(\mathbf{Q}, \omega)$ data in these two different phases.

Finally, in Chapter 12, we summarize and assess the new interpretation of the phonon–maxon–roton dispersion curve based on a condensate-induced hybridization of two different excitation branches. We discuss the history of this scenario and some of its implications, for both normal and superfluid liquid ^{4}He. We briefly sketch the dielectric formalism results for superfluid ^{3}He–^{4}He mixtures and conclude with several suggestions for future work.

Units used in this book

In neutron-scattering data for $S(\mathbf{Q}, \omega)$ which we quote, energies are given interchangeably in milli-electron-volts, kelvins and terahertz. The conversion factors are

$$1 \text{ THz} = 48.0 \text{ K} = 4.14 \text{ meV},$$
$$10 \text{ K} = 0.208 \text{ THz} = 0.863 \text{ meV},$$
$$1 \text{ meV} = 0.24 \text{ THz} = 11.6 \text{ K}.$$

In general, we set $\hbar = 1$ but occasionally we insert it in final formulas or when it will make the discussion more physical.

2

Dynamic response of Helium atoms to thermal neutrons

Much of this book is concerned with the density fluctuation spectrum of superfluid ^{4}He. As background, in this chapter, we review some standard material concerning the dynamic structure factor $S(\mathbf{Q},\omega)$ and give an introductory summary of the extensive inelastic neutron-scattering data on liquid ^{4}He (see also Section 7.1).

A systematic formulation of neutron scattering as a microscopic probe of atomic motions in condensed phases was first developed in the early 1950's by Placzek and van Hove. This led to the epochal work of van Hove (1954), who showed that the inelastic scattering cross-section (in the first Born approximation) of thermal neutrons is directly related to the Fourier transform of correlation functions involving local operators at two different space-time points. In liquid ^{4}He, the most important of these are the coherent $S(\mathbf{Q},\omega)$ and incoherent $S_{\mathrm{inc}}(\mathbf{Q},\omega)$ dynamic structure factors. In Section 2.1, we define these correlation functions and discuss the symmetry relations and rigorous "sum rules" which these functions satisfy. In Section 2.2, we discuss the characteristic features exhibited by $S(\mathbf{Q},\omega)$ in superfluid ^{4}He and their traditional interpretations. In Section 2.3, we consider the large Q and ω limit of $S(\mathbf{Q},\omega)$, and introduce the basic physics of the so-called "impulse approximation" (IA).

2.1 Response functions: general properties and sum rules

Neutrons interact only with the nucleus of an ^{4}He atom, which has zero spin. As discussed in standard texts (see Marshall and Lovesey, 1971), a low-energy (thermal) neutron impinging on a sample of liquid Helium sees a total scattering potential

$$V(\mathbf{r}) = \frac{2\pi\hbar^2}{m}\sum_j b\delta(\mathbf{r}-\mathbf{r}_j) \equiv \frac{2\pi\hbar^2 b}{m}\hat{\rho}(\mathbf{r}), \tag{2.1}$$

where $\mathbf{r}_j$ is the position operator of the jth ^{4}He nucleus (mass m) and b is the "bound" scattering length describing the neutron–Helium-nucleus interaction at the thermal energies of interest. Treating this potential in the first Born approximation, one finds that the double differential scattering cross-section per target atom is given by (van Hove, 1954)

$$\frac{d^2\sigma}{d\Omega dE} = \frac{b^2}{\hbar}\frac{k_1}{k_0}S(\mathbf{Q},\omega), \tag{2.2}$$

where the dynamic structure factor is defined as

$$S(\mathbf{Q},\omega) \equiv \frac{1}{2\pi N}\int_{-\infty}^{\infty} dt\, e^{i\omega t}\langle\hat{\rho}(\mathbf{Q},t)\hat{\rho}^{+}(\mathbf{Q},0)\rangle. \tag{2.3}$$

Here $k_0(k_1)$ is the incoming (outgoing) neutron wavevector and

$$\left.\begin{aligned} \hbar\mathbf{Q} &= \hbar\mathbf{k}_0 - \hbar\mathbf{k}_1\ , \\ \hbar\omega &= \frac{\hbar^2k_0^2}{2m} - \frac{\hbar^2k_1^2}{2m} \end{aligned}\right\} \tag{2.4}$$

are, respectively, the momentum and energy transfer to the fluid which take place in the scattering process; also

$$\hat{\rho}(\mathbf{Q}) \equiv \sum_j e^{-i\mathbf{Q}\cdot\mathbf{r}_j} = \hat{\rho}^{+}(-\mathbf{Q}) \tag{2.5}$$

is the Fourier component of the ^{4}He number-density operator

$$\hat{\rho}(\mathbf{r}) = \sum_j \delta(\mathbf{r}-\mathbf{r}_j)\ . \tag{2.6}$$

It is sometimes convenient to introduce the time-dependent "intermediate" scattering function

$$S_{\text{coh}}(\mathbf{Q},t) \equiv \frac{1}{N}\langle\hat{\rho}(\mathbf{Q},t)\hat{\rho}(-\mathbf{Q},0)\rangle\ . \tag{2.7}$$

At high momentum transfers, $S(\mathbf{Q},\omega)$ in (2.3) may be approximated by the "incoherent" structure factor

$$S_{\text{inc}}(\mathbf{Q},\omega) \equiv \frac{1}{2\pi}\int_{-\infty}^{\infty} dt\, e^{i\omega t}\langle e^{-i\mathbf{Q}\cdot\mathbf{r}_j(t)}e^{i\mathbf{Q}\cdot\mathbf{r}_j(0)}\rangle\ . \tag{2.8}$$

This describes the contributions to $S(\mathbf{Q},\omega)$ from the terms $k = j$. $S_{\text{inc}}(\mathbf{Q},\omega)$ in (2.8) is the same for all atoms (i.e., it is independent of the label j). This correlation function describes the motion of a single atom and is thus of independent interest in the theory of liquids (Hansen and McDonald, 1986).

The static or instantaneous structure factor is defined as

$$S(\mathbf{Q}) \equiv S(\mathbf{Q}, t = 0) = \frac{1}{N} \sum_{i,j} \langle e^{-i\mathbf{Q}\cdot\mathbf{r}_i} e^{i\mathbf{Q}\cdot\mathbf{r}_j} \rangle \ . \tag{2.9}$$

This is related to the static pair correlation function $g(\mathbf{r})$ through the expression (p. 371, Marshall and Lovesey, 1971)

$$S(\mathbf{Q}) = 1 + n \int d\mathbf{r}\, e^{i\mathbf{Q}\cdot\mathbf{r}} [g(\mathbf{r}) - 1] \ , \tag{2.10}$$

where $g(\mathbf{r})$ describes the instantaneous correlations between He atoms. It is proportional to the probability that, if there is an atom at the origin, then there is simultaneously another atom at position $\mathbf{r}$. Measurements of $S(\mathbf{Q})$ can be deconvoluted using (2.10) to give $g(\mathbf{r})$. An equivalent definition of $S(\mathbf{Q})$ is

$$S(\mathbf{Q}) = \int_{-\infty}^{\infty} d\omega\; S(\mathbf{Q}, \omega) \ , \tag{2.11}$$

the dynamic structure factor integrated over all energy transfers.

We also note explicitly that, with the definitions (2.8) and (2.9),

$$S_{\rm inc}(\mathbf{Q}) \equiv \int_{-\infty}^{\infty} d\omega\; S_{\rm inc}(\mathbf{Q}, \omega) = 1 \ . \tag{2.12}$$

This is often used to determine how high the momentum transfer must be before $S(\mathbf{Q}, \omega)$ can be approximated by $S_{\rm inc}(\mathbf{Q}, \omega)$ in (2.8). The oscillations observed in $S(\mathbf{Q})$ in (2.9) are associated with the terms $i \neq j$. In liquid ^{4}He, these oscillations have disappeared for $Q \gtrsim 8$ Å^{-1} (Svensson *et al.*, 1980) and this is interpreted to mean that the $i \neq j$ terms are also not important in $S(\mathbf{Q}, \omega)$ for large Q.

We can express $S(\mathbf{Q}, \omega)$ explicitly in terms of the exact many-body eigenstates $|n\rangle$ of the effective Hamiltonian $\hat{H}' = \hat{H} - \mu\hat{N}$, where μ is the chemical potential. Defining

$$\hat{H}'|n\rangle = E'_n|n\rangle \equiv (E_n - \mu N_n)|n\rangle \ , \tag{2.13}$$

we obtain (see pp. 148ff, Mahan, 1990)

$$S(\mathbf{Q}, \omega) = \frac{1}{N} \sum_{n,m} \frac{e^{-\beta E'_n}}{Z} |\langle m|\hat{\rho}^+(\mathbf{Q})|n\rangle|^2 \delta[\omega - (E'_m - E'_n)] \ . \tag{2.14}$$

Here Z is the grand canonical partition function and $\beta = 1/k_B T$. In (2.14), the states $|n\rangle$ and $|m\rangle$ have the same number of atoms since the effect of the operator $\hat{\rho}^+(\mathbf{Q})$ does not involve changing the number of atoms ($N_n = N_m$). At $T = 0$, the sum over $|n\rangle$ reduces to the ground state corresponding to the correct value of the chemical potential μ. The exact spectral representation (2.14) is useful in deriving various general properties of $S(\mathbf{Q},\omega)$. In addition, it emphasizes that $S(\mathbf{Q},\omega)$ involves the contribution of all energetically possible transitions $|n\rangle \rightarrow |m\rangle$, each weighted by the absolute square of the matrix element $\langle m|\hat{\rho}^{+}(\mathbf{Q})|n\rangle$. Physically, this says that the density operator $\hat{\rho}^{+}(\mathbf{Q})$ acting on a state $|n\rangle$ must produce a state with a *finite* overlap with the many-body state $|m\rangle$ *if* these states are to be important in the summation involved in (2.14).

In the isotropic fluid case, one can introduce operators which create or destroy atoms of specified momentum. In this second quantized language, (2.5) is equivalent to

$$\hat{\rho}^{+}(\mathbf{Q}) = \sum_{\mathbf{k}} \hat{a}^{+}_{\mathbf{k}+\mathbf{Q}}\hat{a}_{\mathbf{k}} \,. \tag{2.15}$$

This form naturally leads to an interpretation in terms of destroying an atom of momentum $\hbar\mathbf{k}$ and creating one with momentum $\hbar\mathbf{k}+\hbar\mathbf{Q}$, that is, a particle $(k+Q)-$ hole (k) excitation. This language is familiar in describing interacting Fermi systems (Pines and Nozières, 1966). It is equally useful in discussing density fluctuations in Bose fluids and we shall often use it in this book. Thus $S(\mathbf{Q},\omega)$ in (2.3) may be viewed as related to the propagator of particle–hole (p–h) pairs created at $t=0$ and destroyed at time t.

The elementary excitations are described by single-particle Green's functions. It is important to understand the distinction between these functions and the density correlation function in (2.3) which is measured by inelastic neutron scattering. A generic example of a single-particle correlation function is given by (see Section 3.2)

$$A(\mathbf{Q},t) = \langle \hat{a}_{Q}(t)\hat{a}^{+}_{Q}(0)\rangle \ . \tag{2.16}$$

In contrast to $S(\mathbf{Q},t)$ defined in (2.7), this correlation function describes how a single *atom* of momentum $\hbar\mathbf{Q}$ propagates over the time interval t. In terms of an exact eigenstate representation used in deriving (2.14), one obtains (see p.142 of Lifshitz and Pitaevskii, 1980)

$$A(\mathbf{Q},\omega) = \sum_{m,n} \frac{e^{-\beta E'_n}}{Z} |\langle m|\hat{a}^{+}_{Q}|n\rangle|^2 \delta\left[\omega - \mu - (E'_m - E'_n)\right] \ . \tag{2.17}$$

Clearly only states $|m\rangle$ which contain one more atom than $|n\rangle$ can contribute to the summation in (2.17): $N_m = N_n + 1$. Thus the excited states $|m\rangle$ which contribute to $A(\mathbf{Q},\omega)$ will, in general, be different from those which contribute significantly to $S(\mathbf{Q},\omega)$ in (2.14). In more technical language, the single-particle spectrum does *not* usually overlap strongly with the p–h or density fluctuation spectrum. However, we shall see (Chapters 3 and 5) that in the presence of a Bose broken symmetry, these two functions are closely related to each other. Thus in superfluid ^{4}He, we will be interested in $A(\mathbf{Q},\omega)$ and $S(\mathbf{Q},\omega)$ as inter-related correlation functions. At high energies, these functions are in turn coupled into the two-particle spectrum discussed in Chapter 10.

In explicit calculations of the correlation function $S(\mathbf{Q},\omega)$, it is convenient to work with the associated density-response function

$$\chi_{nn}(\mathbf{Q},t) \equiv \frac{-i\theta(t)}{\Omega}\langle[\hat{\rho}(\mathbf{Q},t),\hat{\rho}(-\mathbf{Q},0)]\rangle \ , \tag{2.18}$$

where $\theta(t)$ is the step function and Ω is the sample volume. The square bracket is the commutator $[A,B] \equiv AB-BA$. $\chi_{nn}(\mathbf{Q},t)$ is interchangeably called a retarded Green's function, a dynamic susceptibility or a response function. As discussed in standard texts, many-body techniques most naturally give the Fourier components $\chi_{nn}(\mathbf{Q},\omega)$. At finite temperatures, one works with imaginary times in the interval $0 < \tau \leq \beta$ and discrete Matsubara imaginary frequencies $i\omega_n = i2nk_BT$, $n = 0, \pm1, \pm2, \ldots$ (see Chapter 3 of Mahan, 1990).

Evaluating (2.18) in the exact eigenstate representation used in (2.14), one easily finds (η is a positive infinitesimal)

$$\chi_{nn}(\mathbf{Q},\omega+i\eta) = \frac{1}{\Omega}\sum_{m,n}\frac{e^{-\beta E'_n}}{Z}|\langle m|\rho^+(\mathbf{Q})|n\rangle|^2\frac{\left[1-e^{-\beta(E'_m-E'_n)}\right]}{\omega+i\eta-(E'_m-E'_n)} \tag{2.19}$$

and hence

$$\mathrm{Im}\ \chi_{nn}(\mathbf{Q},\omega+i\eta) = -\pi n(1-e^{-\beta\omega})S(\mathbf{Q},\omega) \ , \tag{2.20a}$$

where $n = N/V$ is the density of ^{4}He atoms. The inverted form of (2.20*a*) gives the familiar expression

$$S(\mathbf{Q},\omega) = -\frac{1}{\pi n}[N(\omega)+1]\mathrm{Im}\ \chi_{nn}(\mathbf{Q},\omega+i\eta) \ , \tag{2.20b}$$

where $N(\omega)$ is the Bose distribution

$$N(\omega) = \frac{1}{e^{\beta\omega}-1} \ . \tag{2.21}$$

An important relation between negative and positive energy transfers immediately follows from (2.14), namely

$$S(-\mathbf{Q},-\omega) = \frac{N(\omega)}{N(\omega)+1}S(\mathbf{Q},\omega) = e^{-\beta\omega}S(\mathbf{Q},\omega) \ . \tag{2.22a}$$

This is called the principle of detailed balance. In an isotropic liquid, $S(\mathbf{Q},\omega)$ depends only on the magnitude of $\mathbf{Q}$. In this case, it follows from (2.22*a*) and (2.20*a*) that

$$\text{Im}\ \chi_{nn}(\mathbf{Q},\omega+i\eta) = -\text{Im}\ \chi_{nn}(\mathbf{Q},-\omega+i\eta)\ . \tag{2.22b}$$

In approximate calculations, it is important that (2.22) be satisfied. According to (2.22*a*), the ratio of the intensity for an energy *loss* of $\hbar\omega$ by the system to the scattered neutron (anti-Stokes) to the corresponding energy *gain* of $\hbar\omega$ by the system (Stokes) is given by $\exp(-\beta\hbar\omega)$. At $T = 0$, the anti-Stokes component vanishes, reflecting the fact that the system is in its ground state and thus can only gain energy. To avoid later confusion, note that the single-particle spectral density $A(\mathbf{Q},\omega)$ in (2.17) does not satisfy (2.22*a*). There is no general relation between $A(\mathbf{Q},\omega)$ and $A(\mathbf{Q},-\omega)$.

In the low frequency classical region ($\hbar\omega \ll k_BT$), the detailed balance factor $[N(\omega)+1]$ in (2.20*b*) reduces to $k_BT/\hbar\omega$. In this limit, the $S(\mathbf{Q},\omega)$ line shape can be significantly modified from the underlying Im $\chi_{nn}(\mathbf{Q},\omega)$. This is not a problem in the high-frequency region $\hbar\omega \gg k_BT$.

The dynamic structure factor satisfies various sum rules or constraints which involve frequency moments defined by

$$\langle\omega^n\rangle \equiv \int_{-\infty}^{\infty} d\omega\, \omega^n S(\mathbf{Q},\omega)\ . \tag{2.23}$$

For $n \geq 1$, the main contribution to these moments is from *large* ω, which means that they are related to the *small*-time behaviour of $S(\mathbf{Q},t)$. The well known longitudinal f-sum rule (Placzek, 1952)

$$\langle\omega\rangle \equiv \int_{-\infty}^{\infty} d\omega\, \omega S(\mathbf{Q},\omega) = \frac{Q^2}{2m}\ . \tag{2.24}$$

is derived, for example, on p. 365 of Lifshitz and Pitaevskii (1980). Higher-order moment sum rules can also be derived in a similar manner (Rahman *et al.*, 1962; Puff, 1965). Such sum rules are very useful constraints in experimental studies since the data can be used to compute $\langle\omega^n\rangle$ to see if the results satisfy the sum rules. Sum rules unique to Bose-condensed fluids are discussed in Sections 6.1 and 8.1.

The frequency moments in (2.23) are given directly by a high-frequency expansion of $\chi_{nn}(\mathbf{Q},\omega)$. We recall the spectral relation

$$\chi_{nn}(\mathbf{Q},\omega) = -\int_{-\infty}^{\infty} \frac{d\omega'}{\pi} \frac{\text{Im}\ \chi_{nn}(\mathbf{Q},\omega')}{\omega-\omega'}\ , \tag{2.25}$$

a result which may be explicitly verified using (2.19). Inserting (2.20*a*) into (2.25) and using (2.22), one obtains

$$\chi_{nn}(\mathbf{Q},\omega) = 2n\int_{-\infty}^{\infty} d\omega' \frac{\omega' S(\mathbf{Q},\omega')}{\omega^2-\omega'^2}$$
$$= \frac{2n}{\omega^2}\langle\omega\rangle + \frac{2n}{\omega^4}\langle\omega^3\rangle + \ldots \text{ for } \omega \to \infty \,. \quad (2.26)$$

This high-frequency expansion is a useful constraint in model calculations of $\chi(\mathbf{Q},\omega)$, as we shall see in later chapters.

The zero-frequency limit of (2.24) gives

$$\chi_{nn}(\mathbf{Q},\omega=0) = -2n\int_{-\infty}^{\infty} d\omega' \frac{S(\mathbf{Q},\omega')}{\omega'} \,. \quad (2.27)$$

This inverse frequency moment clearly will be dominated by the *low-frequency* behaviour of $S(\mathbf{Q},\omega)$, i.e., the long-time dynamics of the ^{4}He atoms. The low ω and Q region is usually referred to as the hydrodynamic domain (Section 6.2). One can relate the long-wavelength limit of the zero-frequency response functions to thermodynamic derivatives. In particular, one finds (see p. 136 of Pines and Nozières, 1966)

$$\lim_{Q\to 0}\chi_{nn}(\mathbf{Q},\omega=0) = -\frac{n}{mc^2} \,, \quad (2.28)$$

where c is the isothermal sound velocity. This result is known as the compressibility sum rule.

In connection with the incoherent dynamic structure factor (2.8), it is also useful to define central frequency moments

$$M_n(\mathbf{Q}) \equiv \int_{-\infty}^{\infty} d\omega(\omega-\omega_R)^n S_{\text{inc}}(\mathbf{Q},\omega) \,.$$

These are centred around the free-atom recoil frequency $\omega_R = \hbar Q^2/2m$ and hence are useful at high Q when $S_{\text{inc}}(\mathbf{Q},\omega)$ is peaked near ω_R. These central moments satisfy various sum rules (Rahman *et al.*, 1962). The first central moment sum rule $M_1(\mathbf{Q}) = 0$ is easily seen to be equivalent to the f-sum rule for $S_{\text{inc}}(\mathbf{Q},\omega)$ when (2.12) is taken into account.

2.2 Density fluctuation spectrum of superfluid ^{4}He

In this section, we expand on our introductory remarks in Section 1.1 concerning the density fluctuation spectrum observed in superfluid ^{4}He. A more detailed, critical analysis of recent neutron-scattering data is given in Chapter 7.

As orientation, we first discuss the density-response function of a non-interacting Bose gas. In this case, (2.19) reduces to

$$\chi^0_{nn}(\mathbf{Q},\omega+i\eta)=\frac{1}{\Omega}\sum_{\mathbf{k}}\frac{N(\varepsilon_k-\mu)-N(\varepsilon_{k+Q}-\mu)}{\omega+i\eta-[\varepsilon_{k+Q}-\varepsilon_k]}\ , \tag{2.29}$$

where $\varepsilon_k=k^2/2m$ is the free-particle energy and Ω is the sample volume. This expression is derived in most many-body texts (see, for example, Mahan, 1990; Fetter and Walecka, 1971; Nozières and Pines, 1964, 1990) as well as in Section 5.2 as a limiting case. Inserting (2.29) into (2.20*b*), we obtain that the dynamic structure factor $S^0(\mathbf{Q},\omega)$ for an ideal Bose gas can be written in several equivalent forms:

$$\begin{aligned}
S^0(\mathbf{Q},\omega)&\\
&=\frac{1}{n}\Big[N(\omega)+1\Big]\frac{1}{\Omega}\sum_{\mathbf{k}}\Big[N(\mathbf{k})-N(\mathbf{k}+\mathbf{Q})\Big]\delta\left(\omega-[\varepsilon_{k+Q}-\varepsilon_k]\right) &(2.30a)\\
&=\frac{1}{n}\Big[N(\omega)+1\Big]\frac{1}{\Omega}\sum_{\mathbf{k}}N(\mathbf{k})\left\{\delta\left(\omega-[\varepsilon_{k+Q}-\varepsilon_k]\right)-\delta\left(\omega+[\varepsilon_{k+Q}-\varepsilon_k]\right)\right\} &(2.30b)\\
&=\frac{1}{n}\frac{1}{\Omega}\sum_{\mathbf{k}}N(\mathbf{k})\Big[1+N(\mathbf{k}+\mathbf{Q})\Big]\delta\left(\omega-[\varepsilon_{k+Q}-\varepsilon_k]\right)\ , &(2.30c)
\end{aligned}$$

where $N(\mathbf{k})\equiv N(\varepsilon_k-\mu)$. In the last step, we have used the identity satisfied by the Bose distribution,

$$1+N(\omega)=-N(-\omega)\ . \tag{2.31}$$

We call attention to the appearance of the statistical factors in (2.30*c*) associated with creating and destroying Bosons. Clearly the first term in (2.30*b*) is the Stokes and the second term is the anti-Stokes term. One can also easily verify that the expressions in (2.30) satisfy the f-sum rule (2.24) as well as the detailed-balance condition (2.22).

In the special case of a Bose-condensed gas, (2.30*c*) reduces to

$$\begin{aligned}
S^0(\mathbf{Q},\omega)=&\frac{n_0}{n}\left\{\Big[1+N(\mathbf{Q})\Big]\delta(\omega-\varepsilon_Q)+N(\mathbf{Q})\delta(\omega+\varepsilon_Q)\right\}\\
&+\frac{1}{n}\int\frac{d\mathbf{k}}{(2\pi)^3}N(\mathbf{k})\Big[1+N(\mathbf{k}+\mathbf{Q})\Big]\delta\left(\omega-[\varepsilon_{k+Q}-\varepsilon_k]\right)\ ,
\end{aligned} \tag{2.32}$$

where n_0 is the density of atoms in the $\mathbf{k}=0$ state (the Bose condensate). The second line in (2.32) describes scattering of atoms from state $\mathbf{k}$ to $\mathbf{k}+\mathbf{Q}$. As with an ideal Fermi gas, the latter is a broad *continuum* arising from scattering of incoherent particle–hole (p–h) excitations. In contrast, the first line of (2.32) describes density fluctuations associated with atoms coming out of (or going into) the Bose condensate "reservoir". This is a highly coherent process and gives rise to sharp single-particle peaks at $\pm\varepsilon_Q$, with an overall weight proportional to the condensate fraction n_0/n.

With these results for an ideal Bose gas as background, we now turn to liquid ^{4}He. $S(\mathbf{Q},\omega)$ has been measured by neutron scattering over a wide range of momentum transfers ($0.1 \lesssim Q \lesssim 23$ Å^{-1}) and temperature ($0.35 \leq T \leq 4.2$ K), in both the superfluid and normal phases. In addition, $S(\mathbf{Q},\omega)$ has been studied for pressures up to 25 bar, where liquid ^{4}He solidifies. While recent data have given a more complete picture with better resolution, the general behaviour of $S(\mathbf{Q},\omega)$ was already clear from the extensive data collected at Chalk River in the 1960's (Cowley and Woods, 1971). A later paper by the Chalk River group (Svensson, Martel, Sears and Woods, 1976) also summarizes a wealth of experimental information on $S(\mathbf{Q},\omega)$ for a wide range of values of Q and T. More recent reviews include Glyde and Svensson (1987), which contains a very complete list of references; and Svensson (1989), with emphasis on $S(\mathbf{Q},\omega)$ data at higher temperatures.

Most of the early studies of $S(\mathbf{Q},\omega)$ in superfluid ^{4}He were at low temperatures ($T \sim 1$ K). In addition to the sharp quasiparticle resonance at ω_Q discussed in Section 1.1, $S(\mathbf{Q},\omega)$ exhibits an additional structure at frequencies well above ω_Q, with a broad maximum at an energy ~ 25 K (see Figs. 1.2 and 1.5). This broad peak is referred to as a "multiparticle branch" since it is thought to involve the creation of two (or more) quasiparticles. For $Q \gtrsim 0.8$ Å^{-1}, $S(\mathbf{Q},\omega)$ also exhibits a high-frequency tail extending to very high energies (> 70 K). The general features of this additional scattering are shown in Fig. 1.6. It is customary to take $S(\mathbf{Q},\omega)$ as the sum of a quasiparticle part S_{I} and a multiparticle part S_{II},

$$S(\mathbf{Q},\omega) = S_{\mathrm{I}}(\mathbf{Q},\omega) + S_{\mathrm{II}}(\mathbf{Q},\omega) \; . \qquad (2.33)$$

The existence of these two distinct contributions was first discussed (at $T = 0$) by Miller, Pines and Nozières (1962) in a landmark paper. It is important to emphasize that writing down the two-component expression (2.33) for $S(\mathbf{Q},\omega)$ must be justified theoretically and, as we shall see in Chapter 7, its use prejudices what we mean by a "quasiparticle." At low temperatures, an empirical decomposition into these two parts is easy to make by "eye", if the single-quasiparticle and broad multiparticle background are well separated in energy. As the temperature increases, the quasiparticle resonance broadens and a precise separation of S_{I} and S_{II} becomes more difficult.

At temperatures $T \lesssim 1.7$ K, the quasiparticle resonance at ω_Q is sharp

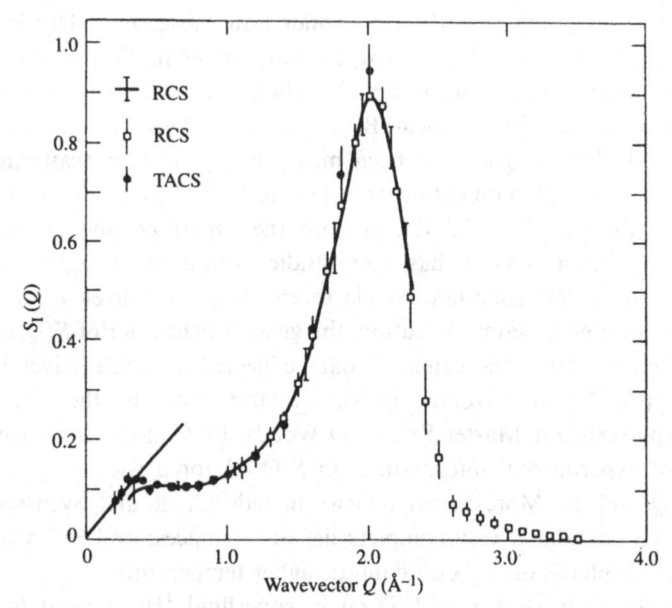

Fig. 2.1. Intensity of the quasiparticle peak in $S(\mathbf{Q}, \omega)$ as a function of Q, at 1.1 K and SVP. This is a compilation of data using different spectrometers [Source: Cowley and Woods, 1971].

and often fitted by (see Appendix of Talbot *et al.*, 1988)

$$S_I(\mathbf{Q}, \omega) = \frac{[N(\omega) + 1]}{\pi} Z(Q) \left[\frac{\Gamma_Q}{(\omega - \omega_Q)^2 + \Gamma_Q^2} - \frac{\Gamma_Q}{(\omega + \omega_Q)^2 + \Gamma_Q^2} \right] , \tag{2.34}$$

where the quasiparticle half-width at half maximum (HWHM) is denoted by Γ_Q. If the damping is negligible, as at low temperature, (2.34) simplifies to

$$S_I(\mathbf{Q}, \omega) = [N(\omega) + 1] \, Z(Q) \left[\delta(\omega - \omega_Q) - \delta(\omega + \omega_Q) \right] . \tag{2.35}$$

Here $Z(Q)$ gives the weight of this excitation in $S(\mathbf{Q}, \omega)$, with the associated static structure factor being

$$S_I(\mathbf{Q}) = Z(Q) \left[2N(\omega_Q) + 1 \right] . \tag{2.36}$$

The double Lorentzian in (2.34) is consistent with detailed balance (2.22); in much of the older literature, only the first (or Stokes) term in (2.34) is used in fitting data. In Fig. 2.1, we plot the relative weight $Z(Q)$ of the so-called "one-phonon" part of $S(\mathbf{Q}, \omega)$ at low temperatures. We note

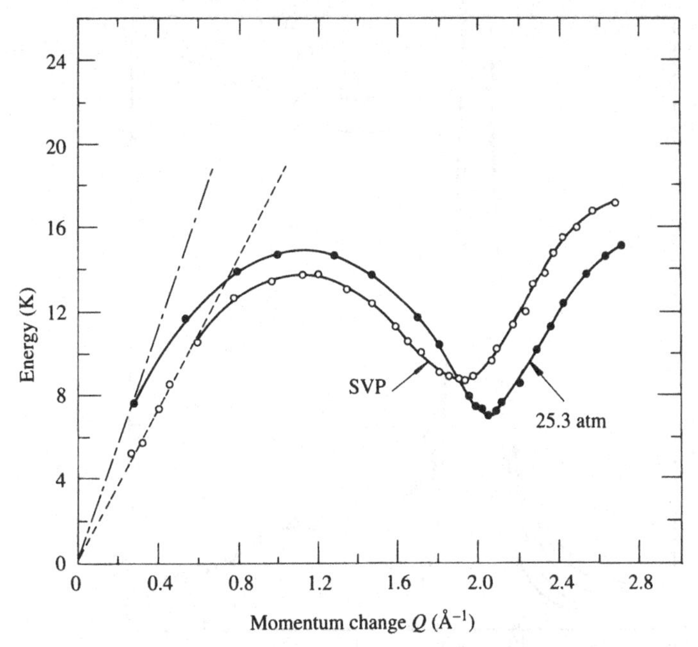

Fig. 2.2. The effect of pressure on the quasiparticle dispersion curve at $T = 1.1$ K. The dashed lines give the measured sound velocity [Source: Woods and Cowley, 1973].

that the full temperature dependence of the static structure factor $S(\mathbf{Q})$ has been measured with high precision (Svensson *et al.*, 1980).

To the extent that $S_{\mathrm{I}}(\mathbf{Q}, \omega)$ can be unambiguously extracted from the full $S(\mathbf{Q}, \omega)$, the quasiparticle peak position and width can be obtained from neutron-scattering data. At $T \lesssim 1.2$ K, the quasiparticle peak width is extremely small and is mainly determined by kinematically restricted decay processes since there are few thermally excited quasiparticles. At $T \lesssim 1$ K, the phonon width is found to decrease abruptly for Q larger than $Q_c \lesssim 0.55$ Å^{-1}. This is consistent with the fact that phonon decay into two phonons is possible only because of (pressure-dependent) anomalous dispersion in the region $Q < Q_c$ (for further discussion see Maris, 1977, and Stirling, 1983). At higher temperatures, the quasiparticle width $2\Gamma_Q$ is apparently mainly due to scattering from thermally excited rotons since it scales roughly with $N_R(T)$, the number of rotons. A width having this temperature dependence was first predicted by Landau and Khalatnikov (1949) for rotons ($Q \sim Q_R$) but it has been found

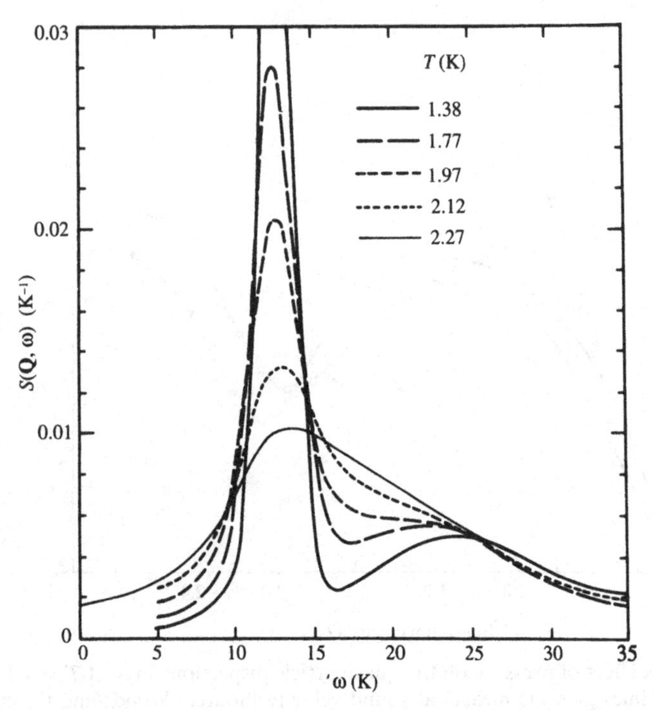

Fig. 2.3. Smoothed scattering data vs. frequency, for $Q = 0.8$ Å^{-1} and SVP, at several temperatures [Source: Woods and Svensson, unpublished; Griffin, 1987].

experimentally to be approximately true for all wavevectors (Cowley and Woods, 1971; Woods and Svensson, 1978; Mezei and Stirling, 1983).

There has been a continuing effort studying the dynamics of superfluid ^{4}He under pressure. In Fig. 2.2, we show the pressure-induced changes to the phonon–maxon–roton curve at low temperatures. As we shall see in Chapters 7 and 10, studies of $S(\mathbf{Q}, \omega)$ as a function of pressure have played an important role in disentangling various contributions and also in understanding the role of the Bose condensate.

There has been increasing interest in how the quasiparticle line shape changes in $S(\mathbf{Q}, \omega)$ as a function of the temperature. For $Q = 0.4$ Å^{-1} (the phonon region), recent high-resolution studies (Stirling and Glyde, 1990) have shown that while the width of the phonon peak steadily increases, there is *no* qualitative change as we go from below to above T_λ. In particular, there is little change in the peak position (see Fig. 1.4). This was first noted in the pioneering study by Woods (1965b). In the roton–maxon wavevector region ($0.8 \lesssim Q \lesssim 2.4$ Å^{-1}), the situation is quite

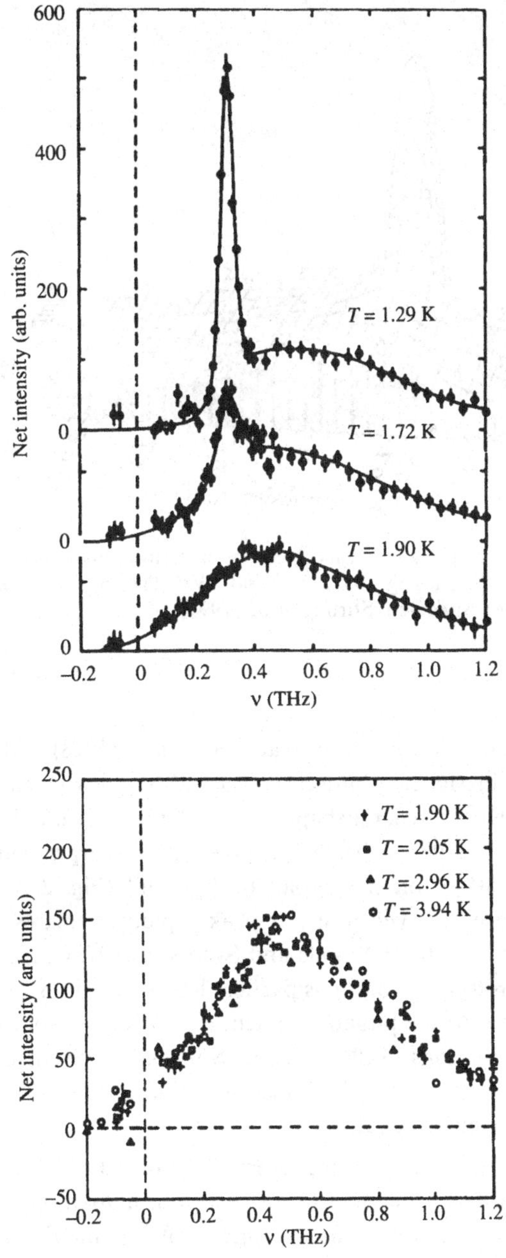

Fig. 2.4. Neutron-scattering intensity vs. frequency, for $Q = 1.13$ Å^{-1} and 20 atm pressure. The top panel shows superfluid-phase data below T_λ while the bottom panel shows data from T_λ (=1.928 K) up to 3.94 K [Source: Talbot, Glyde, Stirling and Svensson, 1988].

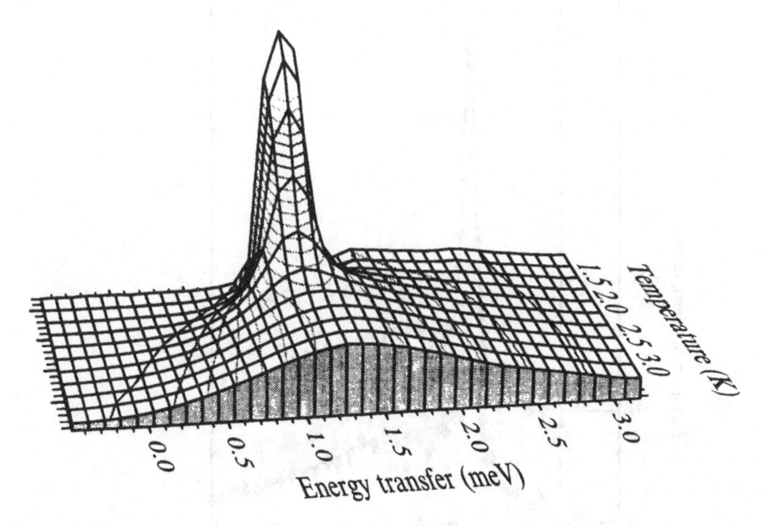

Fig. 2.5. Smoothed scattering intensity given as a contour map of the energy transfer and temperature, for $Q = 1.3$ Å^{-1} and SVP. The maxon peak is on a reduced scale [Source: Andersen, Stirling *et al.*, 1991].

different, as first observed by Woods and Svensson (1978). Above T_λ, $S(\mathbf{Q}, \omega)$ exhibits a broad distribution with a width which increases with wavevector Q but whose general shape is fairly temperature-independent for $T_\lambda \lesssim T \lesssim 3$ K. As we go below T_λ, however, a sharper component seems to appear, sitting on this broad background (Fig. 2.3). As the temperature decreases, the width of this peak rapidly decreases while its weight increases. The data of Woods and Svensson (1978) gave the first evidence that the roton–maxon quasiparticle has weight in $S(\mathbf{Q}, \omega)$ only *below* T_λ, in contrast to the phonon branch. Recent high-resolution data confirm these observations (Talbot, Glyde, Stirling and Svensson, 1988; Stirling and Glyde, 1990). The high-pressure data in Fig. 2.4 probably gives the most striking evidence.

In Figs. 2.5 and 2.6, we show some recent time-of-flight high-resolution ILL data giving the scattering intensity as a contour map of the frequency and temperature at two values of Q. Such plots summarize the huge amount of detailed information which is now available. They provide a real challenge to our understanding of superfluid ^{4}He and have been a major stimulus for the present monograph.

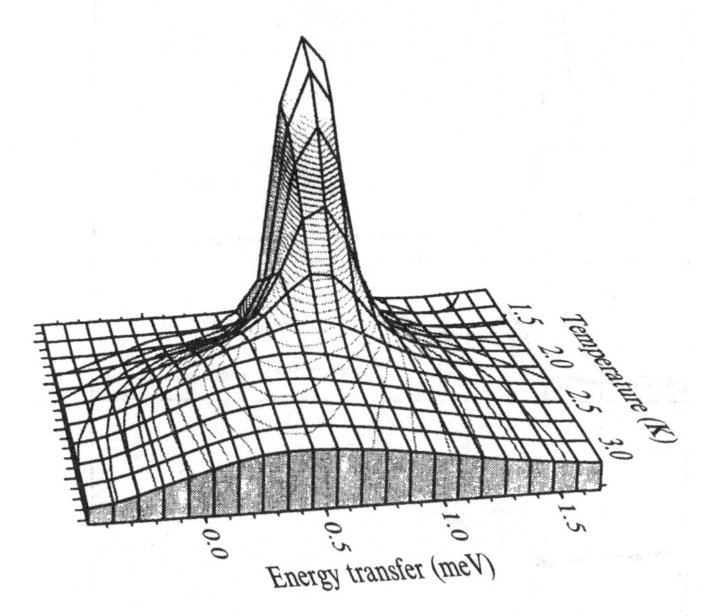

Fig. 2.6. Smoothed scattering intensity plotted as in Fig. 2.5, for Q = 2.0 Å^{-1} and SVP [Source: Andersen, Stirling *et al.*, 1991].

2.3 High momentum transfer and the impulse approximation

For wavevector $Q \gtrsim 2.5$ Å^{-1}, the quasiparticle dispersion relation seems to flatten out abruptly as shown in Fig. 1.3, saturating at about twice the roton energy 2Δ for large Q. There is also a rapid loss in intensity (Fig. 2.1). As can be seen from Figs. 2.7 and 2.8, for Q larger than 2.5 Å^{-1}, the main contribution to the scattering intensity is identified with the multiparticle background S_{II}. In the high-momentum region $Q \gtrsim 3.5$ Å^{-1}, $S(\mathbf{Q}, \omega)$ can be increasingly well described by an expression similar to that of a non-interacting gas of Bosons given by (2.32), with the Doppler-broadened peak centred at free-atom energy $Q^2/2m$ but with a width determined by the momentum distribution $n(\mathbf{p})$ of the ^{4}He atoms. To the extent that this "impulse approximation" for $S(\mathbf{Q}, \omega)$ holds, the momentum distribution $n(\mathbf{p})$ of the ^{4}He atoms can be extracted (see Fig. 2.9) at various temperatures (Sears, Svensson, Martel and Woods, 1982). In turn, this momentum distribution can be used to obtain information about the fraction of atoms $\bar{n}_0$ in the zero-momentum state (the Bose condensate).

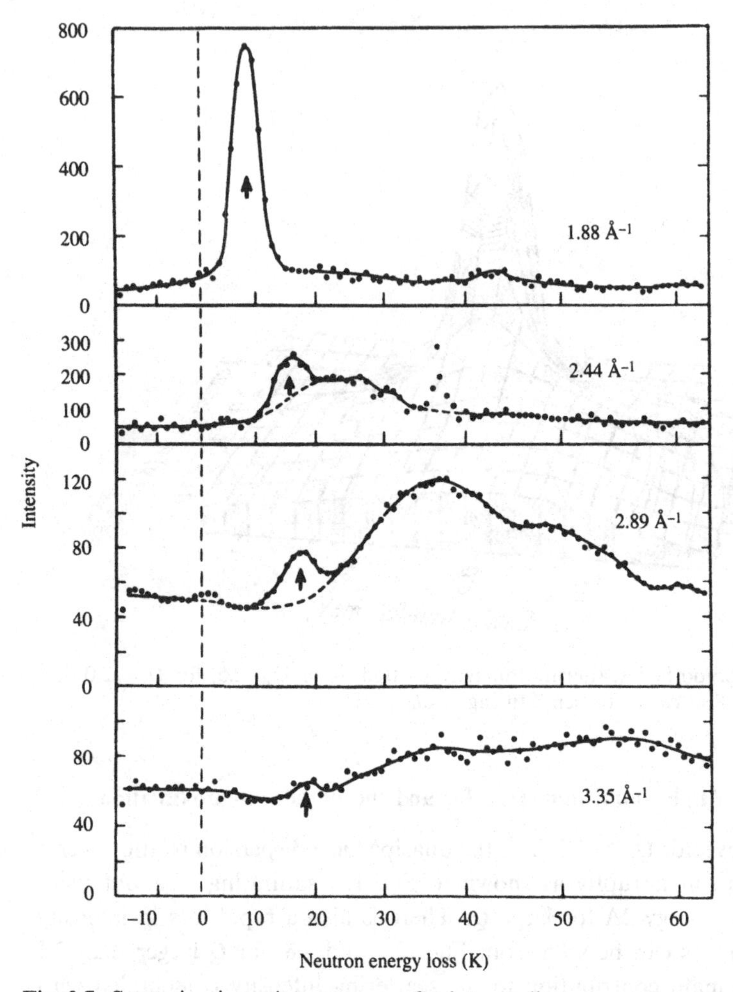

Fig. 2.7. Scattering intensity vs. energy for intermediate momentum transfers, at 1.6 K and SVP. For more recent high-resolution data, see Figs. 7.19 and 7.20 [Source: Woods, 1965a].

It is convenient to introduce here the basic physics which leads to the impulse approximation (IA). The derivation is done in terms of how a specific ^{4}He atom moves (recoils) over small times (see pp. 809ff of Mahan, 1990). An alternative (but physically equivalent) derivation in frequency space is given in Section 4.2. Scattering at high momentum transfers requires a high-momentum neutron. Thus the incident neutron wavelength must be short, even relative to the interatomic spacing. As

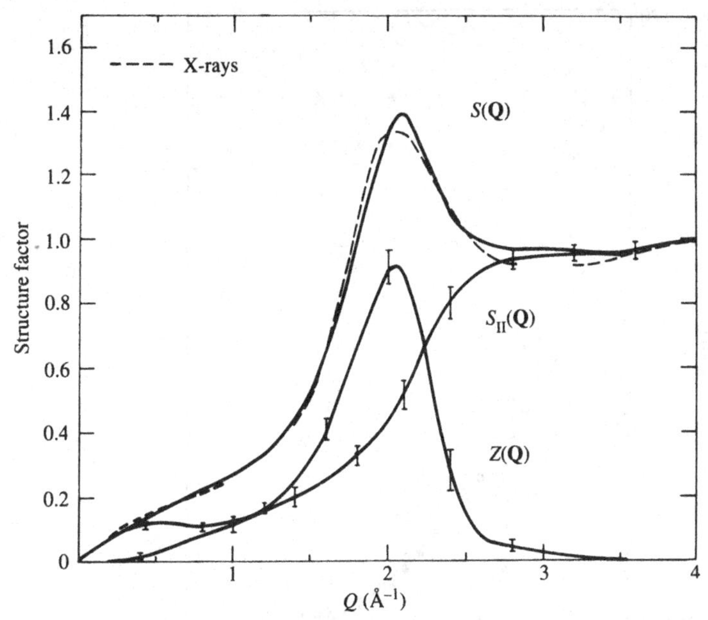

Fig. 2.8. The wavevector dependence of the quasiparticle weight $Z(\mathbf{Q})$ and the full static structure $S(\mathbf{Q})$, at 1.1 K and SVP [Source: Cowley and Woods, 1971].

a result, the neutron can interact with only a single Helium nucleus and consequently we expect $S(\mathbf{Q}, \omega)$ in the scattering intensity (2.2) to be well approximated by $S_{\mathrm{inc}}(\mathbf{Q}, \omega)$ given in (2.8). As we noted after (2.12), this high-Q region is reached for $Q \gtrsim 8$ Å^{-1} in liquid ^{4}He.

At large Q, the scattering time τ_s is very short and the scattered atom can travel only a short distance within this time. This suggests an approximation in which the potential energy of a struck atom does not change appreciably during this scattering time. It is convenient to recast the incoherent intermediate scattering function in terms of a time-ordered correlation function (Rahman, Singwi and Sjölander, 1962)

$$\begin{aligned} S_{\mathrm{inc}}(\mathbf{Q}, t) &= \left\langle e^{-i\mathbf{Q}\cdot\mathbf{r}_j(t)} e^{i\mathbf{Q}\cdot\mathbf{r}_j(0)} \right\rangle \\ &= e^{-i\omega_R t} \left\langle \hat{T} e^{-i\int_0^t dt' \mathbf{p}_j(t')\cdot\mathbf{Q}/m} \right\rangle , \end{aligned} \tag{2.37}$$

where $\omega_R \equiv Q^2/2m$ is the kinetic energy of the recoiling ^{4}He atom. If the important values of t are small, we can approximate $\mathbf{p}_j(t')$ in the exponent by its initial value $\mathbf{p}_j(0) = \mathbf{p}$, in which case (2.37) immediately

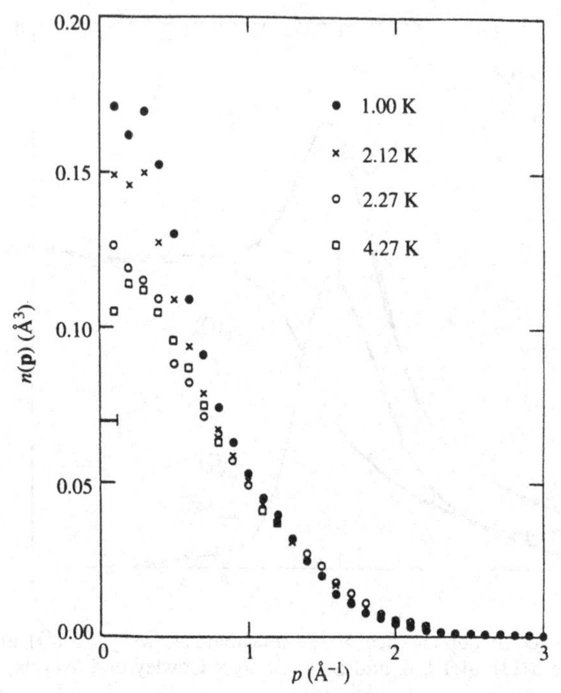

Fig. 2.9. The atomic momentum distribution in the normal and superfluid phases of liquid ^{4}He. These results are obtained from an analysis of high-momentum neutron scattering using a method discussed in Section 4.3 [Source: Sears, Svensson, Martel and Woods, 1982].

reduces to

$$S_{\rm inc}(\mathbf{Q},t) \simeq S_{\rm IA}(\mathbf{Q},t) = e^{-i\omega_R t}\left\langle e^{-i\frac{\mathbf{p}\cdot\mathbf{Q}}{m}t}\right\rangle . \tag{2.38}$$

Here the average is over the thermal equilibrium momentum distribution $n(\mathbf{p})$ of the ^{4}He atoms and thus the Fourier transform of (2.38) leads directly to

$$\begin{aligned} S_{\rm IA}(\mathbf{Q},\omega) &= \int d\mathbf{p}\, n(\mathbf{p})\delta\left(\omega - \omega_R - \frac{\mathbf{p}\cdot\mathbf{Q}}{m}\right) \\ &\equiv \left\langle \delta\left(\omega - \omega_R - \frac{\mathbf{p}\cdot\mathbf{Q}}{m}\right)\right\rangle , \end{aligned} \tag{2.39}$$

where $n(\mathbf{p})$ is normalized to unity. This result is the impulse or independent-particle approximation (IA). In this limit, $S(\mathbf{Q},\omega)$ is seen to depend only on the equilibrium atomic momentum distribution.

Further insight into the physics of the IA result (2.39) as a small-time approximation can be obtained by expanding (2.38) and keeping only terms of order t^2. This gives

$$\left\langle e^{-i\frac{\mathbf{p}\cdot\mathbf{Q}}{m}t}\right\rangle \simeq e^{-\frac{1}{2}\langle\left(\frac{\mathbf{p}\cdot\mathbf{Q}}{m}\right)^2\rangle t^2}$$

$$\equiv e^{-t^2/\tau_s^2} \ . \tag{2.40}$$

In the last line, we have introduced an explicit definition of the scattering time τ_s as the relaxation time of $S_{\mathrm{IA}}(\mathbf{Q},t)$. We see that $\tau_s \propto 1/Q$ and thus the IA should become a better and better approximation as Q increases. Morever we note that if (2.40) is valid, $S_{\mathrm{IA}}(\mathbf{Q},\omega)$ in (2.39) is a Gaussian centred at ω_R,

$$S_{\mathrm{IA}}(\mathbf{Q},\omega) \simeq \frac{1}{\sqrt{2\pi\sigma^2}}e^{-(\omega-\omega_R)^2/2\sigma^2} \ , \tag{2.41}$$

where $\sigma^2 \equiv 2/\tau_s^2$. Parenthetically, we note that if the atomic momentum distribution $n(\mathbf{p})$ were Gaussian, (2.40) and (2.41) would be exact. This can be proven by expanding $\ln\langle e^x\rangle$ and noting that all cumulants higher than the second vanish for a Gaussian average. This is the reason why $S(\mathbf{Q},\omega)$ is often taken to be the Gaussian form (2.41) when analysing high-momentum neutron-scattering data (Sokol, 1987).

One can show that $S_{\mathrm{IA}}(\mathbf{Q},\omega)$ in (2.39) does not depend on Q and ω separately but only on the combination (West, 1975; Gersch and Rodriguez, 1973; Sears, 1985)

$$Y \equiv \left(\frac{m}{Q}\right)(\omega-\omega_R) \ . \tag{2.42}$$

This scaling variable Y is very useful in deep-inelastic scattering studies. One can easily carry out the angular integration in (2.39) to obtain

$$S_{\mathrm{IA}}(\mathbf{Q},\omega) = \frac{m}{Q}J_{\mathrm{IA}}(Y) \ , \tag{2.43}$$

where the "Compton profile" $J(Y)$ for the IA is given by

$$J_{\mathrm{IA}}(Y) = \frac{n_0}{n}\delta(Y) + \int_{|Y|}^{\infty} dp \ pn(p) \ . \tag{2.44}$$

(We have explicitly allowed for a condensate at $\mathbf{p} = 0$ in $n(\mathbf{p})$.) Such Compton profiles are commonly used in the study of electronic systems as well as in nuclear and particle physics (for further references, see Silver and Sokol, 1989). The fact that the IA Compton profile depends only on Y ("Y-scaling") can be traced back to the fact that the peak position in $S_{\mathrm{IA}}(\mathbf{Q},\omega)$ is proportional to Q^2 while its width goes as Q. The Compton

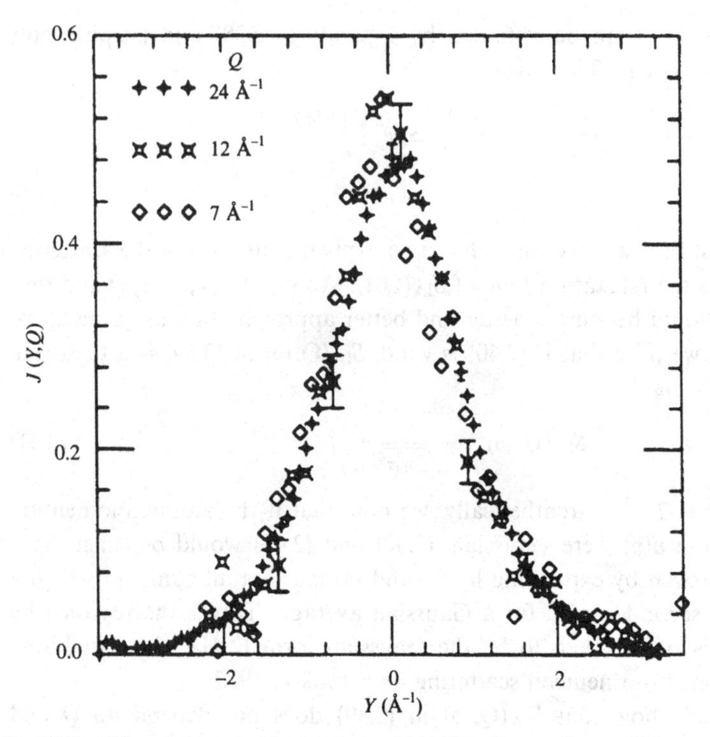

Fig. 2.10. The measured Compton profile $J(Y,Q)$ vs. the scaling variable Y defined in (2.42), for momentum transfers $Q = 7$ and 12 Å^{-1} (at $T = 1.0$ K) and $Q = 24$ Å^{-1} (at $T = 0.32$ K). The results illustrate Y-scaling behaviour [Source: Sosnick, Snow, Silver and Sokol, 1991].

profile of liquid ^{4}He does exhibit Y-scaling fairly well (see Fig. 2.10), which is usually viewed as evidence that the IA is valid. However, we remember that Y-scaling will also occur as long as corrections to the IA are only functions of Y.

It is clear that one needs to understand how large Q must be for the IA to be a sufficiently valid approximation if we use (2.39) to "extract" accurate results for $n(\mathbf{p})$. As a qualitative way of discussing corrections to the IA, one might expect that more generally, (2.39) will be replaced by an expression like

$$S(\mathbf{Q},\omega) = \int d\mathbf{p}\, n(\mathbf{p}) \frac{\Gamma(\mathbf{p};\mathbf{Q},\omega)/\pi}{[\omega - \omega_R - \frac{\mathbf{p}\cdot\mathbf{Q}}{m} - \Delta(\mathbf{p};\mathbf{Q},\omega)]^2 + \Gamma^2(\mathbf{p};\mathbf{Q},\omega)} . \tag{2.45}$$

This includes the collisional broadening described by Γ as well as the *change* in the potential energy Δ of the struck atom between the final and initial states. The presence of Δ in (2.45) emphasizes that the distinction between "initial" and "final" state corrections is not fundamental and we refer to corrections to the IA generically as "Final-State" (FS) effects. In the pioneering study by Hohenberg and Platzman (1966) of the IA in superfluid ^{4}He, Δ was set to zero and the broadening approximated by

$$\Gamma(Q) = n\left(\frac{\hbar Q}{2m}\right)\sigma(Q) \ , \tag{2.46}$$

where $\sigma(Q)$ is the atomic cross-section for two ^{4}He atoms. Martel, Svensson, Woods, Sears and Cowley (1976) later used this approximation to analyse neutron-scattering data at *intermediate* momentum transfers ($Q \sim 5$ Å^{-1}). However, in general, Γ (and Δ) can be expected to be a function of ω also and thus the integrand of (2.45) may deviate significantly from a simple Lorentzian.

The difference of $S_{\mathrm{inc}}(\mathbf{Q}, t)$ in (2.37) from the IA in (2.38) due to FS contributions may be conveniently isolated by introducing a new function R as follows:

$$S_{\mathrm{inc}}(\mathbf{Q}, t) = S_{\mathrm{IA}}(\mathbf{Q}, t)R_{\mathrm{FS}}(\mathbf{Q}, t) \ . \tag{2.47a}$$

In this form, $S_{\mathrm{inc}}(\mathbf{Q}, \omega)$ is given as a convolution over the final-state resolution function $R_{\mathrm{FS}}(\mathbf{Q}, \omega)$,

$$S_{\mathrm{inc}}(\mathbf{Q}, \omega) = \int_{-\infty}^{\infty} d\omega' S_{\mathrm{IA}}(\mathbf{Q}, \omega')R_{\mathrm{FS}}(\mathbf{Q}, \omega - \omega') \ . \tag{2.47b}$$

As an illustration, the Lorentzian form

$$R_{\mathrm{FS}}(\mathbf{Q}, \omega) = \frac{1}{\pi}\frac{\Gamma(Q)}{(\omega - \Delta)^2 + \Gamma^2(Q)} \tag{2.48}$$

gives rise to an integrand of $S_{\mathrm{inc}}(\mathbf{Q}, \omega)$ of the kind assumed in (2.45). The fact that the impulse approximation $S_{\mathrm{IA}}(\mathbf{Q}, \omega)$ already satisfies the first three central moment sum rules ($n = 0, 1, 2$) for $M_n(Q)$ as defined at the end of Section 2.1 means that $R_{\mathrm{FS}}(\mathbf{Q}, \omega)$ must be negative in its high-frequency wings. In particular, this means a simple Lorentzian expression like (2.48) is inconsistent with the $n = 2$ central moment sum rule.

A detailed theory of $R_{\mathrm{FS}}(\mathbf{Q}, \omega)$ was given first by Gersch and Rodriguez (1973), and more recently by Silver (1988, 1989), Rinat (1989), and Carraro and Koonin (1990). Simple approximations to (2.45) based on (2.46) leave out the important short-range spatial correlations present in a liquid (as described by the pair distribution function $g(\mathbf{r})$ defined in

(2.10)). These strongly modify how a recoiling atom moves over atomic distances after it is hit by a high-energy neutron.

FS contributions are especially important if $R_{FS}(\mathbf{Q},\omega)$ is broad compared to the width of any low-frequency peak in $S_{IA}(\mathbf{Q},\omega)$. Specifically, if $S_{IA}(\mathbf{Q},\omega)$ contains a sharp component $n_0\delta(\omega-\omega_R)$ due to a Bose condensate, this component will be broadened to a width given by that of $R_{FS}(\mathbf{Q},\omega)$. This broadening spreads the intensity due to the condensate component into the regions of ω which overlap with the Doppler-broadened contribution from non-condensate atoms. This is a major source of difficulty in extracting information about the condensate fraction. It is clear that determination of $n(\mathbf{p})$ in superfluid ^{4}He using high-momentum neutron-scattering data requires a careful removal of FS effects (see Chapter 4). By way of contrast, the momentum distribution is found to be broad and nearly Gaussian in *normal* liquid ^{4}He, solid Helium, and most classical liquids. The influence of FS corrections to the IA at high Q is relatively less important in these cases.

3
Bose broken symmetry and its implications

In this chapter, we begin our analysis of the dynamical correlation functions in a Bose fluid. Field-theoretic techniques and Green's functions are the most powerful ways of understanding the effect of a Bose broken symmetry. In Section 3.1, we introduce the order parameter associated with this broken symmetry and show how it couples the single-particle excitations and the density fluctuations. In Section 3.2, we review the formal structure of the single-particle Green's functions $G_{\alpha\beta}(\mathbf{Q},\omega)$ and then illustrate this with the well known Bogoliubov approximation. This model approximation really only describes a weakly interacting dilute Bose gas (WIDBG) at low temperatures but it already exhibits characteristic features of superfluid ^{4}He. In Section 3.3, we evaluate the density-response function $\chi_{nn}(\mathbf{Q},\omega)$ in a Bose-condensed fluid within the simple Bogoliubov approximation in order to illustrate these features. Finally, in Section 3.4, we use a simple mean-field approach to illustrate how $G_{\alpha\beta}$ and χ_{nn} share the same poles when there is a Bose condensate. This sets the stage for the more systematic field-theoretic analysis given in Chapter 5.

For orientation, we first summarize the properties of a non-interacting Bose gas (see, for example, pp. 38ff of Fetter and Walecka, 1971). The number of atoms in a free Bose gas with energy ε_k is given by

$$\langle \hat{n}_k \rangle_0 = \frac{1}{e^{\beta(\varepsilon_k - \mu)} - 1} \equiv N(\varepsilon_k - \mu) \tag{3.1}$$

where $\langle \ldots \rangle_0$ is an average in the grand canonical ensemble. The chemical potential μ is defined by the condition

$$N = \sum_{\mathbf{k}} N(\varepsilon_k - \mu) \ . \tag{3.2}$$

One can show that μ must be ≤ 0. If μ is fixed and the temperature decreases, $\langle \hat{n}_k \rangle_0$ decreases and hence the total number of atoms decreases. Indeed, for any finite μ, the value of N must decrease to zero as $T \to 0$. Since N is fixed, as the temperature drops, μ must become zero at some finite temperature T_{BE} given by

$$k_B T_{BE} = 3.31 \frac{\hbar^2 n^{2/3}}{m} , \tag{3.3}$$

where n is the density of atoms. The scenario is therefore as follows: to keep N fixed as the temperature decreases, the chemical potential approaches zero until, at T_{BE}, it reaches zero. For temperatures below T_{BE}, Einstein (1925) first pointed out that extra atoms can go into the $\mathbf{k} = 0$ state, which then becomes macroscopically occupied. We have

$$\langle \hat{a}_0^+ \hat{a}_0 \rangle_0 = N_0 , \tag{3.4}$$

such that $n_0 = N_0/\Omega$ remains finite in the thermodynamic limit (the sample volume $\Omega \to \infty$). Separating the atoms with $\mathbf{k}=0$ and $\mathbf{k} \neq 0$ explicitly, we find that (3.2) takes the form (note that $\mu = 0$ below T_λ)

$$\left. \begin{aligned} N &= N_0 + \tilde{N} , \\ \tilde{N} &= \sum_{\mathbf{k}\neq 0} N(\varepsilon_k) = N \left(\frac{T}{T_{BE}} \right)^{3/2} . \end{aligned} \right\} \tag{3.5}$$

One refers to this macroscopic occupation of the $\mathbf{k} = 0$ state as Bose–Einstein condensation or more simply, Bose condensation. For $T < T_{BE}$, the number of “excited” atoms $\tilde{N}$ decreases as $T^{3/2}$ until at $T = 0$, Bose condensation is complete and there are no excited atoms ($N = N_0, \tilde{N} = 0$).

The prediction of Bose condensation by Einstein (1925) was ignored until London (1938a,b) suggested that this kind of phenomenon might be involved in the then recently discovered superfluid phase of liquid ^{4}He. London noted that using the density of liquid ^{4}He, T_{BE} as given by (3.3) is 3.1 K, very close to the observed lambda transition at 2.17 K (SVP). Since liquid ^{4}He is a strongly interacting system with all the ensuing complications, London’s suggestion that a Bose condensate is involved in the superfluid transition was hard to prove or disprove theoretically (see, however, Feynman, 1953a). It was controversial for many years and, in our view, was only finally settled by the finite-temperature Feynman path-integral Monte Carlo calculations of Ceperley and Pollock (1986), summarized in Section 1.2.

3.1 Bose broken symmetry in a liquid

A formal definition of Bose condensation in an interacting Bose fluid was first provided by Penrose and Onsager (1956), developing earlier work by Penrose (1951). They generalized the criterion for Bose condensation in a gas used earlier by Bogoliubov (1947) to the condition that the one-particle reduced density matrix in a liquid (or equivalently, the equal-time single-particle Green's function) $\rho_1(\mathbf{r},\mathbf{r}') \equiv \rho_1(\mathbf{r}-\mathbf{r}') \equiv \langle\hat{\psi}^+(\mathbf{r})\hat{\psi}(\mathbf{r}')\rangle$ does *not* vanish at large separation $|\mathbf{r}-\mathbf{r}'|$. Here $\hat{\psi}(\mathbf{r})$ and $\hat{\psi}^+(\mathbf{r})$ are the field operators which destroy and create, respectively, ^{4}He atoms at position $\mathbf{r}$. In modern terminology, the average $\langle\ \rangle$ involves a broken symmetry (or restricted ensemble) such that

$$\left.\begin{aligned} \langle\hat{\psi}(\mathbf{r})\rangle &\equiv \Phi(\mathbf{r}) \neq 0\ , \\ \langle\hat{\psi}^+(\mathbf{r})\rangle &\equiv \Phi^*(\mathbf{r}) \neq 0 \end{aligned}\right\} \tag{3.6}$$

below T_λ. Thus this generalized criterion for Bose condensation can be stated in the form

$$\lim_{|\mathbf{r}-\mathbf{r}'|\to\infty} \rho_1(\mathbf{r},\mathbf{r}') = \Phi^*(\mathbf{r})\Phi(\mathbf{r}') \neq 0 \tag{3.7}$$

and $\Phi(\mathbf{r})$ can be interpreted as the "condensate wavefunction" (having amplitude and phase). In the presence of superfluid flow, the superfluid velocity $\mathbf{v}_S$ appears as the gradient of the phase of the condensate wavefunction $\Phi(\mathbf{r})$. For a useful discussion of this condensate wavefunction, we refer to §26 of Lifshitz and Pitaevskii (1980). If the condensate $\Phi(\mathbf{r})$ is spatially uniform, it involves only atoms in the zero-momentum state. This can be seen from a Fourier decomposition, namely

$$\begin{aligned} \langle\hat{\psi}(\mathbf{r})\rangle &= \frac{1}{\sqrt{\Omega}}\sum_{\mathbf{Q}} e^{i\mathbf{Q}\cdot\mathbf{r}}\langle\hat{a}_Q\rangle \\ &= \frac{1}{\sqrt{\Omega}}\langle\hat{a}_0\rangle = \sqrt{n_0}e^{i\phi} \neq 0\ . \end{aligned} \tag{3.8}$$

This follows because $\langle\hat{a}_Q\rangle = 0$ for $Q \neq 0$ and $\langle\hat{a}_0\rangle = \sqrt{N_0}$. With no loss of generality, we can set the phase ϕ of the uniform condensate to zero. The Bose broken symmetry is physically equivalent to "off-diagonal long-range order" (ODLRO) as formulated by Yang (1962). However the usual discussions of ODLRO do not give much insight into the *dynamical* implications of a Bose broken symmetry, which is our major interest in this book.

Penrose and Onsager (1956) had a tremendous influence on further work because they also gave the first numerical estimate of the condensate density $n_0 = |\Phi(\mathbf{r})|^2$, using a crude ground-state variational wavefunction for hard-sphere Bosons originally introduced by Feynman (1953a). For further details, we refer to pp. 313ff of Huang (1987). They concluded that approximately 8% of the ^{4}He atoms are in the zero-momentum state at $T = 0$. However, Penrose and Onsager did not spell out the *precise* relation between the existence of the Bose condensate *and* superfluidity. It was later shown by Bogoliubov (1963, 1970) as well as Hohenberg and Martin (1965) that the Bose broken symmetry described by (3.7) does indeed lead to the two-fluid equations and superfluidity (see Chapter 6 for further discussion). To what extent superfluidity implies the existence of a Bose condensate is, logically, a separate question; but this is clearly of much less interest once a Bose condensate is known to exist in a given Bose fluid.

The condition $\langle\hat{\psi}(\mathbf{r})\rangle \neq 0$ describes a breaking of the gauge symmetry associated with the conservation of particles and is analogous to the broken-symmetry condition $\langle\hat{S}_z\rangle \equiv m \neq 0$ in a ferromagnet. The main difference between (3.6) and the broken-symmetry condition in a ferromagnet is that a state of net magnetization is more easily visualized than a macroscopic wavefunction describing a state having a specific phase but not a fixed number of atoms. From a physical point of view, one can understand (3.6) by noting that the physical average of the phase of the field operator $\hat{\psi}(\mathbf{r})$ is still undefined: Bose condensation corresponds more precisely to enforcing a well defined relation between the phase of $\hat{\psi}$ at $\mathbf{r}$ and that at $\mathbf{r}'$, as given in (3.7). Without the "gauge fixing" or clamping implied by (3.6), the average over all possible phases would result in $\langle\Phi(\mathbf{r})\rangle_{\text{phase}} = 0$ and no Bose condensate. For the expectation value $\langle\hat{\psi}(\mathbf{r})\rangle$ to be non-zero, one must add to the Hamiltonian infinitesimal terms which act as sources and sinks for atoms in the condensate. As long as these terms do not alter the continuity equation, they can be used to fix the phase and hence justify the use of thermal averages over a "restricted" ensemble (Bogoliubov, 1963, 1970; Hohenberg and Martin, 1965). Thus the non-zero expectation values in (3.6) imply a broken gauge symmetry corresponding to the non-conservation of the number of particles.

As we shall discuss in Section 6.3 in more detail, the broken symmetry introduced by a finite value of the order parameter is restored by the appearance of long-wavelength fluctuations of the condensate. This intimate relation between the static and dynamic aspects of the order parameter is in the substance of the Goldstone theorem. The appropriate symmetry operator is the gauge transformation $\exp(i\phi\hat{N})$. For further insight into Bose broken symmetry, we refer the reader to the classic article by Anderson (1966), as well as Chapter 10 of the monograph by Forster (1975).

The introduction of an explicit symmetry-breaking term in the Hamiltonian gives one a "hunting license" to look for a new thermodynamic phase (Bogoliubov, 1963, 1970) within a scheme where we can use the usual many-body techniques of finite-temperature perturbation theory. Landmark papers on the quantum field theory of Bose-condensed fluids include those by Beliaev (1958a,b), Hugenholtz and Pines (1959), Bogoliubov (1963, 1970), Gavoret and Nozières (1964), Hohenberg and Martin (1965) and Ma and Woo (1967). All of these papers treat $T = 0$, but the formal extension to finite temperature is straightforward using the technique of imaginary-time Green's functions (Mahan, 1990).

A fundamental implication of the Bose broken symmetry (3.6) is that the single-particle spectrum appears directly in the density fluctuation spectrum, as we shall now show. In second-quantized form, the number density operator defined in (2.6) is given by

$$\hat{\rho}(\mathbf{r}) = \hat{\psi}^{+}(\mathbf{r})\hat{\psi}(\mathbf{r}) \ . \tag{3.9}$$

In a Bose-condensed system, it is useful to decompose the field operators as follows (Beliaev, 1958a):

$$\left.\begin{aligned} \hat{\psi}(\mathbf{r}) &\equiv \langle\hat{\psi}(\mathbf{r})\rangle + \tilde{\psi}(\mathbf{r}) \ , \\ \hat{\psi}^{+}(\mathbf{r}) &\equiv \langle\hat{\psi}^{+}(\mathbf{r})\rangle + \tilde{\psi}^{+}(\mathbf{r}) \ , \end{aligned}\right\} \tag{3.10}$$

where the $\tilde{\psi}$, $\tilde{\psi}^{+}$ operators only involve atoms *outside* the condensate. Using (3.10) in (3.9), we obtain

$$\begin{aligned} \hat{\rho}(\mathbf{r}) &= |\Phi(\mathbf{r})|^2 + \Phi^{*}(\mathbf{r})\tilde{\psi}(\mathbf{r}) + \Phi(\mathbf{r})\tilde{\psi}^{+}(\mathbf{r}) + \tilde{\psi}^{+}(\mathbf{r})\tilde{\psi}(\mathbf{r}) \\ &= n_0 + \sqrt{n_0}\,[\tilde{\psi}(\mathbf{r}) + \tilde{\psi}^{+}(\mathbf{r})] + \tilde{\psi}^{+}(\mathbf{r})\tilde{\psi}(\mathbf{r}) \ . \end{aligned} \tag{3.11}$$

Clearly the non-zero value of the condensate couples the single-particle operators directly to the density operator. In momentum space, (3.11) is equivalent to ($\mathbf{Q} \neq 0$)

$$\begin{aligned} \hat{\rho}(\mathbf{Q}) &= \sum_{\mathbf{p}} \hat{a}^{+}_{\mathbf{p}}\hat{a}_{\mathbf{p}+\mathbf{Q}} \\ &= \sqrt{N_0}(\hat{a}_{\mathbf{Q}} + \hat{a}^{+}_{-\mathbf{Q}}) + {\sum_{\mathbf{p}}}' \hat{a}^{+}_{\mathbf{p}}\hat{a}_{\mathbf{p}+\mathbf{Q}} \ , \end{aligned} \tag{3.12}$$

where the prime means that the second term (the "normal" density fluctuation operator) involves only atoms outside the condensate. The first term in (3.12) describes density fluctuations involving atoms scattering into and out of the condensate. This is summarized by rewriting (3.12) in the form ($\mathbf{Q} \neq 0$)

$$\hat{\rho}(\mathbf{Q}) = \sqrt{N_0}\hat{A}_{\mathbf{Q}} + \tilde{\rho}_{\mathbf{Q}} \,, \tag{3.13}$$

where

$$\hat{A}_{\mathbf{Q}} \equiv \hat{a}_{\mathbf{Q}} + \hat{a}^{+}_{-\mathbf{Q}} \tag{3.14}$$

describes single-particle excitations. One may think of the separation in (3.13) in analogy with (3.5).

Consider a system of interacting Bosons (in a volume Ω satisfying the boundary conditions) with a uniform Bose condensate described by (3.8), with $\langle \hat{a}_0 \rangle = \sqrt{N_0}$. The second-quantized Hamiltonian is

$$\hat{H} = \sum_{\mathbf{k}} (\varepsilon_k - \mu)\hat{a}^{+}_{k}\hat{a}_{k} + \frac{1}{2\Omega}\sum_{\mathbf{Q}} V(Q)\hat{\rho}(\mathbf{Q})\hat{\rho}(-\mathbf{Q}) \,, \tag{3.15}$$

where $\hat{\rho}(\mathbf{Q})$ is defined in (3.12). As usual, it is convenient to include the chemical potential μ in our effective Hamiltonian as in (3.15) since we work in the grand canonical ensemble. We have not explicitly included the symmetry-breaking perturbation in (3.15). Inserting (3.13) into (3.15), we obtain

$$\begin{aligned}\hat{H} \;=\; & \sum_{\mathbf{k}}{}' (\varepsilon_k - \mu)\hat{a}^{+}_{k}\hat{a}_{k} + \frac{N_0^2}{2\Omega}V(q=0) - \mu N_0 + \frac{N_0}{2\Omega}\sum_{\mathbf{Q}}{}' V(Q)\hat{A}_Q\hat{A}_{-Q} \\ & + \frac{\sqrt{N_0}}{2\Omega}\sum_{\mathbf{Q}}{}' V(Q)(\hat{A}_Q\tilde{\rho}_{-Q} + \tilde{\rho}_Q\hat{A}_{-Q}) + \frac{1}{2\Omega}\sum_{\mathbf{Q}}{}' V(Q)\tilde{\rho}_Q\tilde{\rho}_{-Q},\end{aligned} \tag{3.16}$$

where the prime on the summations again means that atoms with zero momentum are excluded. The first term in the second line of (3.16) shows how the presence of a condensate leads to coupling of the single-particle ($\hat{A}_Q$) and the "normal" density ($\tilde{\rho}_Q$) fluctuations. This coupling will play a crucial role in the subsequent analysis in this and succeeding chapters.

Turning to the dynamic structure factor (see Section 2.1)

$$S(\mathbf{Q},\omega) = (2\pi N)^{-1}\int_{-\infty}^{\infty} dt\, e^{i\omega t}\langle \hat{\rho}(\mathbf{Q},t)\hat{\rho}(-\mathbf{Q})\rangle \,, \tag{3.17}$$

we find (3.13) immediately leads to (Hugenholtz and Pines, 1959)

$$\begin{aligned} S(\mathbf{Q},\omega) &= \frac{1}{2\pi N}\int_{-\infty}^{\infty} dt\, e^{i\omega t}\big[N_0\langle \hat{A}_Q(t)\hat{A}_{-Q}\rangle \\ &\quad + \sqrt{N_0}\{\langle \tilde{\rho}_Q(t)\hat{A}_{-Q}\rangle + \langle \hat{A}_Q(t)\tilde{\rho}_{-Q}\rangle\} + \langle \tilde{\rho}_Q(t)\tilde{\rho}_{-Q}\rangle\big] \\ &\equiv S_1(\mathbf{Q},\omega) + S_{\rm int}(\mathbf{Q},\omega) + \tilde{S}(\mathbf{Q},\omega) \ . \end{aligned} \tag{3.18}$$

S_1 describes the density fluctuations associated with scattering atoms into or out of the Bose condensate and shows the direct role the single-particle excitations play in $S(\mathbf{Q},\omega)$. However, in the presence of a condensate, all terms in (3.18) are coupled and thus exhibit the same poles (as we illustrate in Section 3.4). This motivates the next section, where we introduce the matrix single-particle Green's function $G_{\alpha\beta}(\mathbf{Q},\omega)$.

We note in passing that in the Bardeen–Cooper–Schrieffer theory, superconductivity is associated with the finite value of anomalous correlation functions of the kind $\langle\hat{\psi}^+_\uparrow(\mathbf{r})\hat{\psi}^+_\downarrow(\mathbf{r})\rangle$. This order parameter describes Cooper pairs which are Bose-condensed in the sense that all electron pairs have the same two-particle bound-state wavefunction (see, for example, Schrieffer, 1964). Because both theories invoke a broken gauge symmetry involving non-conservation of particles, there are many similarities between the descriptions of BCS superconductors and Bose-condensed fluids. Indeed, it was probably the BCS theory in 1957 that stimulated a wider appreciation of the original work of Bogoliubov (1947). Among many articles stressing the useful analogies between BCS superconductors and Bose-condensed fluids, we call attention to those of Anderson (1966), Nozières (1966) and Vinen (1969). However, in contrast to superfluid ^{4}He, the single-particle Fermi excitations (quasiparticles) are quite distinct from the collective modes in superfluid ^{3}He. This distinction between Bose and Fermi superfluids is an important one to remember. A full account of the modern theory of superfluid ^{3}He, with emphasis on the broken symmetries involved, is given by Vollhardt and Wölfle (1990).

3.2 Single-particle Green's functions for a Bose-condensed fluid

In this section we discuss the general structure of the single-particle Green's functions taking into account the Bose broken symmetry (3.6), as first worked out by Beliaev (1958a). In addition, we analyse the simplest non-trivial model calculation, for a WIDBG at $T = 0$, due to Bogoliubov (1947). This is discussed in all the standard texts. Finally, we briefly review theories which go past the Bogoliubov approximation.

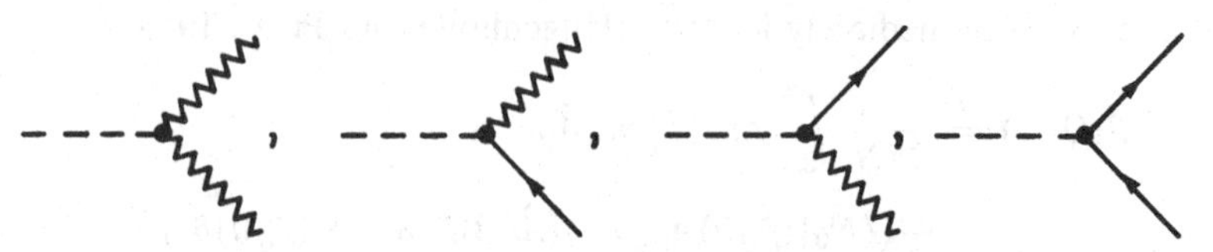

Fig. 3.1. The four different vertices involved in a Bose-condensed fluid. The jagged line represents a condensate atom and the dashed line is the two-particle interaction.

The single-particle spectrum of the non-condensed atoms is given by the poles of the single-particle Green's function, which for real times is given by (for $Q \neq 0$)

$$- iG(\mathbf{Q}, t - t') \equiv \langle T\hat{a}_Q(t)\hat{a}_Q^+(t')\rangle \tag{3.19}$$

where T [...] denotes Wick's time-ordering operator. Since the number of particles is not conserved by the symmetry-breaking term, we also need to allow for the anomalous Green's functions $\langle T\hat{a}^+_{-Q}(t)\hat{a}_Q^+(t')\rangle$ and $\langle T\hat{a}_Q(t)\hat{a}_{-Q}(t')\rangle$ in the Bose-condensed phase. A generalized 2×2 matrix single-particle Green's function can be defined as (see pp. 249ff of Rickayzen, 1980)

$$- iG_{\alpha\beta}(\mathbf{Q}, t) \equiv \begin{pmatrix} -iG_{11} & -iG_{21} \\ -iG_{12} & -iG_{22} \end{pmatrix} = \begin{pmatrix} \langle T\hat{a}_Q(t)\hat{a}_Q^+\rangle & \langle T\hat{a}^+_{-Q}(t)\hat{a}_Q^+\rangle \\ \langle T\hat{a}_Q(t)\hat{a}_{-Q}\rangle & \langle T\hat{a}^+_{-Q}(t)\hat{a}_{-Q}\rangle \end{pmatrix}, \tag{3.20}$$

It is common to introduce the notation (Gavoret and Nozières, 1964)

$$\hat{a}_{Q\alpha} = \begin{cases} \hat{a}_Q & \text{if } \alpha = 1 \text{ or } + , \\ \hat{a}^+_{-Q} & \text{if } \alpha = 2 \text{ or } - . \end{cases} \tag{3.21}$$

We can then write the matrix $G_{\alpha\beta}$ in (3.20) in the compact form $-iG_{\alpha\beta}(\mathbf{Q}, t) = \langle T\hat{a}_{Q\alpha}(t)\hat{a}^+_{Q\beta}\rangle$. In Fig. 3.1, we show the four interaction vertices which are involved in (3.16).

Switching to imaginary time $\tau = it$ $(0 < \tau \leq \beta = 1/k_BT)$, we define the imaginary-time matrix Green's function by (see Chapter 3 of Mahan, 1990)

$$G_{\alpha\beta}(\mathbf{Q}, \tau) = -[\langle T\hat{a}_{Q\alpha}(\tau)\hat{a}^+_{Q\beta}\rangle - \langle\hat{a}_{0\alpha}\rangle\langle\hat{a}^+_{0\beta}\rangle\delta_{Q,0}] , \tag{3.22}$$

where

$$\hat{a}_{Q\alpha}(\tau) = e^{\tau\hat{H}}\hat{a}_{Q\alpha}e^{-\tau\hat{H}} . \tag{3.23}$$

These finite-temperature Green's functions are periodic in τ with period

β and we can expand $G_{\alpha\beta}(Q,\tau)$ as a Fourier series

$$G_{\alpha\beta}(\mathbf{Q},\tau) = k_B T \sum_{\omega_n} e^{-i\omega_n\tau} G_{\alpha\beta}(\mathbf{Q}, i\omega_n) , \tag{3.24}$$

where ω_n is the Bose discrete Matsubara frequency $2\pi n k_B T$ (n is any integer) and the Matsubara Fourier coefficients are given by

$$G_{\alpha\beta}(\mathbf{Q}, i\omega_n) = \int_0^{1/k_B T} d\tau \, e^{i\omega_n\tau} G_{\alpha\beta}(\mathbf{Q},\tau) . \tag{3.25}$$

The Fourier transform of the more physical real-time Green's functions is obtained by the standard technique of analytic continuation of the imaginary Matsubara frequencies to the real frequency axis ($i\omega_n \to \omega + i\eta$). The terms subtracted in (3.22) are only important when dealing with the dynamics of the condensate atoms at $Q = 0$ (see the end of Section 5.1 and Section 6.3).

Finite-temperature many-body perturbation theory can be used to evaluate $G_{\alpha\beta}$. In particular, the equations of motion of $G_{\alpha\beta}$ can be conveniently written in the form of a Dyson equation involving a 2×2 matrix self-energy $\Sigma_{\alpha\beta}$,

$$G_{\alpha\beta} = G^0_{\alpha\beta} + G^0_{\alpha\mu}\Sigma_{\mu\nu}G_{\nu\beta} , \tag{3.26}$$

where the non-interacting Bose Green's function is

$$G^0_{\alpha\beta}(\mathbf{Q}, i\omega_n) = \frac{\delta_{\alpha\beta}}{\alpha i\omega_n - (\varepsilon_Q - \mu)} . \tag{3.27}$$

In (3.26) and elsewhere, repeated Greek *subscripts* (with values 1,2 or +, −) are *summed*. Using rotational invariance, time-reversal invariance, space-inversion symmetry and the fact $\hat{a}_{Q\alpha} = \hat{a}^+_{-Q,-\alpha}$, one finds

$$G_{\alpha\beta}(\mathbf{Q}, i\omega_n) = G_{\alpha\beta}(Q, i\omega_n) = G_{\beta\alpha}(Q, i\omega_n) = G_{-\alpha,-\beta}(Q, -i\omega_n) . \tag{3.28}$$

The self-energies satisfy the same symmetry relations. In particular, one has $\Sigma_{11}(Q, i\omega_n) = \Sigma_{22}(Q, -i\omega_n)$ and $\Sigma_{12} = \Sigma_{21}$. We note that there is no simple relation between $G_{\alpha\beta}(Q, i\omega_n)$ and $G_{\alpha\beta}(Q, -i\omega_n)$. Writing out the Dyson–Beliaev equations (3.26) explicitly for the four components of $G_{\alpha\beta}$ and using (3.28), one can represent them diagrammatically as in Fig. 3.2.

One can easily solve the resulting coupled algebraic equations to obtain (see, for example, pp. 207ff of Fetter and Walecka, 1971; p. 128 of Lifshitz

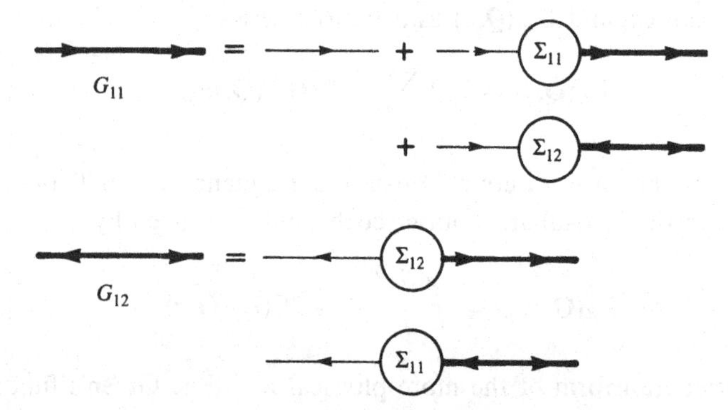

Fig. 3.2. Diagrammatic representation of the Dyson–Beliaev equations in (3.26) describing the single-particle Green's functions of a Bose-condensed fluid.

and Pitaevskii, 1980)

$$\left.\begin{aligned} G_{11}(\mathbf{Q}, i\omega_n) &= \frac{i\omega_n + \varepsilon_Q - \mu + \Sigma_{22}(\mathbf{Q}, i\omega_n)}{D(\mathbf{Q}, i\omega_n)}, \\ G_{22}(\mathbf{Q}, i\omega_n) &= \frac{-i\omega_n + \varepsilon_Q - \mu + \Sigma_{11}(\mathbf{Q}, i\omega_n)}{D(\mathbf{Q}, i\omega_n)}, \\ G_{12}(\mathbf{Q}, i\omega_n) &= G_{21}(\mathbf{Q}, i\omega_n) = -\frac{\Sigma_{12}(\mathbf{Q}, i\omega_n)}{D(\mathbf{Q}, i\omega_n)} . \end{aligned}\right\} \tag{3.29}$$

The poles of $G_{\alpha\beta}$ are given by the zeros of the denominator

$$\begin{aligned} D(\mathbf{Q}, i\omega_n) &= [i\omega_n - (\varepsilon_Q - \mu + \Sigma_{11})][i\omega_n + (\varepsilon_Q - \mu + \Sigma_{22})] + \Sigma_{12}^2 \\ &= [i\omega_n - A_-]^2 - [\varepsilon_Q - \mu + A_+ - \Sigma_{12}][\varepsilon_Q - \mu + A_+ + \Sigma_{12}], \end{aligned} \tag{3.30a}$$

where

$$A_\pm = \frac{1}{2}[\Sigma_{11}(\mathbf{Q}, i\omega_n) \pm \Sigma_{11}(\mathbf{Q}, -i\omega_n)] . \tag{3.30b}$$

The poles of $G_{\alpha\beta}$ give the energies ω_Q of the single-particle (or field-fluctuation) excitations. We emphasize that (3.29) and (3.30) are quite general and apply equally well to Bose gases and liquids.

In this Green's function language, the Bogoliubov (1947) original approximation for a WIDBG at $T = 0$ is easily stated. At $T = 0$, essentially all atoms are in the condensate (i.e., $n_0 \simeq n$), and hence one can ignore dynamical processes which involve only excited atoms. Working to first order in the interaction, we restrict ourselves to self-energies which involve *only* excited atoms interacting with atoms *in* the condensate (see

Fig. 3.3). This defines the Bogoliubov frequency-independent self-energy approximation,

$$\left.\begin{aligned} \Sigma^B_{11}(\mathbf{Q}, i\omega_n) &= \Sigma^B_{22}(\mathbf{Q}, i\omega_n) = n_0 V(\mathbf{p}=0) + n_0 V(\mathbf{Q}) \ , \\ \Sigma^B_{12}(\mathbf{Q}, i\omega_n) &= n_0 V(\mathbf{Q}) \ , \end{aligned}\right\} \tag{3.31}$$

We emphasize that (3.31) describes only a *subset* of the complete Hartree–Fock self-energies, since it ignores the terms arising from interactions *between* excited atoms. The determination of the chemical potential μ within any given approximation is non-trivial but it can be proven that it must satisfy the relation (Hugenholtz and Pines, 1959; also Section 6.1)

$$\mu = \Sigma_{11}(\mathbf{Q} \to 0, \omega = 0) - \Sigma_{12}(\mathbf{Q} \to 0, \omega = 0) \ . \tag{3.32}$$

As can be shown from (3.30), to satisfy the Hugenholtz–Pines "sum rule" (3.32) means that the spectrum of $G_{\alpha\beta}$ in (3.29) must be gapless in the long-wavelength limit $Q \to 0$. Using (3.31) in (3.32) gives

$$\mu^B = n_0 V(\mathbf{p} = 0) \tag{3.33}$$

for the Bogoliubov model approximation. Using (3.31) and (3.33), the denominator (3.30*a*) reduces to

$$D(\mathbf{Q}, i\omega_n) = (i\omega_n)^2 - \omega_Q^2 \ , \tag{3.34}$$

where the Bogoliubov quasiparticle dispersion relation is given by

$$\omega_Q^2 = \varepsilon_Q^2 + 2n_0 V(Q)\varepsilon_Q \ , \tag{3.35}$$

The corresponding single-particle Green's functions are

$$\left.\begin{aligned} G^B_{11}(\mathbf{Q}, i\omega_n) &= \frac{u_Q^2}{i\omega_n - \omega_Q} - \frac{v_Q^2}{i\omega_n + \omega_Q} = G^B_{22}(\mathbf{Q}, -i\omega_n) \ , \\ G^B_{12}(\mathbf{Q}, -i\omega_n) &= -u_Q v_Q \left(\frac{1}{i\omega_n - \omega_Q} - \frac{1}{i\omega_n + \omega_Q} \right) = G^B_{21}(\mathbf{Q}, i\omega_n) \ , \end{aligned}\right\} \tag{3.36}$$

where the Bogoliubov "coherence factors" are

$$\left.\begin{aligned} u_Q^2 &= \frac{\varepsilon_Q + n_0 V(Q)}{2\omega_Q} + \frac{1}{2} \ , \\ v_Q^2 &= \frac{\varepsilon_Q + n_0 V(Q)}{2\omega_Q} - \frac{1}{2} \ . \end{aligned}\right\} \tag{3.37}$$

The expressions in (3.36) shows a general feature which is associated with the presence of a Bose broken symmetry, namely positive and negative energy poles, $\omega = \pm\omega_Q$. This can also be seen from the general structure

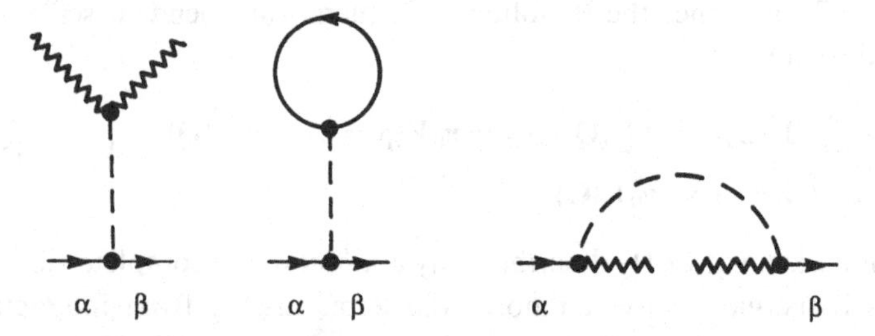

Fig. 3.3. The Bogoliubov approximation (3.31) for the self-energies $\Sigma_{\alpha\beta}(\mathbf{Q},\omega)$. The external propagator lines are not part of the self-energy.

of (3.29) and (3.30). In a free Bose gas, $u_Q=1$ and $v_Q=0$ and hence only the positive-energy pole at $\omega = \varepsilon_Q$ has any weight. Eq. (3.35) shows how the condensate changes the single-particle spectrum from particle-like at high momentum to phonon-like at low momentum.

The Bogoliubov approximation corresponds to keeping only the terms in the first line of (3.16). This simplified Hamiltonian can be diagonalized by the well known Bogoliubov transformation (see p. 314 of Fetter and Walecka, 1971), with results equivalent to (3.36). One can use $G^B_{\alpha\beta}$ in (3.36) as the new unperturbed Green's function in a new renormalized matrix Dyson–Beliaev equation in which the self-energies entirely arise from the terms in the last line of (3.16). The resulting diagrammatic expansion is discussed on pp. 249ff of Rickayzen (1980).

As discussed in the standard texts on many-body problems, one can generalize (3.32) to include multiple scattering of two free Bose atoms within the "ladder approximation." In this approximation, the Hartree–Fock self-energies take the same form with $V(\mathbf{Q})$ replaced by an expression involving $\tilde{f}(\mathbf{p},\mathbf{p}')$, the exact scattering amplitude for two free atoms (see Eq. (11.14) of Fetter and Walecka, 1971). In the limit of low momentum, $V(\mathbf{Q}=0)$ is replaced by $\tilde{f}(\mathbf{p}=0,\mathbf{p}'=0)/m \equiv 4\pi a/m$, where a is the s-wave phase shift. Within this approximation (for hard spheres, a is the diameter), the quasiparticle spectrum in (3.35) has the following limiting behaviour:

$$\omega_Q = c_0 Q, \qquad \text{for } Q \lesssim 2mc_0\,, \tag{3.38}$$

$$\omega_Q = \frac{Q^2}{2m} + \frac{4\pi n_0 a}{m}, \qquad \text{for } Q \gtrsim 2mc_0\,, \tag{3.39}$$

where the Bogoliubov phonon speed is given by $c_0^2 \equiv 4\pi n_0 a/m^2$. In a dilute gas at $T=0$, the small expansion parameter is $(na^3)^{1/2}$, that is the spacing between atoms must be much larger than the interaction range a.

The Bogoliubov approximation results (3.32)–(3.37) are studied at considerable length in standard many-body texts (see especially Chapter 6 of Fetter and Walecka, 1971). This approximation deserves attention since it exhibits the characteristic structure imposed by a Bose broken symmetry on the single-particle fluctuation spectrum. Unfortunately many studies have taken the Bogoliubov approximation (and minor variations of it) as a *realistic* model for a strongly interacting Bose *liquid* like superfluid ^{4}He, which is quite unjustified. This also leads to incorrectly assessing the value of the Bogoliubov approximation (3.36) in terms of how well it reproduces the phonon-roton spectrum of liquid ^{4}He (see, for example, Section 10.1 of Mahan, 1990), rather than for the qualitative insight it gives into the role of the Bose condensate on the excitation spectrum. The structure which the single-particle Green's functions (3.36) exhibit in the Bogoliubov approximation already captures some of the essential features of the exact expression. This is also nicely illustrated, for example, by the renormalization group analysis of the scaling properties of a WIDBG within the Bogoliubov approximation (Weichman, 1988).

Beginning in the late 1950's, there have been many attempts to improve the Bogoliubov approximation by treating the multiple scattering in a self-consistent t-matrix approximation (see, for example, Brueckner and Sawada, 1957; Parry and ter Haar, 1962; Brown and Coopersmith, 1969). That is, the ladder diagrams describing the interaction between excited atoms are calculated using renormalized propagators which include (in an approximate way) the effect of a condensate. In this way, it was hoped that one could include the strong renormalization effects on the excitation spectrum expected in a *liquid*, as well as obtain an estimate of the depletion of the condensate fraction. The limitations of this approach were noted by Hugenholtz and Pines (1959); the dielectric formalism of Chapter 5 pinpoints the key inconsistency. Essentially, the usual t-matrix studies attempt to treat the *diagonal* self-energies Σ_{11} in a way more appropriate to a liquid but only keep the simplest *off-diagonal* self-energy term Σ_{12} appropriate to a *gas*, namely that considered in the Bogoliubov approximation (3.31).

Developing better approximations for the self-energies $\Sigma_{\alpha\beta}$ is not an easy task. Even for a WIDBG, (3.31) is inadequate at higher temperatures where the condensate is thermally depleted ($n_0 \ll n$), since it ignores the interactions *between* excited atoms. In their classic article, Hohenberg and Martin (1965) give a useful classification of various microscopic approximations for a WIDBG (see pp. 348 ff). They emphasize that any self-energy approximation which does not satisfy the Hugenholtz–Pines relation (3.32) will lead to a quasiparticle spectrum which is not phonon-like, i.e., has an energy gap at long wavelengths. This is a well known problem which arises when one uses the *full* self-consistent Hartree–Fock self-energies (Girardeau and Arnowitt, 1959) in place of the Bogoliubov approximation (3.31). In fact, one must include contributions to the self-energy $\Sigma_{\alpha\beta}$ to *second* order in V (Beliaev, 1958b) to obtain the correct "gapless" phonon spectrum for $G_{\alpha\beta}$ at the next level past the Bogoliubov approximation.

The whole subject can be quite confusing to the non-expert since one can still use the full Hartree–Fock self-energy approximation to generate a "conserving approximation" for the density-response function χ_{nn} (see (2.18) or (3.41)) by using the method of functional differentiation (Baym and Kadanoff, 1961; Kadanoff and Baym, 1962). The resulting poles of $S(\mathbf{Q},\omega)$ in the $Q \to 0$ limit are phonon-like, even though the excitations of the single-particle Green's functions used to generate it may have an energy gap at $Q = 0$. In this kind of conserving approximation, the two-particle Green's functions G_2 (such as χ_{nn}) are generated by functional differentiation of the one-particle Green's function G_1. Schematically, we have $G_2 = \delta G_1[W]/\delta W$, where W is some appropriate external field set to zero at the end. More precisely, the equation of motion for G_2 is obtained from functional differentiation of the equation of motion for G_1. Any approximation for the single-particle self-energy Σ defines a G_1 and hence Σ may be viewed as a functional of G_1. One finds that $G_2 = G_1 G_1 + G_1 G_1 \Gamma G_2$, where the interaction vertex is given by $\Gamma \equiv \delta\Sigma[G_1]/\delta G_1$. Thus the self-energy Σ determines G_1 as well as Γ. The two-particle Green's function G_2 given by this procedure is guaranteed to be consistent with various conservation laws but clearly the poles may be quite different from the G_1 used to generate it. Examples of such conserving approximations have been worked out for Bose fluids by Hohenberg and Martin (1965) at $T = 0$ as well as by Cheung and Griffin (1971b) at $T \neq 0$.

The coupling of the single-particle fluctuations with density fluctuations (as exhibited at the end of Section 3.1) means that $G_{\alpha\beta}(\mathbf{Q},\omega)$ and the density response function $\chi_{nn}(\mathbf{Q},\omega)$ share the same singularities, although with different weights. This will be shown using a simple approach in Section 3.4 and in more general terms in Chapter 5. However, while it is clear that an approximate calculation should be consistent with this requirement, it is not easy to satisfy if we simply include more self-energy diagrams in an *ad hoc* manner.

As we mentioned in connection with (3.18), the intimate connection between $G_{\alpha\beta}$ and χ_{nn} was first pointed out by Hugenholtz and Pines (1959). In particular, they showed that in the Beliaev (1958b) approximation, the phonon pole of $G_{\alpha\beta}$ had a velocity corresponding precisely to the compressional sound velocity (as determined by the thermodynamic derivative of the ground-state energy computed in the *same* approximation). This result is of great importance since it justifies the key assumption of Landau and Feynman that "sound waves" play the role of elementary excitations. In this regard, we also call attention to the important field-theoretic results of Gavoret and Nozières (1964). Working to *all* orders of diagrammatic perturbation theory, they evaluated the single-particle Green's function as well as the density and current response functions. At $T = 0$ and in the long-wavelength limit ($Q, \omega \to 0$), they showed explicitly that $G_{\alpha\beta}$ and χ_{nn} exhibit the *same* phonon resonance in a Bose-condensed fluid. This generalized the relation noted by Hugenholtz and Pines (1959). The work of Gavoret and Nozières still provides one of the few rigorous calculations we have for the excitations of a Bose fluid based on an explicit many-body calculation. It is discussed in detail in Sections 5.4 and 6.3.

In Chapter 5, we develop an approach for calculating both $G_{\alpha\beta}$ and χ_{nn} which guarantees from the beginning that they share the *same* spectrum, as they must in the presense of a Bose condensate.

3.3 $S(\mathbf{Q},\omega)$ in the Bogoliubov approximation

It is instructive to evaluate also the dynamic structure factor $S(\mathbf{Q},\omega)$ in the simple Bogoliubov approximation. We recall that $S(\mathbf{Q},\omega)$ is related to the density response function χ_{nn} by the relation (2.20*b*),

$$S(\mathbf{Q},\omega) = -\frac{1}{\pi n}[N(\omega)+1]\mathrm{Im}\,\chi_{nn}(\mathbf{Q},\omega+i\eta)\,. \qquad (3.40)$$

Here χ_{nn} is the usual analytic continuation of the imaginary-frequency Fourier component ($i\omega_n \to \omega + i\eta$)

$$-\chi_{nn}(\mathbf{Q},i\omega_n) = \frac{1}{\Omega}\int_0^\beta d\tau e^{i\omega_n\tau}\left[\langle T\hat{\rho}(\mathbf{Q},\tau)\hat{\rho}(-\mathbf{Q})\rangle - \langle\hat{N}\rangle^2\delta_{Q,0}\right], \qquad (3.41)$$

where $\hat{\rho}(\mathbf{Q})$ is given by (3.12). In the Bogoliubov approximation, we omit the second term on the right hand side of (3.12) since it involves only non-condensate atoms. In the context of (3.18), this means that we can restrict ourselves to the first term on the right side of this equation. One obtains (for $Q \neq 0$)

$$\chi^B_{nn}(\mathbf{Q}, i\omega_n) = \sqrt{n_0} \sum_{\alpha,\beta} G^B_{\alpha\beta}(\mathbf{Q}, i\omega_n) \sqrt{n_0}$$
$$= \frac{2n_0\varepsilon_Q}{(i\omega_n)^2 - \omega_Q^2}, \tag{3.42}$$

where the last line follows from (3.36). This expression was first written down explicitly by Hugenholtz and Pines (1959). In this simple Bogoliubov approximation, we see explicitly how the ω_Q pole of the single-particle Green's function (3.36) appears also in (3.42) as the pole of the density-response function, with a strength directly proportional to n_0. In Chapters 5 and 6, we shall see how this sharing of poles will always occur when $\sqrt{n_0}$ is finite, but *not* in the simple manner shown by (3.42) and (3.36).

The simplest "naive" improvement to the Bogoliubov approximation for χ_{nn} in (3.42) is given by

$$\chi_{nn}(\mathbf{Q}, i\omega_l) = n_0 \sum_{\alpha\beta} G^B_{\alpha\beta}(\mathbf{Q}, i\omega_l) + \chi^0_{\tilde{n}\tilde{n}}(\mathbf{Q}, i\omega_l), \tag{3.43}$$

where the density response of the non-condensate atoms is (see, for example, Fetter, 1970; Griffin and Talbot, 1981)

$$\chi^0_{\tilde{n}\tilde{n}}(\mathbf{Q}, i\omega_l) = -\frac{1}{\beta} \sum_{\omega_n} \int \frac{d\mathbf{p}}{(2\pi)^3} \big[G^B_{11}(\mathbf{p}, i\omega_n) G^B_{11}(\mathbf{p}+\mathbf{Q}, i\omega_n + i\omega_l)$$
$$+ G^B_{12}(\mathbf{p}, i\omega_n) G^B_{12}(\mathbf{p}+\mathbf{Q}, i\omega_n + i\omega_l)\big]. \tag{3.44}$$

The expression (3.44) corresponds to the lowest-order "bubble diagram" using the Bogoliubov approximation (3.36) for the matrix single-particle Green's function $G_{\alpha\beta}$. In physical terms, it is the density response of a non-interacting gas of Bogoliubov quasiparticles as described by the first line of (3.16). In (3.43), no interference terms are included. Computing $S(\mathbf{Q}, \omega)$ using (3.40), a straightforward (but lengthy) calculation gives

$$S(\mathbf{Q}, \omega) =$$
$$\frac{n_0}{n}(u_Q - v_Q)^2 \big\{[N(\omega_Q) + 1]\delta(\omega - \omega_Q) + N(\omega_Q)\delta(\omega + \omega_Q)]\big\}$$
$$+ \frac{1}{n} \int \frac{d\mathbf{p}}{(2\pi)^3} N(\omega_p)[1 + N(\omega_{p+Q})](u_p u_{p+Q} + v_p v_{p+Q})^2 \delta(\omega - [\omega_{p+Q} - \omega_p])$$
$$+ \frac{1}{n} \int \frac{d\mathbf{p}}{(2\pi)^3} \frac{1}{2}(u_p v_{p+Q} + v_p u_{p+Q})^2 \big\{N(\omega_{p+Q}) N(\omega_p) \delta(\omega + [\omega_{p+Q} + \omega_p])$$
$$+ [1 + N(\omega_{p+Q})][1 + N(\omega_p)] \delta(\omega - [\omega_{p+Q} + \omega_p])\big\}. \tag{3.45}$$

We have carried out the standard Bose Matsubara frequency sums using (see pp. 167ff of Mahan, 1990)

$$\left.\begin{aligned} \frac{1}{\beta}\sum_{i\omega_n}\frac{1}{i\omega_n-\omega_p}\frac{1}{i\omega_n+i\omega_l-\omega_{p+Q}} &= \frac{N(\omega_{p+Q})-N(\omega_p)}{i\omega_l-[\omega_{p+Q}-\omega_p]} \equiv R_1(i\omega_l)\ , \\ \frac{1}{\beta}\sum_{i\omega_n}\frac{1}{i\omega_n-\omega_p}\frac{1}{i\omega_n-i\omega_l+\omega_{p+Q}} &= \frac{1+N(\omega_{p+Q})+N(\omega_p)}{i\omega_l-[\omega_{p+Q}+\omega_p]} \equiv R_2(i\omega_l). \end{aligned}\right\} \tag{3.46}$$

In addition, we have used the Bose distribution identities given by (2.31) as well as

$$[N(\omega_2-\omega_1)+1][N(\omega_1)-N(\omega_2)] = N(\omega_1)[1+N(\omega_2)]$$

in simplifying the final expression given in (3.45). The frequency sums R_1 and R_2 in (3.46) will also be needed in later chapters. Factors such as $(u_pu_{p+Q}+v_pv_{p+Q})$ and $(u_pv_{p+q}+u_pv_{p+Q})$ in (3.45) are referred to as Bose coherence factors. They describe the complicated interference effects associated with the mixing of positive and negative energy poles of $G_{\alpha\beta}$ and are the signature of Bose-condensed fluids.

The expression in (3.45) is only an illustration of the *kind* of structure $S(\mathbf{Q},\omega)$ which exhibits when the Bogoliubov approximation is a good starting point. The first line of (3.45) corresponds to S_1 in (3.18) and describes the creation ($N(\omega)+1$) or destruction ($N(\omega)$) of a single Bogoliubov excitation. The remaining terms correspond to $\tilde{S}$ in (3.18). The second line of (3.45) describes the thermal scattering processes destroying an excitation with energy ω_p and creating one with energy ω_{p+Q}. These first two lines in (3.45) have their analogue in a free Bose gas given by (2.32) of Section 2.2.

The third line of (3.45) describes the creation (or destruction) of *two* quasiparticles, with total energy $\omega_p+\omega_{p+Q}$. This term is associated with the existence of negative and positive energy poles of $G^B_{\alpha\beta}$ in (3.36). As with the first line of (3.45), this multiparticle term disappears when n_0=0 (since v_Q vanishes) and thus it is characteristic of a Bose-condensed fluid. The "two-phonon" or multiphonon contribution in (3.45) will give rise to a broad frequency spectrum, in contrast to the "one-phonon" terms which are sharp resonances at $\pm\omega_Q$. At $T=0$, all the Bose occupation factors $N(\omega)$ vanish, in which case we can only *create* one or two quasiparticles. There are no thermal scattering terms at $T=0$. The two-excitation or pair spectrum in liquid ^{4}He is discussed in detail in Chapter 10.

Eq. (3.45) is, at best, only appropriate in the "weak-coupling" limit.

In using (3.43), one has completely ignored any collective (zero sound) density fluctuations, as well as the interference terms coupling such fluctuations into the single-particle terms described by the first line in (3.45). To include such effects, we need a more systematic procedure for calculating χ_{nn} past the Bogoliubov approximation.

In Chapters 5 and 6, we show quite generally that χ_{nn} can always be separated into two parts, one of which is directly proportional to the single-particle Green's function:

$$\chi_{nn}(\mathbf{Q},\omega) = \sum_{\alpha,\beta} \Lambda_\alpha(\mathbf{Q},\omega) G_{\alpha\beta}(\mathbf{Q},\omega) \Lambda_\beta(\mathbf{Q},\omega) + \chi^R_{nn}(\mathbf{Q},\omega) \ . \tag{3.47}$$

The result in (3.43) may be viewed as an illustration of this. The Bose vertex functions $\Lambda_\alpha(\mathbf{Q},\omega)$ in (3.47) determine the strength with which the single-particle excitations appear in the density-response function. These symmetry-breaking vertex functions vanish with n_0, while χ^R_{nn} goes over into the full response function of the normal Bose fluid. Such rigorous decompositions of χ_{nn} into what one might interpret as condensate and normal contributions were first derived (at $T = 0$) by Gavoret and Nozières (1964) and Hohenberg and Martin (1965). Formulas such as (3.47) will play a central role in future chapters.

3.4 Mean-field analysis

All three contributions in (3.18) are strongly modified by the effect of the $\tilde{\rho}$–$\hat{A}$ coupling terms in the second line of (3.16). They are all related, with the result that the single-particle spectrum of $G_{\alpha\beta}(\mathbf{Q},\omega)$ contains the density fluctuation spectrum of $S(\mathbf{Q},\omega)$ and vice versa. This is the key feature of a Bose-condensed fluid, as is already apparent from (3.13). To understand the details of the resulting hybridization of the spectra of $S(\mathbf{Q},\omega)$ and $G_{\alpha\beta}(\mathbf{Q},\omega)$ requires a fairly sophisticated diagrammatic analysis in terms of proper, irreducible diagrams. This is called the "dielectric formalism" and is the subject of Chapter 5. As an introduction to the basic physics involved, in this section we calculate both $\chi_{nn}(\mathbf{Q},\omega)$ and $G_{\alpha\beta}(\mathbf{Q},\omega)$ using a simple mean-field analysis (Griffin, 1991).

We consider the linear response of a Bose fluid to an *external* scalar potential $\delta\phi^0$ which couples into the variable $\tilde{\rho}_Q$ and an *external* broken-symmetry potential $\delta\eta^0$ which couples into the variable $\hat{A}_Q$ (see (3.13) and (3.14)). This perturbing Hamiltonian is given by

$$\hat{V}_{\text{ex}} = \int \frac{d\mathbf{Q}}{(2\pi)^3} \delta\phi^0(\mathbf{Q},\omega)\tilde{\rho}_Q + \int \frac{d\mathbf{Q}}{(2\pi)^3} \delta\eta^0(\mathbf{Q},\omega)\hat{A}_Q \ . \tag{3.48}$$

Thus we have

$$\delta\tilde{n}(\mathbf{Q},\omega) = \chi_{\tilde{n}\tilde{n}}(\mathbf{Q},\omega)\delta\phi^0(\mathbf{Q},\omega) + \chi_{\tilde{n}A}(\mathbf{Q},\omega)\delta\eta^0(\mathbf{Q},\omega)$$
$$\simeq \chi^0_{\tilde{n}\tilde{n}}(\mathbf{Q},\omega)\left[\delta\phi^0(\mathbf{Q},\omega) + V(\mathbf{Q})\delta\tilde{n}(\mathbf{Q},\omega) + \sqrt{n_0}V(\mathbf{Q})\delta A(\mathbf{Q},\omega)\right], \tag{3.49a}$$

$$\delta A(\mathbf{Q},\omega) = \chi_{AA}(\mathbf{Q},\omega)\delta\eta^0(\mathbf{Q},\omega) + \chi_{A\tilde{n}}(\mathbf{Q},\omega)\delta\phi^0(\mathbf{Q},\omega)$$
$$\simeq \chi^0_{AA}(\mathbf{Q},\omega)\left[\delta\eta^0(\mathbf{Q},\omega) + \sqrt{n_0}V(\mathbf{Q})\delta\tilde{n}(\mathbf{Q},\omega) + n_0 V(\mathbf{Q})\delta A(\mathbf{Q},\omega)\right]. \tag{3.49b}$$

In (3.49), the second lines are written in terms of the *two* self-consistent-fields $\delta\phi$ and $\delta\eta$ which arise from a mean-field approximation to the various terms of the interaction Hamiltonian in (3.16),

$$\hat{V} = \frac{1}{2}\sum_Q V(Q)\left[n_0\hat{A}_Q\hat{A}_{-Q} + \sqrt{n_0}(\hat{A}_Q\tilde{\rho}_{-Q} + \tilde{\rho}_Q\hat{A}_{-Q}) + \tilde{\rho}_Q\tilde{\rho}_{-Q}\right]. \tag{3.50}$$

The "screened" response functions in (3.49) relative to these effective fields are denoted by χ^0. In the simplest approximation we are considering, these are for a non-interacting Bose gas. In this case

$$n_0\chi^0_{AA}(\mathbf{Q},\omega) = \frac{n_0 Q^2/m}{\omega^2 - (Q^2/2m)^2}, \tag{3.51}$$

and $\chi^0_{\tilde{n}A} = \chi^0_{A\tilde{n}} = 0$ (this has already been assumed in the second lines of (3.49*a*) and (3.49*b*)). One can easily solve the coupled equations (3.49*a*) and (3.49*b*) to obtain

$$\chi_{AA}(\mathbf{Q},\omega) = \frac{\chi^0_{AA}}{1 - \dfrac{n_0 V(Q)\chi^0_{AA}}{1 - V(Q)\chi^0_{\tilde{n}\tilde{n}}}}, \tag{3.52a}$$

$$\chi_{\tilde{n}\tilde{n}}(\mathbf{Q},\omega) = \frac{\chi^0_{\tilde{n}\tilde{n}}}{1 - \dfrac{V(Q)\chi^0_{\tilde{n}\tilde{n}}}{1 - n_0 V(Q)\chi^0_{AA}}}, \tag{3.52b}$$

$$\chi_{\tilde{n}A}(\mathbf{Q},\omega) = \chi_{A\tilde{n}}(\mathbf{Q},\omega) = \sqrt{n_0}V(Q)\frac{\chi^0_{AA}\chi^0_{\tilde{n}\tilde{n}}}{\epsilon(\mathbf{Q},\omega)}. \tag{3.52c}$$

We note that the response function χ_{AA} corresponds to $\sum_{\alpha,\beta} G_{\alpha\beta}$ and thus it is directly related to the Beliaev single-particle Green's functions (as given in this mean-field approximation).

As expected, all three response functions $\chi_{\tilde{n}\tilde{n}}$, χ_{AA} and $\chi_{\tilde{n}A}$ in (3.52) are seen to share the *same* denominator

$$\epsilon(\mathbf{Q},\omega) = 1 - V(\mathbf{Q})[n_0\chi^0_{AA} + \chi^0_{\tilde{n}\tilde{n}}] \ , \tag{3.53}$$

and hence they have the same excitation spectrum. Summing up the components given by (3.52), we find that the *total* density-response function in (3.18) is given by

$$\begin{aligned}\chi_{nn}(\mathbf{Q},\omega) &= n_0\chi_{AA} + \sqrt{n_0}\chi_{\tilde{n}A} + \sqrt{n_0}\chi_{A\tilde{n}} + \chi_{\tilde{n}\tilde{n}} \\ &= \frac{n_0\chi^0_{AA} + \chi^0_{\tilde{n}\tilde{n}}}{\epsilon(\mathbf{Q},\omega)} \ .\end{aligned} \tag{3.54}$$

To the extent we approximate the screened response functions by those of a non-interacting Bose gas, the results in (3.52)–(3.54) correspond to the "shielded-potential" approximation (discussed in Section 5.2). The density-response function χ_{nn} as given by (3.53) and (3.54) has the usual RPA structure but (3.52) shows the characteristic hybridization of the single-particle and density fluctuations which results because of the Bose broken symmetry. In the systematic dielectric formalism of Chapter 5, we shall see that these mean-field results correspond to the simplest realization of the hybridization which always occurs in a Bose-condensed fluid.

Even in a dilute Bose gas at $T = 0$, where one can ignore the effect of the non-condensate atoms, the preceding analysis leads to a different interpretation of the Bogoliubov phonon mode. In Sections 3.2 and 3.3, the Bogoliubov excitation (3.35) was viewed as a renormalized pole of the single-particle Green's functions (see (3.36)). The density-response function χ_{nn} was directly proportional to these Green's functions, as in (3.42). The mean-field analysis given above gives an alternative picture in which the Bogoliubov phonon is a particle–hole zero sound mode in χ_{nn}, arising from the dynamic mean-field of the condensate atoms. Thus in a Bose gas at $T = 0$, (3.54) reduces to

$$\chi_{nn}(\mathbf{Q},\omega) = \frac{n_0\chi_{AA}}{1 - V(\mathbf{Q})n_0\chi_{AA}} = \frac{n_0Q^2/m}{\omega^2 - (Q^2/2m)^2 - (n_0V(\mathbf{Q})/m)Q^2} \ . \tag{3.55}$$

This zero sound picture was first emphasized by Pines (1963, 1966) and Ma, Gould and Wong (1971). In a WIDBG at $T = 0$, the single-particle and zero sound pictures are equivalent. However, we shall see in later chapters that the zero sound interpretation is more useful when one deals with finite temperatures and Bose liquids, where the condensate fraction is small.

4

High-momentum scattering and the condensate fraction

As discussed in Chapter 3, our analysis of correlation functions in superfluid ^{4}He is based on the existence of a Bose broken-symmetry and the associated Bose condensate. It seems appropriate to review theories of the momentum distribution of atoms in liquid ^{4}He and measurements of this using neutron scattering at high momentum transfers. Numerical simulations as well as measurements which we review in this chapter both give convincing evidence for a condensate fraction of 9-10% at SVP at $T = 0$, decreasing to zero at T_λ.

If the impulse approximation (IA) discussed in Section 2.3 is valid, then the dynamic structure faction is given by (2.39), i.e.

$$S_{\mathrm{IA}}(\mathbf{Q},\omega) \equiv \int d\mathbf{p}\; n(\mathbf{p})\delta\big(\omega - \frac{Q^2}{2m} - \frac{\mathbf{p}\cdot\mathbf{Q}}{m}\big) , \tag{4.1}$$

where $n(\mathbf{p})$ is the exact atomic momentum distribution (normalized to unity). Thus, in principle, one should be able to extract information about the ^{4}He momentum distribution $n(\mathbf{p})$ from $S(\mathbf{Q},\omega)$ data if Q is large enough. If there is a finite density n_0 of ^{4}He atoms in the $\mathbf{p} = 0$ state, then (4.1) naturally splits into two contributions

$$S_{\mathrm{IA}}(\mathbf{Q},\omega) = \bar{n}_0\delta(\omega - \omega_R) + \int d\mathbf{p}\; \tilde{n}(\mathbf{p})\delta\big(\omega - \omega_R - \frac{\mathbf{p}\cdot\mathbf{Q}}{m}\big) , \tag{4.2}$$

the second term describing the Doppler-broadened distribution (centred at the recoil frequency $\omega_R = Q^2/2m$) of the non-condensate ^{4}He atoms ($\mathbf{p} \neq 0$). In this chapter, we follow the experimental literature on this topic and use a *normalized* excited-atom momentum distribution $\tilde{n}(\mathbf{p}) \equiv \tilde{n}_p/(2\pi)^3 n$ as well as a condensate fraction $\bar{n}_0 \equiv n_0/n$.

The essential simplicity of the experimental procedure mentioned above for measuring the condensate fraction (first suggested by Miller, Pines and Nozières, 1962 and later developed by Hohenberg and Platzman,

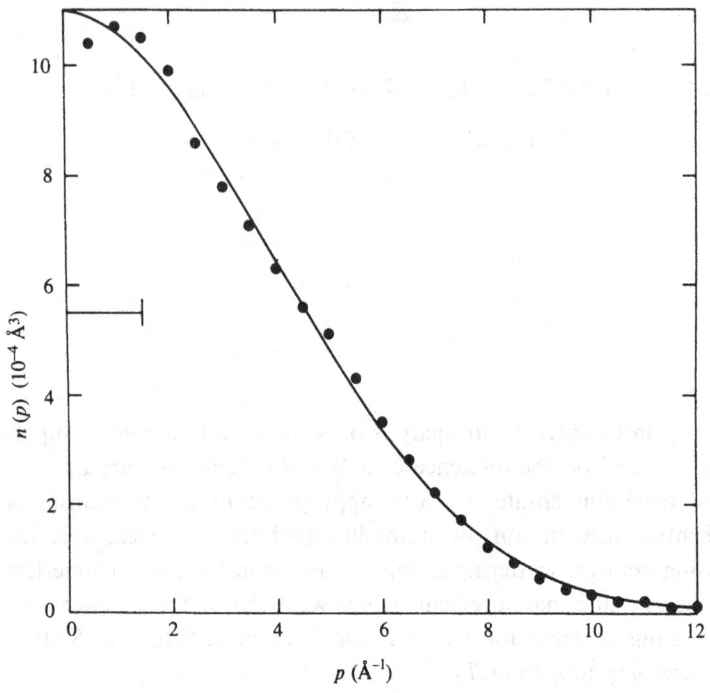

Fig. 4.1. The atomic momentum distribution obtained from $S(\mathbf{Q},\omega)$ data for liquid neon [Source: Sears, 1981].

1966) is very attractive and it has been the basis of a major continuing research effort since the early seventies. A basic problem is the proper way of including the interactions of the recoiling atom (hit by the incoming neutron) with the neighbouring ^{4}He atoms. The simple IA result in (4.1) must be generalized to include such final-state effects (FSE) in some appropriate fashion. Such final-state effects can be expressed as a convolution of $S_{\text{IA}}(\mathbf{Q},\omega)$ with a resolution function as in (2.47*b*), as can the broadening due to finite instrumental resolution. The detailed shape of these resolution functions is important if their widths are comparable to that of the narrow condensate-associated peak in $n(\mathbf{p})$ (the first term in (4.2)). In contrast, in normal liquid ^{4}He as well as in solid ^{4}He, where the atomic momentum distribution is a relatively broad near-Gaussian (see Fig. 4.1), the FS broadening has negligible effect (for $Q \gtrsim 20$ Å^{-1}).

Because it concerns a quantity of such fundamental significance in condensed matter physics, the procedures used to extract or confirm a value of the Bose condensate n_0 from high-momentum scattering data have been subject to intense scrutiny, not to mention a certain amount of scepticism, by the low-temperature community. In particular, in the absence of any direct experimental evidence for a peak due to the first term in (4.2), doubt has been sometimes expressed as to whether the relatively small temperature-dependent changes in the high-momentum neutron-scattering data below T_λ cannot have their explanation in something other than a Bose broken-symmetry. In this connection, it is important to remember that one is analysing high-momentum neutron-scattering data in the context of other experimental studies. There is *independent* direct evidence for the existence of a Bose condensate, including:

(a) The results of Monte Carlo calculations, especially the recent $T \neq 0$ work of Ceperley and Pollock (1986) reviewed in Section 1.2.
(b) The critical exponents at T_λ are those appropriate to a fluid with a complex (two-component) order parameter (see, for example, Ahlers, 1978).
(c) The fact that the sharp maxon–roton quasiparticle resonance in $S(\mathbf{Q}, \omega)$ decreases in intensity and completely vanishes at T_λ, as discussed in Chapter 7. We argue there that this sharp resonance is, in fact, the low-Q equivalent of the free-particle peak in (4.2) at high Q.

Whatever the complications which arise in the detailed analysis of high-momentum scattering data, there is a solid underpinning to this approach. Other methods which have been advanced for obtaining direct information about the condensate fraction (see Glyde and Svensson, 1987) do not appear to have a sound microscopic basis (see Appendix A of Griffin, 1987). This includes the interesting suggestion of Cummings, Hyland and Rowlands (1970), who argue that n_0 can be obtained from the temperature dependence of the pair distribution function $g(\mathbf{r})$.

As general background to this chapter, we recommend the conference proceedings *Momentum Distributions*, edited by Silver and Sokol (1989).

4.1 Condensate-induced changes in the momentum distribution

The atomic momentum distribution of normal liquid ^{4}He is, to a good approximation, a Gaussian. This is consistent with path-integral Monte Carlo (PIMC) calculations of Ceperley and Pollock (1986, 1987) as well as with experimental data. Indeed, such a near-Gaussian momentum distribution is characteristic of most classical liquids as well as quantum solids (see Fig. 4.1). In the superfluid phase, the main difference in $\tilde{n}_p$

appears at low momentum $\lesssim 1$ Å^{-1}. In this section, we discuss the expected changes in $\tilde{n}_p$ at low but finite p which are associated with the Bose broken symmetry.

As we shall see in Section 4.4, this condensate-induced contribution (denoted by δn_c) must be allowed for if we are to extract a reliable value of n_0 from the experimentally determined n_p (Sears *et al.*, 1982; Griffin, 1985). Moreover, we shall see that $\delta n_c(\mathbf{p})$ is a signature of the Bose broken symmetry since it describes how the momentum distribution of ^{4}He *atoms* is modified as a result of the condensate coupling them into the collective density fluctuations (phonon quasiparticles). Our theoretical understanding of $\delta n_c(\mathbf{p})$ is thus closely tied in with the unique nature of the excitation spectrum of a Bose-condensed liquid (see Chapters 5 and 6).

The momentum distribution for atoms with $\mathbf{p} \neq 0$ can be directly related to the single-particle Green's function spectral density defined in (2.17) (see pp. 153ff of Mahan, 1990; Baym, 1969)

$$\begin{aligned}\tilde{n}_p &= \langle \hat{a}_p^+ \hat{a}_p \rangle \\ &= Z^{-1} \sum_{m,n} e^{-\beta E'_m} |\langle m|\hat{a}_p^+|n\rangle|^2 \\ &= \int_{-\infty}^{\infty} \frac{d\omega}{2\pi} N(\omega) A(\mathbf{p}, \omega) \ ,\end{aligned} \tag{4.3}$$

where $N(\omega)$ is the Bose distribution. Considering the $\omega \to \infty$ limit of $G_{11}(\mathbf{p}, \omega)$, one finds the spectral density satisfies two sum rules (see, for example, Griffin, 1984)

$$\int_{-\infty}^{\infty} \frac{d\omega}{2\pi} A(\mathbf{p}, \omega) = 1 \tag{4.4}$$

and

$$\int_{-\infty}^{\infty} \frac{d\omega}{2\pi} \omega A(\mathbf{p}, \omega) = \varepsilon_p - \mu + nV(\mathbf{q} = 0) + n_0 V(\mathbf{p}) + \int \frac{d\mathbf{q}}{(2\pi)^3} V(\mathbf{p} + \mathbf{q}) \tilde{n}_q \ . \tag{4.5}$$

Eq. (4.5) is the well known Wagner (1966) sum rule. In contrast to (4.3), the high-frequency single-particle spectrum will make the most important contributions to the frequency integrals in (4.4) and (4.5). Both (4.3) and (4.5) relate the momentum distribution $\tilde{n}_p$ to frequency integrals over the spectral density $A(\mathbf{p}, \omega)$. We remark that, as written, (4.5) is only correct for a soft interatomic potential, where the Fourier transform $V(\mathbf{p})$ is well defined. A generalized form of (4.5), valid for hard-core potentials, is given in Section 8.1.

The last line in (4.3) reminds us that Bose condensation in a Bose liquid, complicated as it is, involves the same phenomenon as in an ideal Bose gas (see remarks at the beginning of Chapter 3). The macroscopic occupation of the zero-momentum state is related to the presence of the Bose distribution $N(\omega)$ in (4.3). To take advantage of the fact that $N(\omega \to 0) \to \infty$, the spectral density $A(\mathbf{p}, \omega)$ must have a zero-energy pole in the long-wavelength limit. In a free Bose gas, this leads to the requirement that the chemical potential μ vanish below T_λ. More generally, this condition is replaced by the Hugenholtz–Pines relation given in (3.32), which ensures that $G_{\alpha\beta}(\mathbf{p}, \omega)$ in (3.29) has the required zero-energy pole in the $\mathbf{p} \to 0$ limit.

The structure of $A(\mathbf{p}, \omega)$ in the hydrodynamic region described by the two-fluid equations is given in Section 6.2. As discussed there, $A(\mathbf{p}, \omega)$ includes contributions from both first and second sound modes. At $T = 0$ (where the superfluid density $\rho_S = \rho$), the expression in (6.26) is equivalent to the result obtained by Gavoret and Nozières (1964),

$$A(\mathbf{p}, \omega) = 2\pi \frac{n_0}{\rho} m^2 c^2 \delta(\omega^2 - c^2 p^2) \operatorname{sgn} \omega \ , \tag{4.6}$$

where c is the compressional sound velocity. For reasons given in Section 6.3, we believe that (4.6) also gives the correct collisionless low-frequency result at *finite* temperatures in superfluid ^{4}He. At any finite temperature, the transition from the hydrodynamic to the collisionless spectral density $A(\mathbf{p}, \omega)$ involves the second sound contribution disappearing, with the first sound contribution smoothly going over to that of zero sound. In the strict hydrodynamic limit, one has an additional sum rule (see (6.12) in Section 6.1)

$$\lim_{p \to 0} -G_{11}(\mathbf{p}, \omega = 0) \equiv \int_{-\infty}^{\infty} \frac{d\omega}{2\pi} \frac{A(\mathbf{p}, \omega)}{\omega} = \frac{m^2 n_0}{\rho_S p^2} \ . \tag{4.7}$$

Substituting (4.6) into (4.3) gives

$$\tilde{n}_p = \frac{n_0}{n} \frac{mc}{2p} [2N(cp) + 1] \ , \tag{4.8}$$

which has the limits

$$\tilde{n}_p \simeq \begin{cases} \dfrac{n_0}{n} \dfrac{mc}{2p} \ , & cp \gg k_B T \ , \quad (4.9) \\[2ex] \dfrac{n_0}{n} \dfrac{m k_B T}{p^2} \ , & cp \ll k_B T \ . \quad (4.10) \end{cases}$$

The low-temperature limit (4.9) was first derived by Gavoret and Nozières (1964) at $T = 0$. The other limit (4.10) was discussed by Bogoliubov

(1963, 1970) and Hohenberg and Martin (1964). In superfluid ^{4}He, the maximum temperature of ~ 2 K means that the high-temperature limit (4.10) is only appropriate for $p \lesssim 0.1$ Å^{-1}. A heuristic derivation of (4.8) is given on p. 108 of Lifshitz and Pitaevskii (1980), following the approach of Ferrell, Menyhard, Schmidt, Schwabl and Szépfalusy (1968). This approach, based on the effect of the phase fluctuations of the Bose order parameter, has been recently extended to derive the leading-order correction to (4.10) going as p^{-1} (Giorgini, Pitaevskii and Stringari, 1992). Reatto and Chester (1967) were the first to derive the low-temperature limit (4.9) from a Jastrow variational wavefunction by including long-range correlations associated with long-wavelength phonon modes (see Section 9.1). For further discussion of (4.8), we refer to Griffin (1984, 1985). The leading-order corrections to (4.9) and (4.10) are briefly discussed in Section 6.3.

Clearly the momentum distribution (4.8) only includes the effect of the low-frequency, long-wavelength phonons described by (4.6). Is there any weight from high frequencies and what happens at larger values of momentum? Before addressing these questions, we discuss (4.6) in a little more detail.

As illustrated in our discussion of the simple Bogoliubov approximation in Section 3.2, a key feature of the single-particle Green's function in any Bose-condensed fluid is that there is a *negative*-energy pole in addition to the usual positive-energy pole,

$$A(\mathbf{p},\omega) = 2\pi \left[u_p^2\delta(\omega-\omega_p) - v_p^2\delta(\omega+\omega_p)\right] \ . \qquad (4.11)$$

Using this spectral density, one obtains

$$\int_{-\infty}^{\infty}\frac{d\omega}{2\pi}A(\mathbf{p},\omega) = u_p^2 - v_p^2 = 1 \ , \qquad (4.12)$$

$$\int_{-\infty}^{\infty}\frac{d\omega}{2\pi}\omega A(\mathbf{p},\omega) = \omega_p(u_p^2 + v_p^2) \ , \qquad (4.13)$$

$$\tilde{n}_p = \int_{-\infty}^{\infty}\frac{d\omega}{2\pi}N(\omega)A(\mathbf{p},\omega) = N(\omega_p) + v_p^2[2N(\omega_p)+1] \ . \qquad (4.14)$$

At large enough momentum, we expect that only the *positive*-energy pole will survive, with

$$u_p \to 1 \ , \ v_p \to 0 \ . \qquad (4.15)$$

This is illustrated by the Bogoliubov approximation results in (3.37). The effect of the interactions is to take atoms out of the condensate (depletion) and put them into the low-momentum states. This new condensate-induced contribution to $\tilde{n}_p$ is precisely the origin of the second term in (4.14) (see p. 218 of Fetter and Walecka, 1971). At low values of p, the weights u_p^2 and v_p^2 become equal. This is illustrated by (3.37), which gives

$$\lim_{p\to 0} u_p^2 = v_p^2 = \frac{n_0 V(p)}{2\omega_p} = \frac{mc_0}{2p} \ . \tag{4.16}$$

Using (4.16) in (4.11) , we reproduce the WIDBG version of the Gavoret–Nozières result (4.6), with $n_0 \simeq n$. Within the same Bogoliubov approximation, (4.13) gives $\varepsilon_p + n_0 V(p)$. This is consistent with the Wagner sum rule (4.5) since, in this case, $\mu = nV(\mathbf{q} = 0)$ (see (3.33)) and the contribution of the non-condensate atoms (the last term in (4.5)) is negligible.

Outside the phonon region (say $p \gtrsim 0.7$ Å^{-1}), one expects that the condensate-induced negative-energy pole will have negligible weight. However, a simple expression of the kind

$$A(\mathbf{p}, \omega) \simeq 2\pi z_p \delta(\omega - \omega_p) \tag{4.17}$$

is inadequate since using this in (4.3) gives $\tilde{n}_p \sim N(\omega_p)$, which is very temperature-dependent as well as extremely small (since $\omega_p \gg k_B T$). By way of contrast, the *observed* momentum distribution $\tilde{n}_p$ in the region $p \gtrsim 0.8$ Å^{-1} is found to be a remarkably temperature-independent Gaussian (see Fig. 2.9). As with a classical Maxwell–Boltzmann velocity distribution, one can thus write $\tilde{n}_p$ in the form

$$\tilde{n}_p \sim e^{-\frac{3}{2}\frac{p^2}{\bar{p}^2}} \ , \tag{4.18}$$

where

$$\langle K \rangle = \frac{\int d\mathbf{p}(\frac{p^2}{2m})\,\tilde{n}_p}{\int d\mathbf{p}\,\tilde{n}_p} \equiv \frac{\bar{p}^2}{2m} \tag{4.19}$$

is the average kinetic energy of a non-condensate ^{4}He atom. Experimentally one finds $\langle K \rangle \sim 14.5$ K at $T = 0$ at SVP, with only a small increase even for T as high as 4.2 K. Recalling that $N(-\omega) = -[N(\omega) + 1]$, a temperature-independent result like (4.18) must be associated with a spectral density $A(\mathbf{p}, \omega)$ which has significant weight at *negative* frequencies, in addition to the positive-energy quasiparticle peak given by (4.17). This can be seen by rewriting (4.3) in the form (Pitaevskii, 1987)

$$\tilde{n}_p = 2\int_0^\infty \frac{d\omega}{2\pi} N(\omega) A_{\text{ant}}(\mathbf{p}, \omega) - \int_{-\infty}^0 \frac{d\omega}{2\pi} A(\mathbf{p}, \omega) \ , \tag{4.20}$$

where the antisymmetric component $A_{\text{ant}}(\mathbf{p}, \omega) \equiv \frac{1}{2}\,[A(\mathbf{p}, \omega) - A(\mathbf{p}, -\omega)]$

controls the temperature-dependent contribution to $\tilde{n}_p$. Since $\tilde{n}_p$ is the same above and below T_λ in the region $p \gtrsim 0.8$ Å^{-1}, one must conclude that the *second* term in (4.20) is dominant in this momentum region. Moreover, this negative-frequency component of the spectral density must *also* exist in the normal phase and thus cannot be associated with the Bose broken symmetry. We note that in any Bose fluid, one has the rigorous result $A(\mathbf{p}, \omega) \leq 0$ for $\omega < 0$. This can easily be proved using the spectral representation (2.17).

Information about $\tilde{n}_p$ at low momentum is difficult to obtain, either theoretically or experimentally. Monte Carlo studies (either GFMC or PIMC) usually deal with a small number of particles (often as small as $N = 64$) and hence finite-size effects strongly restrict the lowest wavevector at which $\tilde{n}_p$ can be calculated ($p \gtrsim \pi/L$). Moreover, as we discuss in Section 4.3, the impulse approximation expression for the Compton profile is relatively insensitive to the value of $\tilde{n}_p$ at low p. On the other hand, variational many-body calculations (see Section 9.1) have a distinct advantage in that they do give the full momentum distribution including the low-p singular behaviour in (4.9), as shown in Fig. 4.2. For reviews of such variational calculations, see Manousakis (1989) and Clark and Ristig (1989).

We have emphasized that the condensate-induced singularity in $\tilde{n}_p$ at low p is the direct consequence of the coupling of collective density fluctuations into the single-particle Green's function in a Bose-condensed fluid. In this regard, such singularities give a point of contact between the dielectric formalism of Chapter 5 and variational many-body techniques reviewed in Section 9.1. If one has some approximation to the many-body wavefunction $|\mathbf{k}\rangle$ describing liquid ^{4}He when there is a *single* excitation of momentum $\mathbf{k}$, one can calculate the associated atomic momentum distribution $n_p(\mathbf{k})$ from the single-particle density matrix. Manousakis, Pandharipande and Usmani (MPU, 1985) have done this and also argue that the change in the $T = 0$ atomic momentum distribution at finite but low temperatures is given by

$$\delta n_p(T) = \int \frac{d\mathbf{k}}{(2\pi)^3} N(\omega_k)[n_p(\mathbf{k}) - n_p(\mathbf{k} = 0)] \ . \qquad (4.21a)$$

Here $n_p(\mathbf{k} = 0)$ is the atomic momentum distribution in the ground state (no excitations) and ω_k is the energy of the excited state $|\mathbf{k}\rangle$ (relative to the ground-state energy). By explicit calculations based on correlated basis functions using Feynman–Cohen eigenstates (see Section 9.1), MPU

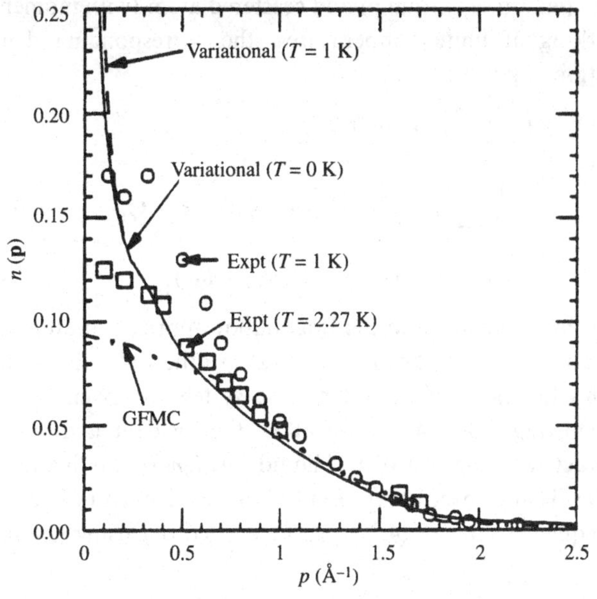

Fig. 4.2. The atomic momentum distribution $n(\mathbf{p})$ as given by three methods: variational; Green's function Monte Carlo GFMC (Whitlock and Panoff, 1987); and high-momentum $S(\mathbf{Q},\omega)$ data as analysed by Sears *et al.*, (1982) [Source: Manousakis, 1989].

show that the expression in (4.21*a*) leads to

$$\begin{aligned} n_p(T) &= n_p(T=0) + \delta n_p(T) \\ &= \frac{1}{2}\frac{n_0}{n}\frac{mc}{p} + \frac{n_0}{n}\frac{mc}{p}N(cp) \ , \end{aligned} \tag{4.21b}$$

in precise agreement with the result in (4.8).

4.2 Impulse approximation using a Green's function formulation

We now derive the impulse approximation formula (4.2) directly from a Green's function approach, working in momentum and frequency space. This allows us to see explicitly how the unique coherence factors associated with Bose condensation enter into the analysis which ultimately leads to the IA. This approach complements the usual derivation given in Section 2.3.

In the IA, the starting assumption is that the scattering cross-section can be approximated by the product of the single-particle propagators of the initial atom (with momentum $\mathbf{p}$) and scattered atom (with momentum $\mathbf{p}+\mathbf{Q}$). Working at finite temperatures, the corresponding density-response function is given by

$$\begin{aligned}\chi_{nn}(\mathbf{Q}, i\omega_l) &= n_0 \sum_{\alpha,\beta} G_{\alpha\beta}(\mathbf{Q}, i\omega_l)\\ &-\frac{1}{\beta}\sum_{\omega_n}\int \frac{d\mathbf{p}}{(2\pi)^3}\big[G_{11}(\mathbf{p}, i\omega_n)G_{11}(\mathbf{p}+\mathbf{Q}, i\omega_n + i\omega_l)\\ &+G_{12}(\mathbf{p}, i\omega_n)G_{12}(\mathbf{p}+\mathbf{Q}, i\omega_n + i\omega_l)\big] \ . \end{aligned} \quad (4.22)$$

This form is formally identical to the Bogoliubov approximation in (3.43) and (3.44), except that now the $G_{\alpha\beta}$ in (4.22) are the full Beliaev single-particle Green's functions of the interacting system, as given by (3.29). Basically, in starting from (4.22), we allow the incident and scattered atoms to interact with the rest of the liquid but *ignore* any interactions with each other. Using spectral representations (as defined in (6.25)), the Matsubara frequency sums in (4.22) can be carried out using (3.46), and we obtain

$$\begin{aligned}\chi_{nn}(\mathbf{Q}, i\omega_l) &= n_0 \sum_{\alpha,\beta} G_{\alpha\beta}(\mathbf{Q}, i\omega_l)\\ &+\int \frac{d\mathbf{p}}{(2\pi)^3}\int_{-\infty}^{\infty}\frac{d\omega'}{2\pi}\int_{-\infty}^{\infty}\frac{d\omega''}{2\pi}\left[\frac{N(\omega')-N(\omega'')}{i\omega_l-(\omega''-\omega')}\right]\\ &\times\big[A_{11}(\mathbf{p},\omega')A_{11}(\mathbf{p}+\mathbf{Q},\omega'')+A_{12}(\mathbf{p},\omega')A_{12}(\mathbf{p}+\mathbf{Q},\omega'')\big] \ . \end{aligned} \quad (4.23)$$

Here $A_{11} \equiv A$ is the spectral density of G_{11} and A_{12} is the spectral density of G_{12}. Making the change of variable $\mathbf{p}+\mathbf{Q} \to -\mathbf{p}'$ and $\omega' \leftrightarrow \omega''$, one can rewrite the first factor in the integrand of (4.23) as

$$N(\omega')\left\{\frac{1}{i\omega_l-(\omega''-\omega')}-\frac{1}{i\omega_l+(\omega''-\omega')}\right\} \ . \quad (4.24)$$

The characteristic simplifying feature of *high* momentum transfers is that for the scattered atom, we can use the approximations

$$\left.\begin{aligned}A_{11}(\mathbf{p}+\mathbf{Q},\omega'') &\simeq 2\pi\delta(\omega''-\varepsilon_{p+Q}) \ ,\\ A_{12}(\mathbf{p}+\mathbf{Q},\omega'') &\simeq 0 \ .\end{aligned}\right\} \quad (4.25)$$

That is to say, the Bose coherence factors disappear at high values of momentum and so does the anomalous Beliaev Green's function G_{12}. Inserting (4.25) into (4.23), we obtain

$$\chi_{nn}(\mathbf{Q}, i\omega_l) = n_0\big[G^0_{11}(\mathbf{Q}, i\omega_l)+G^0_{22}(\mathbf{Q}, i\omega_l)\big]$$

$$+\int \frac{d\mathbf{p}}{(2\pi)^3} \int_{-\infty}^{\infty} \frac{d\omega'}{2\pi} A(\mathbf{p},\omega')N(\omega')$$
$$\times \left\{ \frac{1}{i\omega_l - (\varepsilon_{p+Q} - \omega')} - \frac{1}{i\omega_l + (\varepsilon_{p+Q} - \omega')} \right\} . \quad (4.26)$$

If the dominant contribution to the ω' integration in (4.26) is from frequencies much less that ε_{p+Q} (see Section 4.1), we can use

$$\varepsilon_{p+Q} - \omega' \simeq \varepsilon_{p+Q} \simeq \omega_R + \frac{\mathbf{p}\cdot\mathbf{Q}}{m} . \quad (4.27)$$

Within this approximation, (4.26) finally reduces to (after the usual analytic continuation $i\omega_l \to \omega + i\eta$)

$$\chi_{nn}(\mathbf{Q},\omega) = n_0 \left(\frac{1}{\omega - \omega_R} - \frac{1}{\omega + \omega_R} \right)$$
$$+ \int \frac{d\mathbf{p}}{(2\pi)^3} \tilde{n}_p \left\{ \frac{1}{\omega - (\omega_R + \frac{\mathbf{p}\cdot\mathbf{Q}}{m})} - \frac{1}{\omega + (\omega_R + \frac{\mathbf{p}\cdot\mathbf{Q}}{m})} \right\} , \quad (4.28)$$

where $\tilde{n}_p$ is the *exact* excited-atom momentum distribution as defined in the last line of (4.3). Using this last expression in (3.40) leads directly to the IA in (4.2).

This kind of derivation (which can be easily extended to liquid ^{3}He and solid He) shows how, once we accept (4.22), the IA follows from specific properties of the single-particle spectral densities. In essence, it is a Green's function reformulation of an approach first used by Platzman and Tzoar (1965). The present derivation suggests an obvious extension to include final-state effects due to lifetime broadening of the scattered atom, namely by using a Lorentzian for G_{11} instead of (4.25). On the other hand, in starting from (4.22), we are precluded from consideration of any final-state effects due to correlations between the incident and scattered states.

4.3 Measurement of the atomic momentum distribution

As we noted in Section 2.3 and in the introductory remarks to this chapter, using the impulse approximation expression (4.2) to obtain information about the momentum distribution $n(\mathbf{p})$ is complicated because the *observed* $S(\mathbf{Q},\omega)$ contains

(a) Corrections to the impulse approximation: While small in the limit of large Q, these may be important for ω close to the recoil frequency

ω_R. These include both initial-state and final-state (FS) effects, the latter being associated with collisional effects on the recoiling atom.

(b) Instrumental-energy-resolution broadening effects.

In the experimental literature on superfluid ^{4}He, many different procedures have been used to carry out this comparison between the underlying momentum distribution and the observed $S(\mathbf{Q},\omega)$ data, with the ultimate goal of estimating the condensate density n_0. These procedures have become increasingly sophisticated and, in our opinion, increasingly convincing.

The first wave of experimental studies was in the early 1970's, including those of Cowley and Woods (1968, 1971), Harling (1970, 1971), Mook, Scherm and Wilkinson (1972) and Mook (1974). This stimulated several theoretical studies, especially concerning the final-state corrections. In particular, we call attention to Puff and Tenn (1970) and Gersch and Rodriguez (1973). It became clear that the values of n_0 obtained from the analysis of the high-momentum scattering data were quite model-dependent and, as such, the data might even be consistent with $n_0 = 0$ (as suggested by Jackson, 1974).

The next stage was initiated by the work of Martel, Svensson, Woods, Sears and Cowley (1976), who emphasized the need to remove instrumental-resolution broadening as well as final-state effects in a well defined manner. $S(\mathbf{Q},\omega)$ can be divided into symmetric and antisymmetric contributions, relative to $\omega-\omega_R$. $S_{\mathrm{IA}}(\mathbf{Q},\omega)$ in (2.39) is clearly symmetric. Sears (1969) showed that if the interatomic potential does not have a hard core, the leading-order additive correction S_1 to the impulse approximation, as defined by

$$S(\mathbf{Q},\omega) = S_{\mathrm{IA}}(\mathbf{Q},\omega) + S_1(\mathbf{Q},\omega) \ + \dots \ , \qquad (4.29)$$

is of order $1/Q$, antisymmetric and entirely due to final-state effects. Consequently the Chalk River group (Martel *et al.*, 1976; Woods and Sears, 1977; Sears *et al.*, 1982) argued that the final-state effects can be largely removed from the $S(\mathbf{Q},\omega)$ data by first symmetrizing it with respect to $\omega-\omega_R$. The remaining part is identified with the instrumentally broadened $S_{\mathrm{IA}}(\mathbf{Q},\omega)$ and $n(\mathbf{p})$ is extracted from this by direct numerical differentiation with respect to ω. The adequacy of this procedure in the case of superfluid ^{4}He has become the subject of some recent debate (see Sosnick, Snow, Silver and Sokol, 1991). However, at the time it was a distinct improvement over earlier studies and allowed Woods and Sears (1977) to obtain more trustworthy results for $n(\mathbf{p})$ from reactor data in

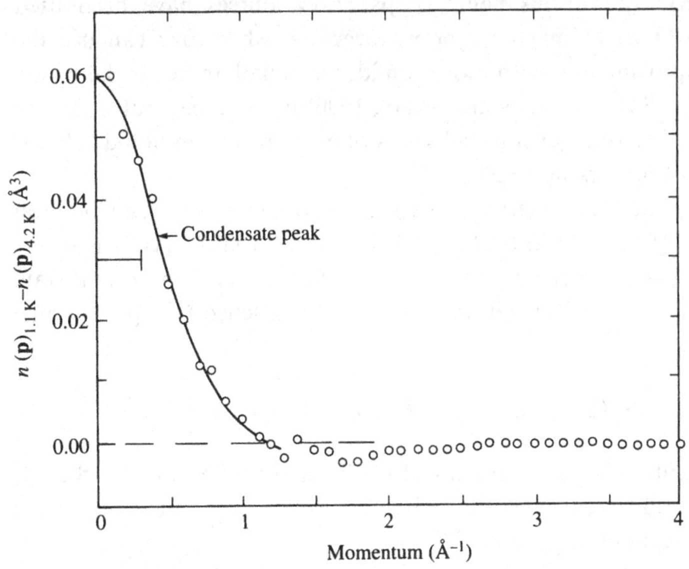

Fig. 4.3. Momentum distribution difference in liquid ^{4}He at $T = 1.1$ and 4.2 K [Source: Woods and Sears, 1977].

the range $6 \leq Q \leq 8$ Å^{-1} at 1.1 and 4.2 K (see Fig. 4.3). This led to an important paper by Sears, Svensson, Martel and Woods (1982), which was based on a similar analysis of data in the range $5 \leq Q \leq 7$ Å^{-1} at a series of temperatures above and below T_λ. This was the first data analysis which tried to incorporate the condensate-induced low-momentum singularity in $n(\mathbf{p})$ discussed in Section 4.1. Their final results for $n(\mathbf{p}, T)$ are shown in Fig. 2.9.

As mentioned in Section 2.2, the final-state corrections to the IA can be incorporated either using a convolution with an FS resolution function as in (2.47b) or as an additive correction, as in (4.29). The latter procedure is very convenient computationally (Glyde, 1992b) but it is only really appropriate when one is dealing with a broad momentum distribution at low p (as in normal liquid ^{4}He, where $n(\mathbf{p})$ is very close to a Gaussian). In contrast, when one is looking for anomalous behaviour in $n(\mathbf{p})$ at low momentum as in the case of superfluid ^{4}He, the convolution approach (2.47b) to FS effects is best. Of course, in both approaches, one must also fold in the inevitable broadening due to the finite energy resolution of the neutron spectrometers.

More recently, various neutron spallation sources have been used to study $S(\mathbf{Q},\omega)$ at much larger wavevectors, where one can use the impulse approximation with more confidence. Sokol and coworkers (see, in particular, Sosnick, Snow and Sokol, 1990) have carried out what one might call a "second generation" series of experiments on liquid ^{4}He for momentum transfers up to 23 Å^{-1}.

Following the recent literature, we shall work in terms of the Compton profile $J(Y,Q)$ defined in Section 2.3. For a given $n(\mathbf{p})$, $J_{\rm IA}(Y)$ is determined by (2.44). Expressing the IA corrections in terms of a final-state broadening function $R(Y,Q)$, the final-state broadened Compton profile is given by

$$J_{\rm FS}(Y,Q) \equiv \int_{-\infty}^{\infty} dY' R_{\rm FS}(Y-Y',Q) J_{\rm IA}(Y') \tag{4.30}$$

This convolution is the equivalent of (2.47*a*) and (2.47*b*). Finally, the $J_{\rm FS}$ given by (4.30) is broadened by finite *instrumental* resolution to give the "observed" Compton profile

$$J_{\rm obs}(Y,Q) \equiv \int_{-\infty}^{\infty} dY' I(Y-Y',Q) J_{\rm FS}(Y',Q) \; . \tag{4.31}$$

What we shall call Method I attempts to find $J_{\rm IA}(Y)$ from $J_{\rm obs}(Y,Q)$ in (4.31) by effectively deconvoluting (or otherwise removing) the effects of the instrumental and final-state broadening. One then numerically differentiates the resulting $J_{\rm IA}(Y)$ to obtain the atomic momentum distribution,

$$n(p) = -\frac{1}{2\pi p}\frac{dJ_{\rm IA}(Y)}{dY}\bigg|_{|Y|=p} \; . \tag{4.32}$$

In a broad sense, this is the procedure originally used by the Chalk River group in 1982 with the results shown in Fig. 2.9, as well as in most earlier determinations of the atomic momentum distribution. Method I inevitably leads to errors since it involves deconvolutions and numerical differentiation of raw data.

Sosnick, Snow and Sokol (SSS, 1990) suggested a more modest procedure, which we shall call Method II. This is based on starting with an *assumed* form for $n(\mathbf{p})$ and then, as described above, calculating $J_{\rm IA}, J_{\rm FS}$ and finally $J_{\rm obs}$ using (2.44), (4.30) and (4.31), respectively. One then compares this "theoretical" value of $J_{\rm obs}(Y,Q)$ with the experimental Compton profile data.

SSS have used Method II with several different starting assumptions about the form of $n(\mathbf{p})$. Before we discuss these, we note that (4.31)

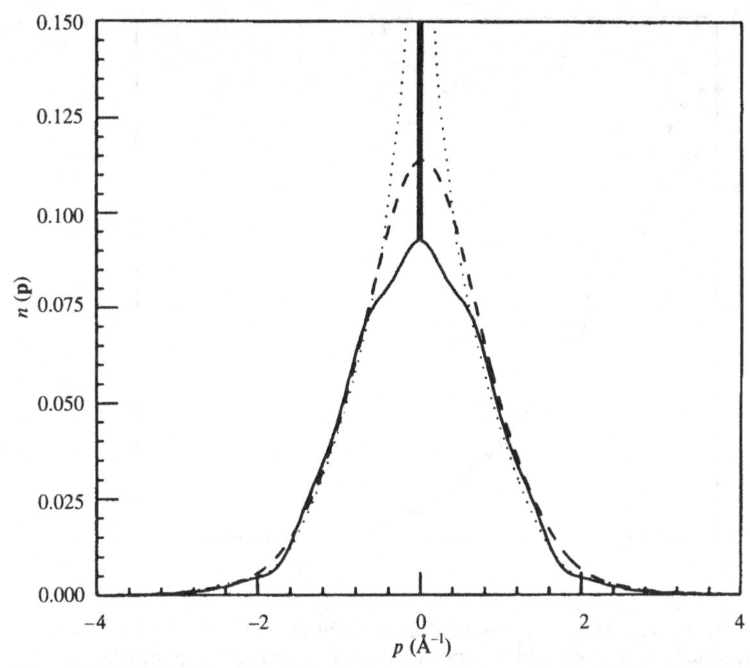

Fig. 4.4. Theoretical results for $n(\mathbf{p})$ at $T = 0$ given by GFMC (solid line: Whitlock and Panoff, 1987) and variational (dotted line: Manousakis, Pandharipande and Usmani, 1985) techniques. For comparison, $n(\mathbf{p})$ is shown for normal liquid ^{4}He as given by PIMC (dashed line: Ceperley and Pollock, 1986) [Source: Sokol, Sosnick and Snow, 1989].

already indicates that J_{IA} (and hence J_{obs}) will not be very dependent on the low-p behaviour of $\tilde{n}(\mathbf{p})$ because of the factor p in the integrand of $J_{\mathrm{IA}}(Y)$ in (2.44). This insensitivity is shown dramatically by the similarity of $J_{\mathrm{IA}}(Y)$ which results from using the three theoretical momentum distributions shown in Fig. 4.4. On the other hand, it is precisely the $p = 0$ and small-p region (see Section 4.1) where $\tilde{n}(\mathbf{p})$ is altered with the appearance of a Bose-condensate. Thus the Compton profile in liquid ^{4}He will only exhibit relatively small changes in the region $Y \simeq 0$ when we go from above to well below T_λ. The task at hand is to make sure that the analysis of the scattering data is sufficiently good so the small changes observed in $J_{\mathrm{obs}}(Y, Q)$ near $Y \simeq 0$ are significant. It is an experimental *tour de force* that this has now been accomplished in a convincing manner.

In using Method II, it is important to remember that the form of

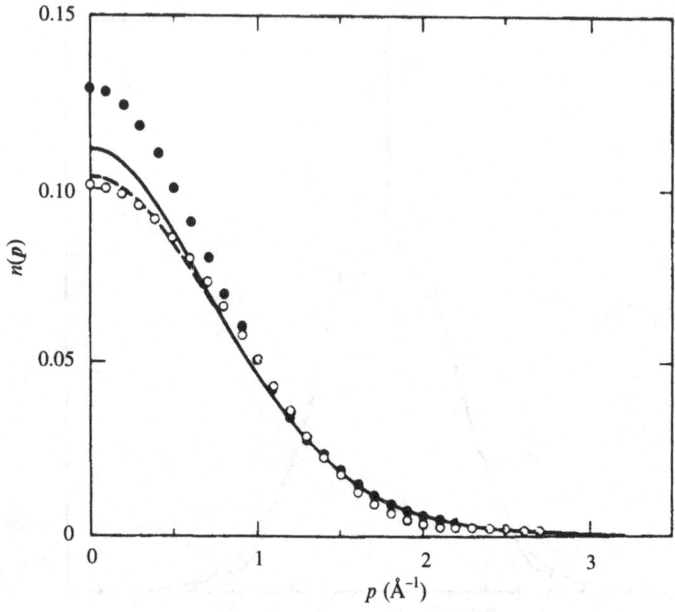

Fig. 4.5. PIMC results for the momentum distribution at 3.33 K (solid curve), 2.22 K (dashed curve), and 1.18 K (open circles). The excited-atom distribution at low momentum is being depleted, forming the condensate. For comparison, the solid circles show results at 2.22 K if one ignores Bose symmetrization [Source: Ceperley and Pollock, 1986].

the momentum distribution $n(\mathbf{p})$ for the non-condensate atoms is *not* independent of the condensate value. Both are interdependent in the real world as well as in any satisfactory microscopic calculation! Accurate numerical calculations of $n(\mathbf{p})$ (see Fig. 4.4) and the associated condensate density n_0 are now available. At $T = 0$, these include variational calculations based on the Hypernetted-Chain-Scaling HCN/S approximation (Manousakis, Pandharipande and Usmani, 1985) as well as Green's function Monte Carlo (GFMC) results (Whitlock and Panoff, 1987). Both calculations give a $T = 0$ condensate fraction of 9.2%. In Fig. 4.5, we show the PIMC results of Ceperley and Pollock (1986, 1987) for $T \gtrsim 1$ K (including above T_λ).

With these theoretical results for $n(\mathbf{p})$, SSS (1990) have used Method II and computed $J_{\rm obs}(Y, Q)$ with (2.44), (4.30) and (4.31), for comparison with data at $Q = 23$ Å^{-1} taken at the Intense Pulsed Neutron Source (IPNS) at the Argonne National Laboratory. In Figs. 4.6 and 4.7, we compare the predicted $J_{\rm obs}(Y, Q)$ with the experimental Compton profiles

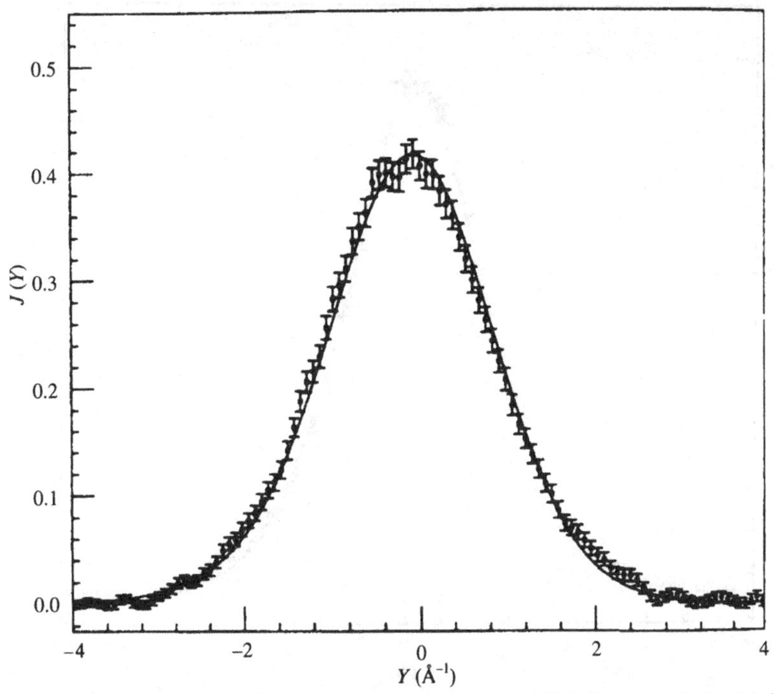

Fig. 4.6. Observed neutron scattering (Compton profile) in normal liquid ^{4}He at $T = 3.5$ K. The line is the impulse approximation based on the PIMC $n(\mathbf{p})$ results of Ceperley and Pollock (1986), broadened by instrumental-resolution and final-state effects. The latter effects are treated following Silver (1988, 1989). The resolution functions used here are shown in Fig. 4.8 [Source: Sosnick, Snow and Sokol, 1990].

at $T = 3.5$ K (normal ^{4}He) and $T = 0.35$ K (superfluid). The final-state $R_{\text{FS}}(Y)$ and resolution-broadening $I(Y)$ functions used in (4.30) and (4.31) are shown in Fig. 4.8. The final-state broadening function which is used is that recently developed by Silver (1988, 1989), whose work may be viewed as a generalized version of the pioneering study by Gersch and Rodriguez (1973). For more recent theoretical work on final-state effects, see Carraro and Koonin (1990) and Carraro and Rinat (1992). For a detailed experimental study of various theories of final-state effects see Sosnick *et al.* (1991) and Herwig *et al.* (1991).

In a second application of Method II, SSS (1990) introduce a modified expression for $n(\mathbf{p})$, based on slightly altering the number of atoms in the condensate. The non-condensed atomic distribution is assumed to

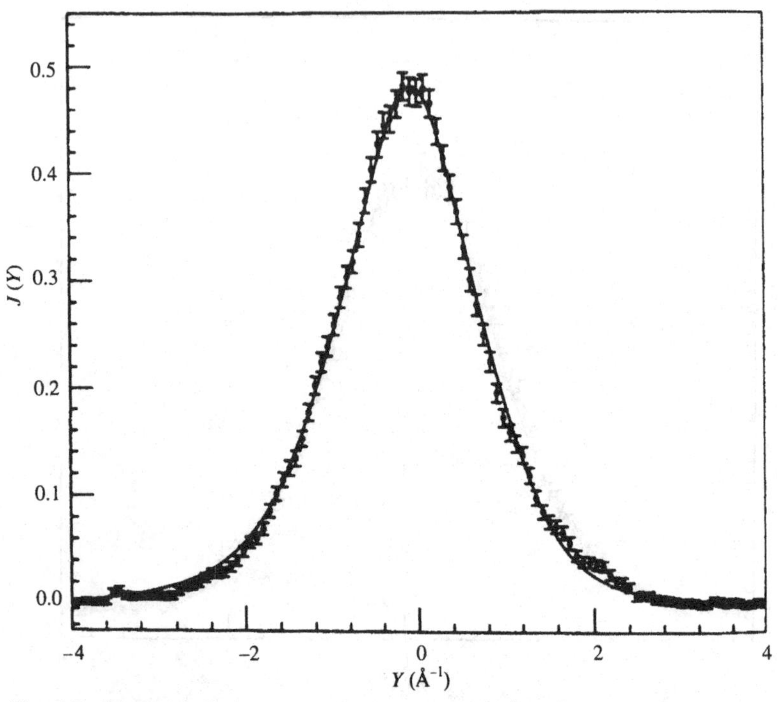

Fig. 4.7. Observed Compton profile in superfluid ^{4}He at $T = 0.35$ K. The line is the theoretical result based on the $T = 0$ GFMC momentum distribution of Whitlock and Panoff (1987), broadened as described in Fig. 4.6 [Source: Sosnick, Snow and Sokol, 1990].

have the *same* shape it had when $\bar{n}_0 = 0.092$ but is now normalized to contain only a fraction $1 - \bar{n}_0$ of the atoms. As shown in Fig. 4.9, this *ad hoc* model expression for $n(\mathbf{p})$ leads to a $J_{\text{obs}}(Y, Q)$ via Method II which agrees with the observed Compton profile for an assumed condensate fraction $\bar{n}_0$ of 10%, within a range $\pm 2\%$. As SSS note, while this analysis is based on a model expression for $n(\mathbf{p})$ which does *not* correspond to any microscopic calculation, it does give an estimate of the range of uncertainty in the value of $\bar{n}_0$ obtained in these studies.

At $T = 3.5$ K, the PIMC $n(\mathbf{p})$ is very close to a Gaussian, as is the momentum distribution of solid Helium and most normal liquids (see Figs. 4.1 and 4.5). When the low-p behaviour of $n(\mathbf{p})$ is a smooth, broad function like this, the final-state broadening (the effect of which is concentrated at small Y) is not expected to have much effect. Indeed, SSS find $J_{\text{IA}}(Y)$ and $J_{\text{FS}}(Y)$ are almost identical at all temperatures *above* T_λ.

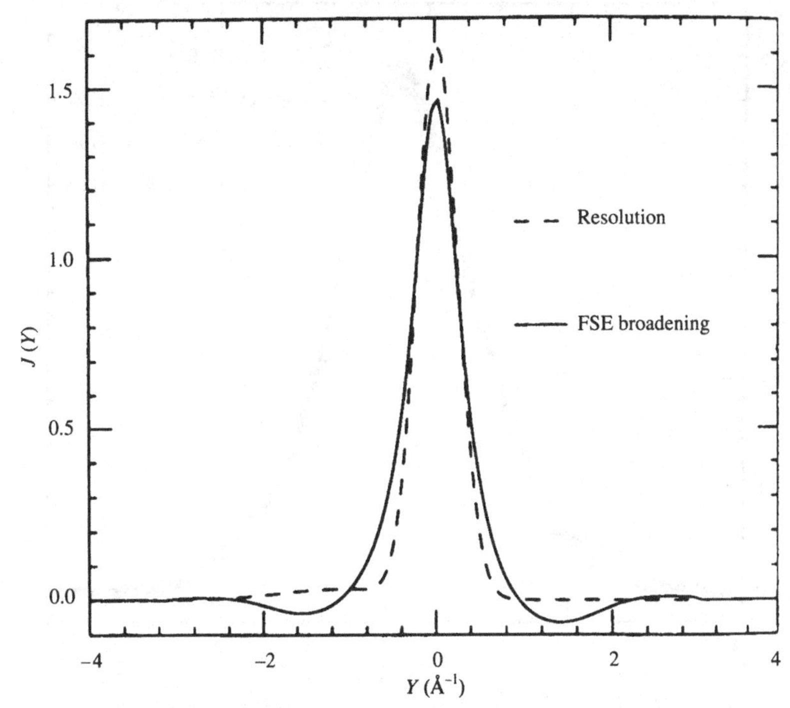

Fig. 4.8. Resolution functions at $Q = 23$ Å^{-1} for instrumental broadening $I(Y)$ and final-state broadening $R_{\mathrm{FS}}(Y)$ as given by Silver's theory [Source: Sosnick, Snow and Sokol, 1990].

In contrast, the *quantitative* agreement with the calculated and measured Compton profiles $J_{\mathrm{obs}}(Y, Q)$ shown in Fig. 4.7 for $T = 0.35$ K disappears when the FS broadening is not included (see Fig. 4.10). As mentioned earlier, in order to treat the small (but all-important) corrections near $Y \simeq 0$, it seems preferable to include the FSE via a broadening function as in (4.30) rather than as an additive correction (4.29).

At a more quantitative level, the PIMC calculations of Ceperley and Pollock (1986), as well as experiment (Herwig *et al.*, 1990) show that $n(\mathbf{p})$ in the normal phase exhibits non-Gaussian behaviour in the *high*-momentum tails. Glyde (1992*b*) has emphasized that in the normal phase, it is more useful to view the final-state corrections as additive corrections to the impulse approximation, as in (4.29). The constraints imposed by the central moment sum rules (see end of Section 2.1) can then be used to extract information about the non-Gaussian behaviour of $n(\mathbf{p})$ at large momentum.

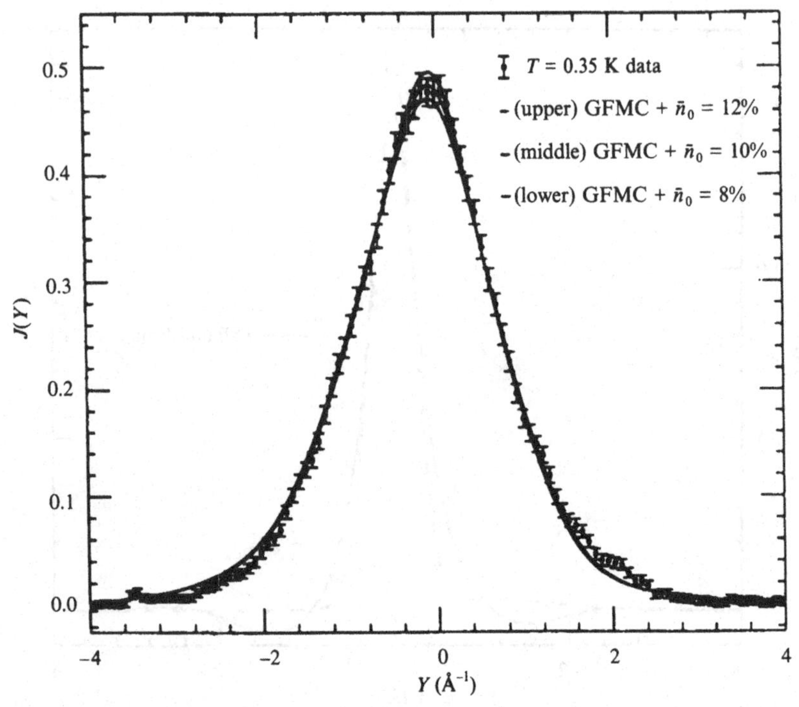

Fig. 4.9. Observed Compton profile in superfluid ^{4}He at $T = 0.35$ K (see Fig. 4.7) compared with theoretical renormalized GFMC results for three different values of $\bar{n}_0$ [Source: Sosnick, Snow and Sokol, 1990].

4.4 Extraction of the condensate fraction

Recent work has made use of the ingenious procedure introduced by Sears, Svensson, Martel and Woods (1982) to extract n_0 from a given atomic distribution $n(\mathbf{p})$. In their original work, Sears *et al.* (1982) worked with the momentum distribution given in Fig. 2.9 which they obtained using Method I (see Section 4.3), the final-state effects having being removed by averaging and antisymmetrizing the $S(\mathbf{Q}, \omega)$ data.

We now summarize a third version of Method II (as defined in the preceding section) used by Sokol *et al.* (1989) to directly extract n_0 from the $S(\mathbf{Q}, \omega)$ data. They introduce a model form for $J_{\text{IA}}(Y)$ as sum of *two* Gaussians,

$$J_{\text{model}}(Y) \equiv \sum_i \frac{a_i}{(2\pi\sigma_i^2)^{1/2}} e^{-(Y-Y_c)^2/2\sigma_i^2} . \tag{4.33}$$

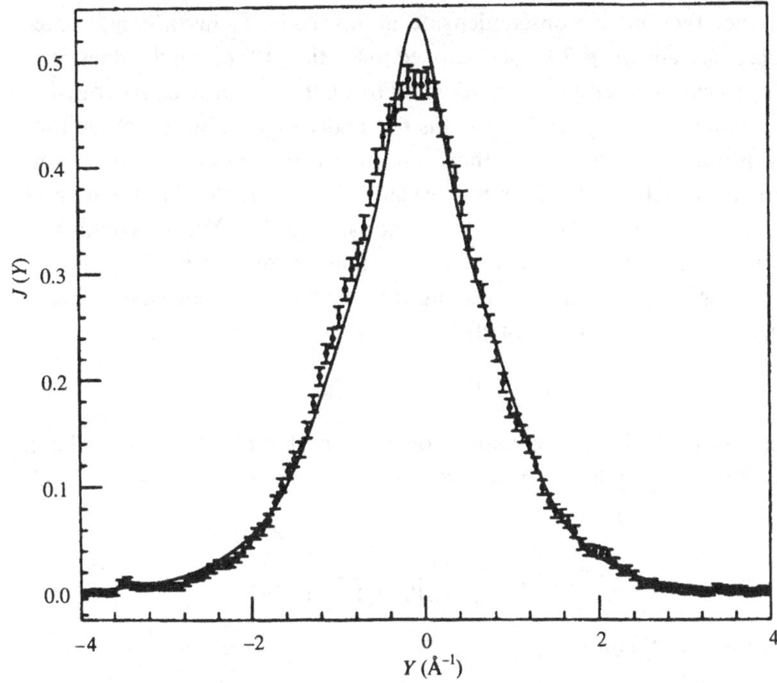

Fig. 4.10. Same as Fig. 4.7 except that the calculated Compton profile is not broadened by final-state effects [Source: Sosnick, Snow and Sokol, 1990].

This model function is convoluted using (4.30) and (4.31) to obtain $J_{\text{obs}}(Y,Q)$. The Gaussian amplitudes, widths and (common) centre Y_c are then adjusted to give the best agreement between J_{obs} and the experimental Compton profile. It was found that excellent fits could be made at all temperatures using only *two* Gaussians in (4.33), one of which is considerably narrower than the other. With this momentum distribution, Sokol *et al.* (1989) used the "Chalk River" procedure (Sears *et al.*, 1982) for extracting a value of n_0. It is assumed that $n(\mathbf{p})$ in superfluid ^{4}He can be written in the form

$$n(\mathbf{p},T) = \bar{n}_0(T)\delta(\mathbf{p}) + \big(1-\bar{n}_0(T)\big)n(\mathbf{p},T_\lambda) + \delta n_c(\mathbf{p},T) \ , \tag{4.34}$$

where we recall that $n(\mathbf{p})$ is normalized to unity:

$$\int d\mathbf{p}\ n(\mathbf{p},T) = 1 \ . \tag{4.35}$$

In (4.34), the non-condensate atom distribution $\tilde{n}(\mathbf{p},T)$ in the superfluid is assumed to be the same as in the *normal* liquid at T_λ (but scaled with the fraction of non-condensate atoms $1 - \bar{n}_0$) plus a condensate-induced change $\delta n_c(\mathbf{p}, T)$. As discussed in Section 4.1, at *small* values of p, $\delta n_c(\mathbf{p}, T)$ can be adequately represented by (4.9). How δn_c behaves outside this phonon region ($p \gtrsim 0.7$ Å^{-1}) is not really known, but we argue that it vanishes, based on the idea that, *above* a certain cross-over wavevector, the quasiparticle spectrum is intrinsically single-particle-like (Glyde and Griffin, 1990). The associated atomic momentum distribution would then *not* be affected by the presence of a condensate (see Section 7.2).

In using (4.34), a further assumption is that the condensate-induced change is of the form (see (4.9))

$$\delta n_c(\mathbf{p}, T) = \bar{n}_0(T)\delta n^*(\mathbf{p}, T) \ , \tag{4.36}$$

where $\delta n^*(\mathbf{p}, T)$ is *not* dependent on n_0. Combining (4.36) with (4.34), and integrating up to some wavevector p_c, one obtains (Sears *et al.*, 1982; Sokol *et al.*, 1989)

$$\bar{n}_0(T, p_c) = \frac{\alpha(T, p_c) - \alpha(T_\lambda, p_c)}{1 - \alpha(T, p_c) + \gamma(T, p_c)} \ , \tag{4.37}$$

where we have defined

$$\left.\begin{aligned} \alpha(T, p_c) &\equiv \int_0^{p_c} dp \ 4\pi p^2 n(p, T) \ , \\ \gamma(T, p_c) &\equiv \int_0^{p_c} dp \ 4\pi p^2 \delta n^*(p, T) \ . \end{aligned}\right\} \tag{4.38}$$

The value of the condensate fraction $\bar{n}_0(T)$ given by (4.37) should be independent of the value of the cutoff p_c chosen; this is not the case, as shown in Fig. 4.11. However, the decrease in the value of $\bar{n}_0(p_c)$ exhibited in Fig. 4.11 at large and small values of p_c is expected. The low-p behaviour of $n(\mathbf{p})$ in the superfluid phase is not correctly approximated by the two-Gaussian functional form (4.33) used for $n(\mathbf{p}, T)$. At high p, the difference between the superfluid and normal phase momentum distributions is very small. We agree with the argument of Sokol *et al.* (1989) that the peak value of $\bar{n}_0$ as a function of p_c gives the most accurate estimate. This always occurs at p_c in the region 0.6–0.8 Å^{-1}, where (4.9) and (4.36) should be valid (Griffin, 1985). Using these, one finds $\gamma(T, p_c) = 0.85p_c^2$, where p_c is measured in Å^{-1}.

Fig. 4.12 shows the values for $\bar{n}_0(T)$ obtained by Sokol, Sosnick and Snow (1989) by the procedure described above. For comparison, the Monte Carlo path-integral results of Ceperley and Pollock (1986) are also plotted. Earlier estimates by Sears *et al.* (1982) and Mook (1983)

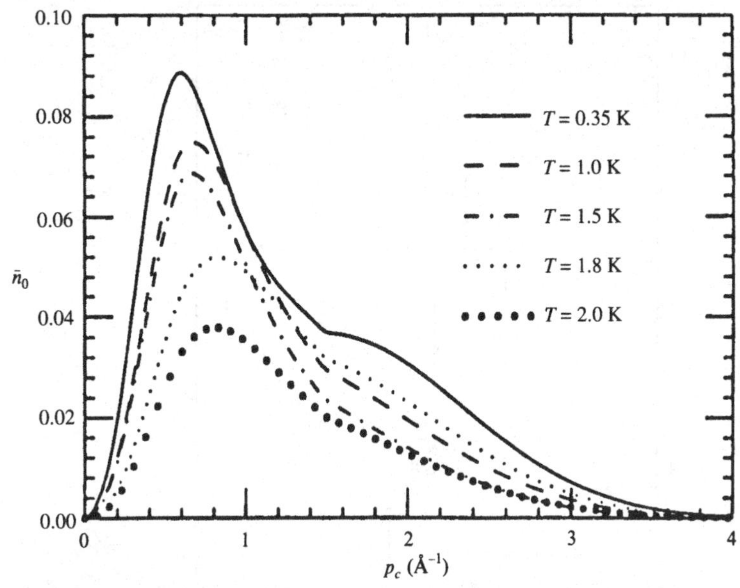

Fig. 4.11. Inferred values of $\bar{n}_0$ at several temperatures as a function of the cutoff momentum p_c. See discussion after (4.38) [Source: Sokol, Sosnick and Snow, 1989].

used an inadequate treatment of the key $\gamma(T, p_c)$ contribution in (4.37), which led to an overestimate of $\bar{n}_0$ (Griffin, 1985).

We have given a careful analysis of the expression in (4.37) because it was the basis of the first adequate analysis of $S(\mathbf{Q}, \omega)$ data to extract the condensate density. However, (4.37) is based on the form (4.34), which assumes that the non-condensate atom momentum distribution in superfluid ^{4}He has a component which scales with that in the normal phase. An improved version which does *not* rely on this assumption has been developed by Sokol and Snow (1991). They assume that $n(\mathbf{p})$ is given by

$$n(\mathbf{p}) = \left\{ \begin{array}{ll} \bar{n}_0[\delta(\mathbf{p}) + f(p)] + A_1 e^{-p^2/2\sigma_1^2} & \text{for } p \leq p_c\,, \\ A_2 e^{-p^2/2\sigma_2^2} & \text{for } p > p_c\,, \end{array} \right\} \tag{4.39}$$

where $\bar{n}_0 f(p)$ is the condensate-induced low-p anomaly in $n(\mathbf{p})$. This form for $n(\mathbf{p})$ incorporates what is known rigorously from microscopic theory (see Section 4.1): it is positive, normalized to unity, continuous at p_c and symmetric about $p = 0$. Fortunately, the parameters in (4.39) are mainly

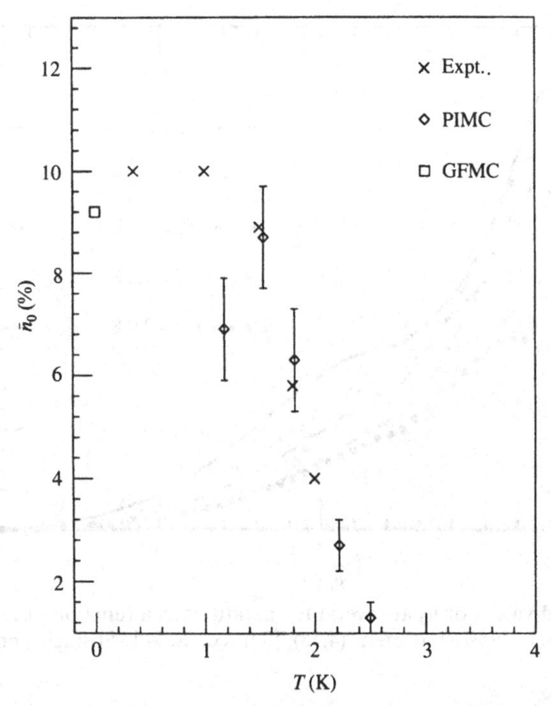

Fig. 4.12. Temperature dependence of $\bar{n}_0$ (crosses) based on (4.37) and (4.38). The theoretical results given by GFMC (square) and PIMC (diamonds) are shown for comparison [Source: Snow, 1990].

sensitive to different wavevector regions. The choice of p_c is subject to the considerations discussed after (4.38). As in Section 4.3, (4.39) is used to calculate the equivalent $J_{\mathrm{IA}}(Y)$, which is then broadened by convolution with the instrumental and FS resolution functions. Finally, the parameters in (4.39) are determined by a fit to the observed Compton profile data. Snow, Wang and Sokol (1992) find a value of $\bar{n}_0 \simeq 10 \pm 1.25\%$ at $T = 0.75$ K at the density 0.14 g/cm^3, in good agreement with earlier results. In principle, this new method of analysis should give the best extracted values for $\bar{n}_0(T)$.

Sokol and Snow (1991) (see also Snow, 1990) have also used such procedures to determine $\bar{n}_0$ as a function of the density at a fixed temperature of 0.75 K (see Fig. 4.13). While $\bar{n}_0$ is found to decrease as the first-order superfluid–solid phase transition is approached, the data indicates that $\bar{n}_0$ remains large and finite in the superfluid phase and then abruptly vanishes in the solid phase. More studies of condensate fraction in this

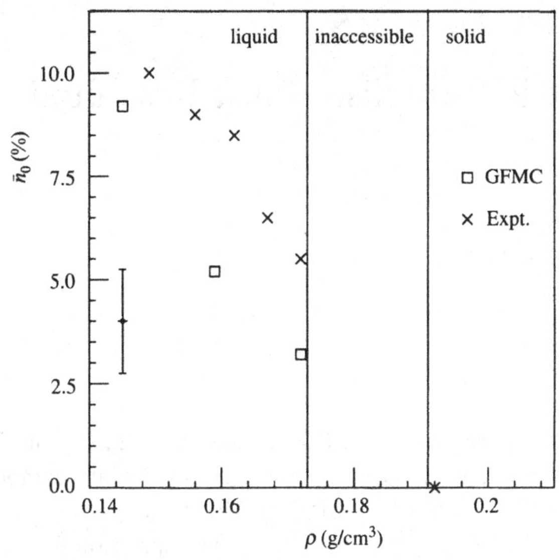

Fig. 4.13. Condensate fraction $\bar{n}_0$ as a function of the density at 0.75 K. The theoretical predictions based on GFMC are also shown [Source: Snow, 1990].

phase boundary region would be useful. This application emphasizes the utility of having tested techniques for determining $n(\mathbf{p})$ and $\bar{n}_0$ in more exotic situations, such as superfluid ^{4}He in porous media and in ^{3}He–^{4}He mixtures, where numerical simulations are not as well developed.

In concluding this section, we note that neither the numerical nor experimental results for $n_0(T)$ are yet accurate enough to determine the precise temperature dependence near T_λ (see Fig. 4.12). In the critical region, $\sqrt{n_0(T)}$ and $\rho_S(T)$ are predicted to vanish as

$$\left.\begin{aligned} \sqrt{n_0} &\sim (T_\lambda - T)^\beta \ , \\ \rho_S &\sim (T_\lambda - T)^{(2-\alpha)/3} \ , \end{aligned}\right\} \qquad (4.40)$$

based on a two-component order parameter. Since one finds that $\beta \simeq \frac{1}{3}$ and $\alpha \simeq 0$, we see that both n_0 and ρ_S should vanish approximately as $(T_\lambda - T)^{\frac{2}{3}}$ in bulk superfluid ^{4}He (see p. 111 of Lifshitz and Pitaevskii, 1980).

5
Dielectric formalism for a Bose fluid

In Chapter 3, we have shown how a Bose condensate leads to the single-particle Green's function $G_{\alpha\beta}(\mathbf{Q},\omega)$ appearing as a distinct component of the the dynamic structure factor $S(\mathbf{Q},\omega)$. This feature is a direct consequence of (3.13) and is explicitly shown by the expression for $S(\mathbf{Q},\omega)$ in (3.18). In the present chapter, we review the many-body diagrammatic analysis which specifies the precise relation between $G_{\alpha\beta}(\mathbf{Q},\omega)$ and $S(\mathbf{Q},\omega)$ in a Bose-condensed fluid ($n_0 \neq 0$). The power of this "dielectric formalism" lies in showing that both functions mirror each other, as a result of the condensate-induced intermixing of single-particle and density excitations. At the same time, the formalism emphasizes the *continuity* of these excitations as the fluid passes through T_λ, although these excitations have very temperature-dependent weights in $G_{\alpha\beta}(\mathbf{Q},\omega)$ and $S(\mathbf{Q},\omega)$. While it is the poles of the single-particle Green's function which describe the elementary excitations, experimentally it is difficult to probe directly this field-fluctuation spectrum. In particular, it is only in a Bose-condensed fluid that one can obtain information about $G_{\alpha\beta}(\mathbf{Q},\omega)$ from inelastic neutron-scattering data on $S(\mathbf{Q},\omega)$. In the absence of a condensate, there is little relation between the density-fluctuation modes and the single-particle excitations in a Bose fluid.

After setting up the dielectric formalism in Section 5.1, the next two sections consider some model approximations at finite temperatures. These models are most appropriate to a dilute weakly interacting Bose gas (WIDBG), but we believe they also give crucial insight into various aspects of superfluid ^{4}He. In Section 5.4, we summarize the famous study of Gavoret and Nozières (1964) and relate it to the analysis given in Section 5.1. Their explicit conclusions at $T = 0$ in the limit of small Q and ω are discussed at length in Section 6.3.

The Bose condensate also leads to a coupling of the two-particle (or pair) spectrum with the single-particle and density excitations. While we allude to this briefly in this chapter, a detailed analysis of the two-particle spectrum is deferred to Chapter 10.

Sections 5.1–5.4 are somewhat technical but the basic results we need in connection with superfluid ^{4}He are simple. In Section 5.5, we summarize these key features of the dielectric formulation and give a physical interpretation in two important limits: the collective zero sound (ZS) and the single-particle (SP) regions.

As we noted in Section 1.2, the dielectric formalism had its beginnings in the seminal work of Gavoret and Nozières (1964) and Gavoret (1963), and was later formalized by Ma and Woo (1967) and Kondor and Szépfalusy (1968). It was then developed by Wong, Gould and Ma as an efficient way of doing perturbative calculations for a WIDBG that manifestly fulfilled conservation laws. Their work concentrated on $T = 0$ (Wong and Gould, 1974) or at low temperatures (Ma, Gould and Wong, 1971; Wong and Gould, 1976), when most of the atoms are in the condensate ($n_0 \simeq n$). However, the dielectric formalism is valid at all temperatures and can be used to understand the finite-temperature behaviour of $G_{\alpha\beta}(\mathbf{Q},\omega)$ and $S(\mathbf{Q},\omega)$ right through T_λ. This aspect of the dielectric formalism was first emphasized by Griffin and Cheung (1973) as well as Szépfalusy and Kondor (1974).

5.1 Dielectric formalism

In diagrammatic perturbation theory, it is useful to introduce *irreducible* quantities. These are defined as diagrammatic contributions which cannot be split into two parts by cutting a single *interaction* line. For example, the density-response function χ_{nn} in (3.41) can be written as (see Fig. 5.1)

$$\chi_{nn} = \bar{\chi}_{nn} + \bar{\chi}_{nn} V(Q) \bar{\chi}_{nn} + \bar{\chi}_{nn} V(Q) \bar{\chi}_{nn} V(Q) \bar{\chi}_{nn} + \dots , \tag{5.1}$$

where $\bar{\chi}_{nn}$ is the irreducible component of χ_{nn}. This can be immediately summed up as a geometric series to give

$$\chi_{nn} = \frac{\bar{\chi}_{nn}}{1 - V(Q)\bar{\chi}_{nn}} \equiv \frac{\bar{\chi}_{nn}}{\epsilon(\mathbf{Q},\omega)} , \tag{5.2}$$

where the denominator defines the "dielectric function" $\epsilon(\mathbf{Q},\omega)$. An irreducible contribution to a correlation function will hereafter be denoted by a bar. One can further separate out diagrammatic contributions which are *proper*, i.e. those which cannot be separated into two parts by cutting

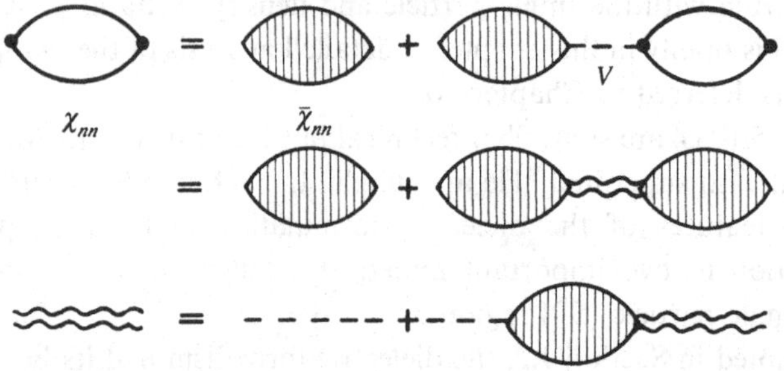

Fig. 5.1. Diagrammatic representation of the density-response function (5.1) in terms of the irreducible contribution $\bar{\chi}_{nn}$.

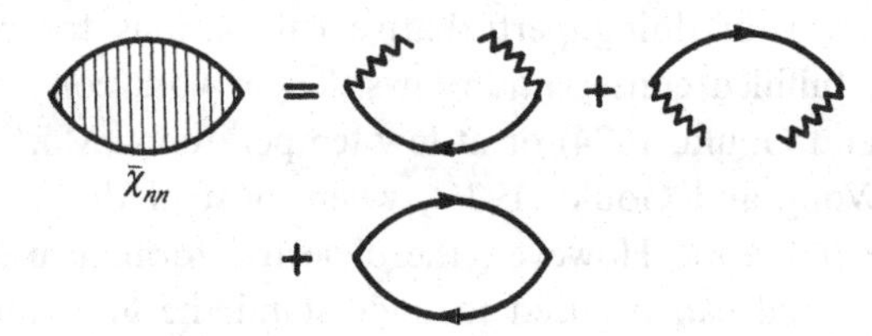

Fig. 5.2. Lowest-order improper and proper contributions to the irreducible density-response function $\bar{\chi}_{nn}$ for $Q \neq 0$ (see Fig. 3.1 for notation).

one line representing a single-particle propagator. Some examples of "improper" contributions to $\bar{\chi}_{nn}$ are shown in Fig. 5.2. A proper, irreducible contribution will be referred to as a "regular" part and will be denoted by a superscript R (in many-body literature, we caution, the meanings of the terms "proper" and "irreducible" are sometimes interchanged). As we shall see, the characteristic structure of the various correlation functions in a Bose-condensed fluid is clarified when they are expressed in terms of "regular" functions, as defined here.

We first need to define some additional correlation functions involving the single-particle operators $\hat{a}_{Q\alpha}$ in (3.21), the number density $\hat{\rho}(\mathbf{Q})$ in (3.12) and the current-density operator defined by

$$\begin{aligned} \mathbf{J}(\mathbf{Q}) &= \sum_{\mathbf{p}} \left(\mathbf{p} + \frac{\mathbf{Q}}{2}\right) \hat{a}_p^+ \hat{a}_{p+Q} \\ &= \sum_{\mathbf{p}\neq 0,-\mathbf{Q}} \left(\mathbf{p} + \frac{\mathbf{Q}}{2}\right) \hat{a}_p^+ \hat{a}_{p+Q} + \frac{\sqrt{N_0}}{2}\mathbf{Q}(\hat{a}_Q - \hat{a}_{-Q}^+) \ . \end{aligned} \tag{5.3}$$

Besides (3.20) and (3.41), the following correlation functions will be needed:

$$\left.\begin{aligned}
-\chi_{n\alpha}(\mathbf{Q},\tau) &\equiv \tfrac{1}{\sqrt{\Omega}}\langle T\hat{\rho}(\mathbf{Q},\tau)\hat{a}^{+}_{\mathbf{Q}\alpha}\rangle \ , \\
-\chi_{i\alpha}(\mathbf{Q},\tau) &\equiv \tfrac{1}{\sqrt{\Omega}}\langle T\hat{J}_i(\mathbf{Q},\tau)\hat{a}^{+}_{\mathbf{Q}\alpha}\rangle \ , \\
-\chi_{in}(\mathbf{Q},\tau) &\equiv \tfrac{1}{\Omega}\langle T\hat{J}_i(\mathbf{Q},\tau)\hat{\rho}(-\mathbf{Q})\rangle \ , \\
-\chi_{ij}(\mathbf{Q},\tau) &\equiv \tfrac{1}{\Omega}\langle T\hat{J}_i(\mathbf{Q},\tau)\hat{J}_j(-\mathbf{Q})\rangle \ ,
\end{aligned}\right\} \qquad (5.4)$$

where the indices i, j represent Cartesian components of the current operator in (5.3). As with $G_{\alpha\beta}$ in (3.29), these correlation functions can be shown to satisfy the symmetry relations (Wong and Gould, 1974; Talbot, 1983)

$$\left.\begin{aligned}
\chi_{nn}(\mathbf{Q},\omega) &= \chi_{nn}(Q,\omega) = \chi_{nn}(Q,-\omega) \ , \\
\chi_{n\alpha}(\mathbf{Q},\omega) &= \chi_{n\alpha}(Q,\omega) = \chi_{n,-\alpha}(Q,-\omega) \ , \\
\chi_{in}(\mathbf{Q},\omega) &= \tfrac{Q_i}{Q}\chi^{\ell}_{Jn}(Q,\omega) = -\chi_{in}(\mathbf{Q},-\omega) \ , \\
\chi_{i\alpha}(\mathbf{Q},\omega) &= \tfrac{Q_i}{Q}\chi^{\ell}_{J\alpha}(Q,\omega) = -\chi_{i,-\alpha}(\mathbf{Q},-\omega) \ , \\
\chi_{ij}(\mathbf{Q},\omega) &= \tfrac{Q_iQ_j}{Q^2}\,\chi^{\ell}_{JJ}(Q,\omega) + (\delta_{ij} - \tfrac{Q_iQ_j}{Q^2})\chi^{t}_{JJ}(Q,\omega) \\
&= \chi_{ij}(\mathbf{Q},-\omega) \ .
\end{aligned}\right\} \qquad (5.5)$$

Here χ^{ℓ}_{Jn}, $\chi^{\ell}_{J\alpha}$ and χ^{ℓ}_{JJ} are correlation functions involving the longitudinal current while χ^{t}_{JJ} is the transverse component of the current–current correlation function. We shall find that $\chi_{n\alpha}$ and $\chi^{\ell}_{J\alpha}$ *vanish* in the absence of a Bose condensate. They involve a mixing or interference between the single-particle fluctuations and the density (and current) fluctuations, as in the second line of (3.18). These mixed (broken-symmetry) correlation functions play a key role in the subsequent analysis.

As with χ_{nn} in (5.1), other correlation functions can be expressed in terms of irreducible contributions (denoted by a bar). These can also be resummed as geometric series to give

$$\left.\begin{aligned}
\chi_{n\alpha} &= \frac{\bar{\chi}_{n\alpha}}{\epsilon} \ , \\
\chi^{\ell}_{JJ} &= \bar{\chi}^{\ell}_{JJ} + \frac{\bar{\chi}^{\ell}_{Jn}V(Q)\bar{\chi}^{\ell}_{nJ}}{\epsilon} \ , \qquad \chi^{t}_{JJ} = \bar{\chi}^{t}_{JJ} \ , \\
\chi^{\ell}_{Jn} &= \frac{\bar{\chi}^{\ell}_{Jn}}{\epsilon} \ , \\
\chi^{\ell}_{J\alpha} &= \bar{\chi}^{\ell}_{J\alpha} + \frac{\bar{\chi}^{\ell}_{Jn}V(Q)\bar{\chi}_{n\alpha}}{\epsilon} \ ,
\end{aligned}\right\} \qquad (5.6)$$

where ϵ has been defined in (5.2). This decomposition into irreducible parts is illustrated in Figs. 5.1 and 5.3 for χ_{nn} and χ^{ℓ}_{JJ}. The diagrams

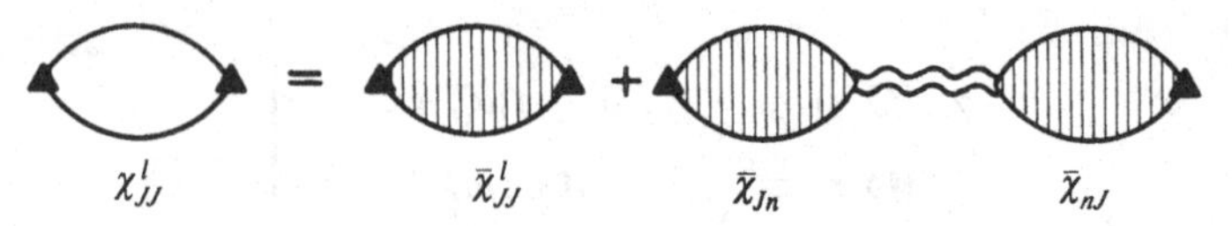

Fig. 5.3. The longitudinal current-response function in (5.6) given in terms of the irreducible contributions. The triangle represents the momentum-current vertex $(\mathbf{p} + \frac{1}{2}\mathbf{Q})$ (see (5.3)). The effective interaction (double wiggly line) is defined in Fig. 5.1.

contributing to the single-particle Green's function can be similarly decomposed,

$$G_{\alpha\beta} = \bar{G}_{\alpha\beta} + \frac{\bar{\chi}_{\alpha n} V(Q) \bar{\chi}_{n\beta}}{\epsilon} \ . \tag{5.7}$$

The irreducible matrix single-particle Green's function $\bar{G}_{\alpha\beta}$ has the same structure as $G_{\alpha\beta}$, except that it only involves the *irreducible* self-energies $\bar{\Sigma}_{\alpha\beta}$ (see also (5.12)). In a Bose-condensed fluid, $G_{\alpha\beta}$ has additional *reducible* diagrammatic contributions (the second term on the r.h.s. of (5.7)) because of the finite value of the broken-symmetry correlation functions $\bar{\chi}_{\alpha n}$. As we shall see, when $n_0 \neq 0$, the poles of $G_{\alpha\beta}$ are completely determined by the zeros of $\epsilon(\mathbf{Q}, \omega)$ and not by the poles of $\bar{G}_{\alpha\beta}$.

The irreducible contributions may be further decomposed into *proper* and *improper* parts (relative to cutting a *single* internal propagator line). We first note that the diagrams for the broken-symmetry density-field correlation function $\chi_{n\alpha}$ always contain a single propagator $G_{\alpha\beta}$ as a separate factor, and thus we can write $(\alpha, \beta = +, -)$

$$\chi_{n\alpha} = \Lambda_\beta G_{\beta\alpha} \quad , \quad \chi_{\alpha n} = G_{\alpha\beta} \Lambda_\beta \ . \tag{5.8}$$

The vertex function Λ_β is, by definition, proper. It can be expressed in terms of an associated irreducible vertex function $\bar{\Lambda}_\beta$ by (see Fig. 5.4)

$$\Lambda_\beta = \frac{\bar{\Lambda}_\beta}{\epsilon^R} \ , \tag{5.9}$$

where we have defined

$$\epsilon^R = 1 - V(Q) \bar{\chi}^R_{nn} \ . \tag{5.10}$$

The regular function $\bar{\chi}^R_{nn}$ describes the subclass of proper diagrams contributing to the irreducible function $\bar{\chi}_{nn}$.

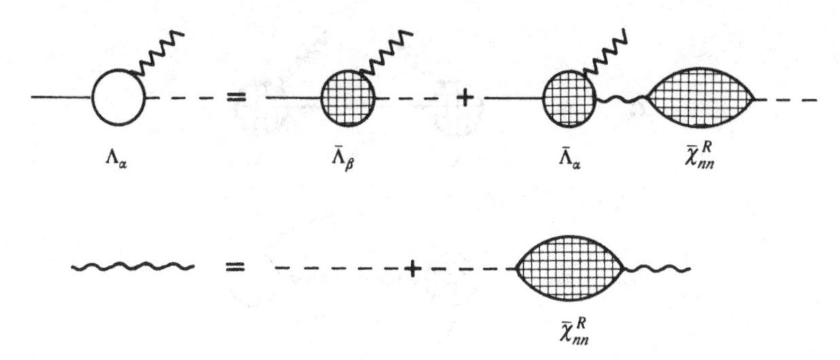

Fig. 5.4. Diagrammatic representation of irreducible vertex function Λ_α. The interaction and propagator external lines are not part of the functions.

Diagrammatically, it is clear that the irreducible contributions to $\bar{\chi}_{n\alpha}$ (see (5.6)) always have an irreducible $\bar{G}_{\alpha\beta}$ as a separate factor,

$$\bar{\chi}_{n\alpha} = \bar{\Lambda}_\beta \bar{G}_{\beta\alpha} \quad , \quad \bar{\chi}_{\alpha\beta} = \bar{G}_{\alpha\beta}\bar{\Lambda}_\beta \ , \tag{5.11a}$$

where $\bar{\Lambda}_\beta$ is the irreducible subclass of the proper diagrams contributing to Λ_β. Using the definition (5.5), one may verify that $\bar{\Lambda}_\beta(\mathbf{Q},\omega) = \bar{\Lambda}_{-\beta}(\mathbf{Q},-\omega)$. All diagrams for the irreducible density response function $\bar{\chi}_{nn}$ can be decomposed into proper and improper contributions. Making use of (5.11*a*), we see that

$$\begin{aligned}\bar{\chi}_{nn} &= \bar{\chi}^R_{nn} + \bar{\Lambda}_\alpha \bar{G}_{\alpha\beta}\bar{\Lambda}_\beta \\ &\equiv \bar{\chi}^R_{nn} + \bar{\chi}^c_{nn} \ .\end{aligned} \tag{5.11b}$$

Similarly, the proper single-particle self-energies can also be split into irreducible and reducible contributions

$$\begin{aligned}\Sigma_{\alpha\beta} &= \bar{\Sigma}_{\alpha\beta} + \frac{\bar{\Lambda}_\alpha V(Q)\bar{\Lambda}_\beta}{\epsilon^R} \\ &\equiv \bar{\Sigma}_{\alpha\beta} + \Sigma^c_{\alpha\beta} \ ,\end{aligned} \tag{5.12}$$

where ϵ^R has been defined in (5.10). In Fig. 5.5, we show the diagrammatic structure of both $\bar{\chi}^c_{nn}$ and $\Sigma^c_{\alpha\beta}$. In (5.12), the second term arises because of the condensate (recall that $\bar{\Lambda}_\alpha$ vanishes if $n_0 = 0$). It mixes or hybridizes the non-condensate density fluctuations (described by the zeros of ϵ^R) into the reducible single-particle self-energy denoted by $\Sigma^c_{\alpha\beta}$.

As with the case of correlation functions involving the density, one can carry out the analogous decomposition of the correlation functions involving the momentum current (Ma and Woo, 1967; Wong and Gould,

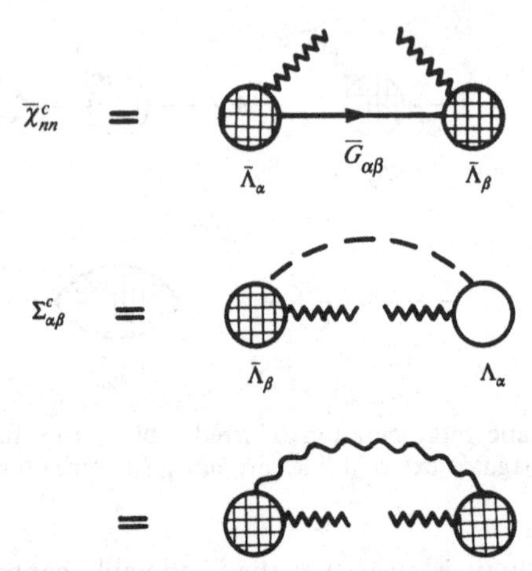

Fig. 5.5. Diagrammatic representation of the improper part $\bar{\chi}^c_{nn}$ and the irreducible self-energy $\Sigma^c_{\alpha\beta}$. The effective interaction is shown in Fig. 5.4.

1974). As with χ_{nn}, one finds that a momentum Bose vertex function Λ^ℓ naturally arises,

$$\left.\begin{aligned} \chi^\ell_{J\alpha} &= \Lambda^\ell_\beta G_{\beta\alpha} \,, \\ \bar{\chi}^\ell_{J\alpha} &= \bar{\Lambda}^\ell_\beta \bar{G}_{\beta\alpha} \,. \end{aligned}\right\} \tag{5.13a}$$

The irreducible functions in (5.6) can be decomposed into proper and improper diagrams

$$\left.\begin{aligned} \bar{\chi}^\ell_{JJ} &= \bar{\chi}^{\ell R}_{JJ} + \bar{\Lambda}^\ell_\alpha \bar{G}_{\alpha\beta} \bar{\Lambda}^\ell_\beta \,, \\ \bar{\chi}^\ell_{Jn} &= \bar{\chi}^{\ell R}_{Jn} + \bar{\Lambda}^\ell_\alpha \bar{G}_{\alpha\beta} \bar{\Lambda}_\beta \,, \\ \bar{\chi}^t_{JJ} &= \bar{\chi}^{tR}_{JJ} \,. \end{aligned}\right\} \tag{5.13b}$$

Using (5.5), one finds the symmetry-breaking longitudinal momentum vertex functions satisfy $\bar{\Lambda}^\ell_\alpha(\mathbf{Q}, \omega) = -\bar{\Lambda}^\ell_{-\alpha}(\mathbf{Q}, -\omega)$. We shall see shortly that it is most convenient to calculate $\bar{\chi}^\ell_{JJ}$ first and then use it to find $\bar{\chi}_{nn}$ and hence χ_{nn}. In addition, the results in (5.6) and (5.13*b*) for the transverse component of the momentum current-response function will be needed in Section 6.1 in defining what is meant by the superfluid density in the low-frequency two-fluid limit.

A remark about notation is appropriate here. Strictly speaking, we should use $\bar{\Sigma}^R_{\alpha\beta}, \bar{\Lambda}^R_\alpha$ and $\bar{\Lambda}^{\ell R}_\alpha$ in (5.12), (5.9) and (5.13*a*) but for simplicity, we drop the *R* superscript. These quantities are, by definition, proper and the bar means that we are dealing with the irreducible parts.

The continuity equation

$$\frac{\partial \hat{\rho}(\mathbf{Q},\tau)}{\partial \tau} = \frac{\mathbf{Q}}{m} \cdot \mathbf{J}(\mathbf{Q},\tau) \tag{5.14}$$

leads to important relations between the correlation functions defined in (5.4). Three of these are (Wong and Gould, 1974)

$$\omega \chi_{n\alpha}(Q,\omega) = \frac{Q}{m} \chi^{\ell}_{J\alpha}(Q,\omega) + \alpha\sqrt{n_0} \ , \tag{5.15a}$$

$$\omega \chi_{nn}(Q,\omega) = \frac{Q}{m} \chi^{\ell}_{Jn}(Q,\omega) \ , \tag{5.15b}$$

$$\omega \chi^{\ell}_{Jn}(Q,\omega) = \frac{Q}{m} [\chi^{\ell}_{JJ}(Q,\omega) + \rho] \ . \tag{5.15c}$$

It is crucial that approximations are consistent with these exact relations. The *irreducible* functions also satisfy the same identities as in (5.15), with all quantities replaced by the corresponding barred functions. We note that (5.15*b*) and (5.15*c*) can be combined to give the "continuity equation"

$$\chi_{nn} = \frac{Q^2}{m^2\omega^2} \ [\chi^{\ell}_{JJ} + \rho] \ . \tag{5.16a}$$

The irreducible version of these equations gives the same relation between $\bar{\chi}_{nn}$ and $\bar{\chi}^{\ell}_{JJ}$,

$$\bar{\chi}_{nn} = \frac{Q^2}{m^2\omega^2} [\bar{\chi}^{\ell}_{JJ} + \rho] \ . \tag{5.16b}$$

This last relation enables one to compute $\bar{\chi}_{nn}$ (and hence χ_{nn}) from $\bar{\chi}^{\ell}_{JJ}$ in such a way that the f-sum rule (2.24) is obeyed.

Using these relations in conjunction with (5.8)–(5.13), we see that the regular (proper and irreducible) functions satisfy the following important identities (Wong and Gould, 1974):

$$\begin{aligned} \omega \bar{\Lambda}_{\alpha}(Q,\omega) &= \frac{Q}{m} \bar{\Lambda}^{\ell}_{\alpha}(Q,\omega) + \sqrt{n_0} \ [\omega - \alpha(\varepsilon_Q - \mu)] \\ &\quad -\sqrt{n_0} \ [\bar{\Sigma}_{1,\alpha}(Q,\omega) - \bar{\Sigma}_{2,\alpha}(Q,\omega)] \ , \end{aligned} \tag{5.17}$$

$$\omega \bar{\chi}^{R}_{nn}(Q,\omega) = \frac{Q}{m} \bar{\chi}^{\ell R}_{Jn}(Q,\omega) - \sqrt{n_0} \ [\bar{\Lambda}_1(Q,\omega) - \bar{\Lambda}_2(Q,\omega)] \ , \tag{5.18}$$

$$\omega \bar{\chi}^{\ell R}_{Jn}(Q,\omega) = \frac{Q}{m} (\bar{\chi}^{\ell R}_{JJ}(Q,\omega) + \rho) - \sqrt{n_0} \ [\bar{\Lambda}^{\ell}_1(Q,\omega) - \bar{\Lambda}^{\ell}_2(Q,\omega)] \ . \tag{5.19}$$

In deriving these results, we have used Dyson's equation for the *inverse* irreducible matrix Green's function (see (3.26))

$$\bar{G}^{-1}_{\alpha\beta} = \delta_{\alpha\beta} [\alpha\omega - (\varepsilon_Q - \mu)] - \bar{\Sigma}_{\alpha\beta} \tag{5.20}$$

and the matrix identity $\bar{G}^{-1}_{\alpha\beta}\bar{G}_{\beta\delta} = \delta_{\alpha\delta}$. Eqs. (5.17)–(5.19) are referred to as generalized Ward identities (Hohenberg and Martin, 1965; Wong and Gould, 1974; Talbot and Griffin, 1983), by analogy with the Ward identity for currents in field theory (see also Nozières, 1964). These identities show how the regular correlation functions $\bar{\chi}^R_{nn}$ and $\bar{\chi}^{RR}_{JJ}$ are connected to the regular single-particle self-energies $\bar{\Sigma}_{\alpha\beta}$ via the Bose symmetry-breaking vertex functions $\bar{\Lambda}_\alpha$ and $\bar{\Lambda}^{\ell}_\alpha$. Unless these identities are satisfied, one cannot be sure that the poles of $G_{\alpha\beta}$ and χ_{nn} will be shared, as required in a Bose-condensed fluid.

We next analyse the structure of both $G_{\alpha\beta}$ and χ_{nn} to show more explicitly how the condensate hybridizes the two spectra. Defining $\bar{G}_{\alpha\beta} \equiv \bar{N}_{\alpha\beta}/\bar{D}$, and using (5.11*b*) in (5.2), we find

$$\chi_{nn} = \frac{\bar{\chi}^R_{nn}\bar{D} + \bar{\Lambda}_\alpha \bar{N}_{\alpha\beta}\bar{\Lambda}_\beta}{C(Q,\omega)} , \tag{5.21}$$

where the denominator is defined as

$$\begin{aligned} C(\mathbf{Q},\omega) \equiv \bar{D}\epsilon &= \bar{D}\left[\epsilon^R - V(Q)\frac{\bar{\Lambda}_\alpha \bar{N}_{\alpha\beta}\bar{\Lambda}_\beta}{\bar{D}}\right] \\ &= \bar{D}\epsilon^R - V(Q)\bar{\Lambda}_\alpha \bar{N}_{\alpha\beta}\bar{\Lambda}_\beta . \end{aligned} \tag{5.22}$$

Turning to the denominator of $G_{\alpha\beta} \equiv N_{\alpha\beta}/D$, one can separate out the irreducible (proper) self-energy $\bar{\Sigma}_{\alpha\beta}$ from the reducible proper self-energy as in (5.12). After some calculation, one finds

$$\begin{aligned} D &= \bar{D} - \bar{N}_{\alpha\beta}\Sigma^c_{\beta\alpha} \\ &= \bar{D} - V(Q)\frac{\bar{\Lambda}_\alpha \bar{N}_{\alpha\beta}\bar{\Lambda}_\beta}{\epsilon^R} . \end{aligned} \tag{5.23}$$

Comparing (5.22) and (5.23), we arrive at an important identity relating ϵ and D, namely $\epsilon^R D = \epsilon \bar{D}$; this shows that the poles of *both* $G_{\alpha\beta}$ and χ_{nn} are given by the zeros of $C(\mathbf{Q},\omega)$ in (5.22). The crucial role of the condensate is manifest through the Bose vertex functions $\bar{\Lambda}_\alpha$ in (5.22) and (5.23). In the absence of a Bose broken symmetry ($\bar{\Lambda}_\alpha = 0$), the pole $\bar{\omega}_2$ of $G_{\alpha\beta}$ (given by $\bar{D}(\mathbf{Q},\bar{\omega}_2) = 0$) and the pole $\bar{\omega}_1$ of χ_{nn} (given by $\epsilon^R(\mathbf{Q},\bar{\omega}_1) = 0$) are uncoupled. "Turning on" the condensate ($\bar{\Lambda}_\alpha \neq 0$) hybridizes these modes, as shown by (5.22). The renormalized zeros of $C(\mathbf{Q},\omega)$ will be denoted by ω_1 and ω_2. (In Griffin and Cheung, 1973, the definitions of ω_i and $\bar{\omega}_i$ are interchanged.)

From (3.12), we see that in the presence of a condensate, the density-response function χ_{nn} will contain contributions which are proportional to a single-particle propagator $G_{\alpha\beta}$ as a *separate* factor (i.e., as an intermediate state). Gavoret and Nozières (1964) refer to these contributions as the "singular" part of χ_{nn}. Within the dielectric formalism, the Gavoret–Nozières decomposition of χ_{nn} (and other correlation functions) into "singular" and "regular" parts corresponds to splitting diagrammatic contributions into *improper* and *proper* parts (see Fig. 5.6)

$$\chi_{nn} = \frac{\bar{\Lambda}_\alpha}{\epsilon^R} G_{\alpha\beta} \frac{\bar{\Lambda}_\beta}{\epsilon^R} + \frac{\bar{\chi}^R_{nn}}{\epsilon^R} , \tag{5.24}$$

where the first term is equivalent to the Gavoret–Nozières singular term. This key formula will now be derived. First of all, (5.2) can be split as follows:

$$\begin{aligned} \chi_{nn} = \frac{\bar{\chi}_{nn}}{\epsilon} &= \frac{(\bar{\chi}^c_{nn} + \bar{\chi}^R_{nn})(1 - V\bar{\chi}^R_{nn})}{\epsilon\epsilon^R} \\ &= \frac{\bar{\chi}^c_{nn} + \bar{\chi}^R_{nn}(1 - V\bar{\chi}^c_{nn} - V\bar{\chi}^R_{nn})}{\epsilon\epsilon^R} \\ &= \frac{\bar{\chi}^c_{nn}}{\epsilon^R\epsilon} + \frac{\bar{\chi}^R_{nn}}{\epsilon^R} , \end{aligned} \tag{5.25}$$

where $\bar{\chi}_{nn}$ has been defined in (5.11). To prove that the first terms in (5.24) and (5.25) are identical, we note that the irreducible parts of the numerator $N_{\alpha\beta}$ of $G_{\alpha\beta}$ given in (3.29) can be separated out, giving

$$\begin{aligned} N_{\alpha\beta} &= \bar{N}_{\alpha\beta} - \alpha\beta(\Sigma_{-\alpha,-\beta} - \bar{\Sigma}_{-\alpha,-\beta}) \\ &= \bar{N}_{\alpha\beta} - \alpha\beta\bar{\Lambda}_{-\alpha}\frac{V(Q)}{\epsilon^R}\bar{\Lambda}_{-\beta} \end{aligned} \tag{5.26}$$

(recall that summation is only over repeated Greek *subscripts*, which do not occur in (5.26)). Using the trivial identity

$$\sum_\alpha \alpha\bar{\Lambda}_\alpha\bar{\Lambda}_{-\alpha} = 0 , \tag{5.27}$$

one easily obtains from (5.26) the surprising result $\bar{\Lambda}_\alpha N_{\alpha\beta}\bar{\Lambda}_\beta = \bar{\Lambda}_\alpha\bar{N}_{\alpha\beta}\bar{\Lambda}_\beta$. Note that there is no contribution from the second term in (5.26). Using this fact and the key identity $\epsilon^R D = \epsilon\bar{D}$ derived above, the first term in (5.25) can be transformed as follows:

$$\frac{\bar{\Lambda}_\alpha\bar{G}_{\alpha\beta}\bar{\Lambda}_\beta}{\epsilon^R\epsilon} = \frac{\bar{\Lambda}_\alpha}{\epsilon^R}\frac{N_{\alpha\beta}}{D}\frac{\bar{\Lambda}_\beta}{\epsilon^R} = \frac{\bar{\Lambda}_\alpha}{\epsilon^R}G_{\alpha\beta}\frac{\bar{\Lambda}_\beta}{\epsilon^R} , \tag{5.28}$$

which completes the proof of (5.24).

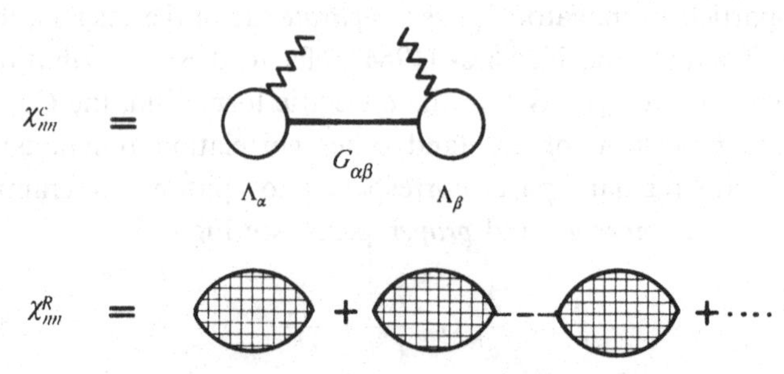

Fig. 5.6. Diagrammatic structure of the proper and improper contributions to χ_{nn} as given by (5.24). See Fig. 5.4 for the Bose vertex function Λ_α.

A result analogous to (5.24) can also be derived for the density single-particle correlation function $\chi_{\alpha n}$. Starting from (5.6), we find

$$\chi_{\alpha n} = \frac{\bar{\chi}_{\alpha n}}{\epsilon} = \frac{\bar{\Lambda}_\beta}{\epsilon}\frac{\bar{N}_{\beta\alpha}}{\bar{D}} = \frac{\bar{\Lambda}_\beta N_{\beta\alpha}}{\epsilon^R D} = \frac{\bar{\Lambda}_\beta}{\epsilon^R} G_{\beta\alpha} \; , \tag{5.29}$$

using (5.26) and (5.27), as well as the identity given after (5.23). Recalling (5.9), the last result is seen to be equivalent to (5.8), i.e., $\chi_{\alpha n} = \Lambda_\beta G_{\beta\alpha}$.

We have shown how the density-response function $\chi_{nn}(\mathbf{Q}, \omega)$ naturally splits into two parts as in (5.24), with a "singular" or "condensate" part having $G_{\alpha\beta}$ as a separate factor. The analogous decomposition of the longitudinal momentum current-response function $\chi^\ell_{JJ}(\mathbf{Q}, \omega)$ in (5.6) into improper and proper parts is given by Eqs. (2.5) and (2.6) of Talbot and Griffin (1984b). Such decompositions are useful because the proper term which is separated off may be loosely viewed as the "normal" contribution. This terminology is justified because, above T_λ, only the second term in (5.24) remains. However, below T_λ, $G_{\alpha\beta}$ and χ_{nn} have singularities given only by the zeros of $C(\mathbf{Q}, \omega)$ in (5.22) and *not* by the zeros of ϵ^R or $\bar{D}$. This fact emphasizes that one should not automatically think of the two contributions in (5.24) as distinct. This inter-relation is already clear when one realizes that they can always be summed to give the expression in (5.2). The key point is that ϵ^R enters *both* terms in (5.24) explicitly, as well as implicitly in $G_{\alpha\beta}$ via the reducible self-energy in (5.12). Talbot and Griffin (1984a) have shown there are cancellations between the two contributions in (5.24) *when* there is significant structure arising from zeros of $\epsilon^R(\mathbf{Q}, \omega)$. Whether we start from (5.2) or (5.24) is central to the interpretation of the structure exhibited by $S(\mathbf{Q}, \omega)$ in superfluid ^{4}He, as we discuss at the end of Section 5.5 and in Section 7.2.

The Dyson–Beliaev matrix equations in (3.26) can also be rewritten with $\bar{G}_{\alpha\beta}$ playing the role of $G^0_{\alpha\beta}$ (Szépfalusy and Kondor, 1974). One finds

$$G_{\alpha\beta} = \bar{G}_{\alpha\beta} + \bar{G}_{\alpha\delta}\Sigma^c_{\delta\gamma}G_{\gamma\beta} \ , \tag{5.30}$$

where the reducible matrix self-energy $\Sigma^c_{\delta\gamma}$ is defined by (5.12). Making use of (5.26) and (5.27) again, one can reduce (5.30) to

$$G_{\alpha\beta} = \bar{G}_{\alpha\delta}\frac{\bar{\Lambda}_\delta V(Q)\bar{\Lambda}_\gamma}{\epsilon}\bar{G}_{\gamma\beta} + \bar{G}_{\alpha\beta} \ . \tag{5.31}$$

Recalling (5.11*a*), the result (5.31) is completely equivalent to (5.7). It shows explicitly how $G_{\alpha\beta}$ has the same poles as $\chi_{nn} = \bar{\chi}_{nn}/\epsilon$, with a weight controlled by the Bose broken-symmetry vertex functions $\bar{\Lambda}_\alpha$. It is the equivalent of (5.24) for χ_{nn} in that we have separated out the "normal" part $\bar{G}_{\alpha\beta}$ from a "condensate" contribution, the latter arising only in the superfluid phase.

In the dielectric formulation we have been discussing, a given approximation to χ_{nn} *and* $G_{\alpha\beta}$ is determined by our choice of the *regular* functions. As emphasized by Wong and Gould (1974), it is best to start with a specific diagrammatic approximation to the three regular functions $\bar{\Sigma}_{\alpha\beta}$, $\bar{\chi}^{\ell R}_{JJ}$ and $\bar{\Lambda}^\ell_\alpha$ and *then* use the Ward identities in (5.17)–(5.19) to determine $\bar{\Lambda}_\alpha$, $\bar{\chi}^{\ell R}_{Jn}$, $\bar{\chi}^R_{nn}$. Starting from some choice for $\bar{\Lambda}_\alpha$ and $\bar{\chi}^R_{nn}$ is less preferable because (5.17) and (5.18) do not give any information about $\bar{\Lambda}_\alpha$ or $\bar{\chi}^R_{nn}$ at zero frequency. In contrast, (5.17)–(5.19) give exact constraints on the zero-frequency limit of $\bar{\Lambda}^l_\alpha$ and $\bar{\chi}^{lR}_{JJ}$ (see Section 6.1). A convenient way of summarizing this section is to give the explicit procedure for calculating $G_{\alpha\beta}$ and χ_{nn}, starting from a specific approximation to the "building blocks" $\bar{\Lambda}^\ell_\alpha$, $\bar{\chi}^{\ell R}_{JJ}$ and $\bar{\Sigma}_{\alpha\beta}$. This algorithm is as follows:

(a) Use (5.13*b*) to find $\bar{\chi}^\ell_{JJ}$.
(b) Use $\bar{\chi}^\ell_{JJ}$ in (5.16*b*) to find $\bar{\chi}_{nn}$.
(c) Use $\bar{\chi}_{nn}$ in (5.2) to find ϵ and hence χ_{nn}.
(d) Use (5.17) to find $\bar{\Lambda}_\alpha$.
(e) Use $\bar{\Lambda}_\alpha$ to find $\bar{\chi}^R_{nn}$ from (5.18) and (5.19) and hence obtain ϵ^R.
(f) Use $\bar{\Lambda}_\alpha$ and ϵ^R in (5.12) to find $\Sigma_{\alpha\beta}$ and hence $G_{\alpha\beta}$.

By construction, the resulting expressions for $G_{\alpha\beta}$ and χ_{nn} will share the same poles, albeit with different weights in the two functions.

In Sections 5.2 and 5.3, we illustrate this procedure with two choices for the "building blocks" $\bar{\Lambda}^\ell_\alpha$, $\bar{\chi}^{\ell R}_{JJ}$ and $\bar{\Sigma}_{\alpha\beta}$. These model approximations are really most appropriate for a WIDBG, but they illustrate behaviour which mimics much of the essential physics of superfluid ^{4}He. The dielectric formalism can also be used to understand the consequences of the condensate-induced coupling of the two-particle spectra into the single-particle self-energy. In Section 10.2, we also show how singularities in the two-roton spectra (bound states) can show up in the irreducible self-energy $\bar{\Sigma}_{\alpha\beta}$ (Pitaevskii, 1959; Ruvalds and Zawadowski, 1970).

We conclude this section with some technical remarks concerning the treatment of the Bose broken symmetry in field-theoretic calculations. Almost all theoretical papers on interacting Bose fluids simply treat the condensate atom operators $\hat{a}_0$ and $\hat{a}_0^+$ as c-numbers equal to $\sqrt{N_0}$. Since the number of atoms N_0 in the condensate is macroscopic, $[\hat{a}_0, \hat{a}_0^+] = 1 \ll N_0$ and hence one should be able to ignore condensate fluctuations resulting from the non-commuting nature of $\hat{a}_0$ and $\hat{a}_0^+$. The c-number treatment should not be confused with the specific model approximation Bogoliubov used within this general scheme. For simplicity, we have used this c-number prescription in our discussion so far. However, this procedure, strictly speaking, invalidates the continuity equation. Within the simple "Bogoliubov prescription" for treating the condensate part of $\hat{\rho}(\mathbf{Q})$ and $\hat{\mathbf{J}}(\mathbf{Q})$, calculation shows there is a "spurious" contribution on the r.h.s. of (5.14) given by

$$[nV(\mathbf{q}=0)-\mu]\ \sqrt{N_0}[\hat{a}_Q(\tau)-\hat{a}^+_{-Q}(\tau)]\ . \tag{5.32}$$

As a result, one does *not* obtain the important relations given in (5.15). This is a serious problem since these relations are the basis of the generalized Ward identities, which ensure that $G_{\alpha\beta}$ and χ_{nn} share the same spectrum. Most authors (for example, Wong and Gould, 1974) simply impose the relations in (5.15) by fiat and proceed from there.

A more satisfactory procedure is that of Szépfalusy (1965), who introduces the new zero-momentum operators

$$\left.\begin{aligned} \hat{b}_0 &\equiv \hat{a}_0 - \sqrt{N_0}\ , \\ \hat{b}_0^+ &= \hat{a}_0^+ - \sqrt{N_0}\ , \end{aligned}\right\} \tag{5.33}$$

with N_0 chosen so that $\langle \hat{a}_0 \rangle = \langle \hat{a}_0^+ \rangle = \sqrt{N_0}$. Keeping the operator nature of $\hat{a}_0$ and $\hat{a}_0^+$, but letting them have a non-zero thermal average, allows one to preserve the continuity equation (Talbot, 1983; Talbot and Griffin, 1983). Talbot and Griffin (1984b) and Talbot (1983) give the Ward identities (5.17)–(5.19) with the symmetry-breaking terms explicitly exhibited. The use of this approach is crucial when calculating response functions at $\mathbf{Q} = 0$ as well as thermodynamic derivatives (see Section 6.3 for more details). Talbot (1983) and Weichman (1988) have given a careful discussion of these functions and also the method of determining the thermodynamic parameters μ and n_0 in the Szépfalusy scheme. For a given value of $n_0 = N_0/\Omega$, $\langle \hat{a}_0 \rangle$ determines a "physical value" for the chemical potential μ. In turn, for each n_0 and $\mu(n_0)$, we have a definite density n, as determined from either the single-particle Green's function or the zero-frequency longitudinal current correlation function. One then chooses the value of n_0 to give the desired density n. Talbot (1983) has proved that this procedure gives the same result as requiring that the thermodynamic potential $A(\Omega, T, \mu, n_0)$ be a minimum. Talbot has also shown this procedure to be equivalent to that used at $T = 0$ by Hugenholtz and Pines (1959) as well as Gavoret and Nozières (1964), namely that the free energy $F(\Omega, T, n, n_0) \equiv A + \mu N$ is minimized at the correct value of n_0.

5.2 RPA or zero-loop approximation

We now illustrate the structure of the dielectric formalism results of Section 5.1 by taking the regular functions to be those of a Bose gas. Keeping the Hartree self-energy, this simple approximation corresponds to

$$\left.\begin{aligned} \bar{\Sigma}_{\alpha\beta} &= nV(0)\delta_{\alpha\beta} \ , \\ \bar{\chi}^{\ell R}_{JJ} &= \bar{\chi}^{\ell R0}_{JJ}, \ \bar{\Lambda}^{\ell}_{\alpha} = \alpha\tfrac{Q}{2}\sqrt{n_0}, \ \mu = nV(0) \ , \end{aligned}\right\} \tag{5.34}$$

where the rules of diagrammatic perturbation theory give (Wong and Gould, 1974)

$$\begin{aligned} \bar{\chi}^{\ell R0}_{JJ}(\mathbf{Q}, i\omega_\ell) &= -\frac{1}{2\beta}\sum_{\omega_m}\int\frac{d\mathbf{p}}{(2\pi)^3}\left[\mathbf{Q}\cdot\left(\mathbf{p}+\frac{\mathbf{Q}}{2}\right)\right]^2 \\ &\qquad \times G_0(\mathbf{p}, i\omega_m)G_0(\mathbf{p}+\mathbf{Q}, i\omega_m + i\omega_\ell) \ , \\ &= -\int\frac{d\mathbf{p}}{(2\pi)^2}\left[\mathbf{Q}\cdot\left(\mathbf{p}+\frac{\mathbf{Q}}{2}\right)\right]^2 \frac{N^0(\varepsilon_{p+Q}) - N^0(\varepsilon_p)}{i\omega_\ell - (\varepsilon_{p+Q} - \varepsilon_p)}, \end{aligned} \tag{5.35}$$

with $G_0 = G^0_{11}$. We recall that excitation energies are defined with respect to μ and hence $E_Q = \varepsilon_Q + nV(0) - \mu = \varepsilon_Q$. Using (5.34) in the Ward identities (5.17)–(5.19), we find

$$\begin{aligned} \bar{\Lambda}_\alpha &= \sqrt{n_0} \ , \\ \omega^2\bar{\chi}^{R0}_{nn} &= \frac{Q^2}{m^2}[\bar{\chi}^{\ell R0}_{JJ} + (n - n_0)m] \ . \end{aligned} \tag{5.36}$$

Using (5.35) in (5.36), one obtains the expected result (after some calculation)

$$\bar{\chi}_{nn}^{R0}(\mathbf{Q}, i\omega_l \to \omega) = \int \frac{d\mathbf{p}}{(2\pi)^3} \frac{N(\varepsilon_p) - N(\varepsilon_{p+Q})}{\omega - (\varepsilon_{p+Q} - \varepsilon_p)} . \tag{5.37}$$

The improper or "condensate" part of $\bar{\chi}_{nn}$ is given by the second term in (5.11*b*). In the Hartree Bose gas, $\bar{\Lambda}_\alpha = \sqrt{n_0}$ and hence

$$\bar{\chi}_{nn}^{c0}(\mathbf{Q}, \omega) = \sqrt{n_0} \frac{2\varepsilon_Q}{\omega^2 - \varepsilon_Q^2} \sqrt{n_0} . \tag{5.38}$$

This is the density response arising from processes in which atoms are excited out of (or into) a Bose gas condensate.

Adding (5.37) and (5.38), the *sum* can be written in the form

$$\bar{\chi}_{nn}(\mathbf{Q}, \omega) = \int \frac{d\mathbf{p}}{(2\pi)^3} \frac{n_p^0 - n_{p+Q}^0}{\omega - (\varepsilon_{p+Q} - \varepsilon_p)} \equiv \chi_{nn}^0 , \tag{5.39}$$

where the momentum distribution is

$$n_p^0 = n_0 (2\pi)^3 \delta(\mathbf{p}) + N(\varepsilon_p) . \tag{5.40}$$

$\bar{\chi}_{nn}$ in (5.39) is simply the *total* density-response function χ_{nn}^0 of a free Bose gas including the contribution of atoms in the condensate. This result was quoted earlier in (2.29).

In conjunction with (5.2), (5.39) gives

$$\chi_{nn}(\mathbf{Q}, \omega) = \frac{\chi_{nn}^0}{1 - V(Q)\chi_{nn}^0} , \tag{5.41}$$

which is precisely the well known random phase approximation (RPA) result for the density-response function. As for the single-particle self-energies in (5.12), we have

$$\Sigma_{\alpha\beta}(\mathbf{Q}, \omega) = n_0 V(0) \delta_{\alpha\beta} + \frac{n_0 V(Q)}{\epsilon^R} , \tag{5.42}$$

where

$$\epsilon^R(\mathbf{Q}, \omega) = 1 - V(Q) \bar{\chi}_{nn}^{R0}(\mathbf{Q}, \omega) , \tag{5.43}$$

with $\bar{\chi}_{nn}^{R0}$ given by (5.37). The result (5.42) should be compared with the Bogoliubov approximation (3.31). One can interpret the second term in (5.42) as the exchange self-energy with a dynamically screened interaction defined by

$$V_S(\mathbf{Q}, \omega) = \frac{V(Q)}{\epsilon^R(\mathbf{Q}, \omega)} . \tag{5.44}$$

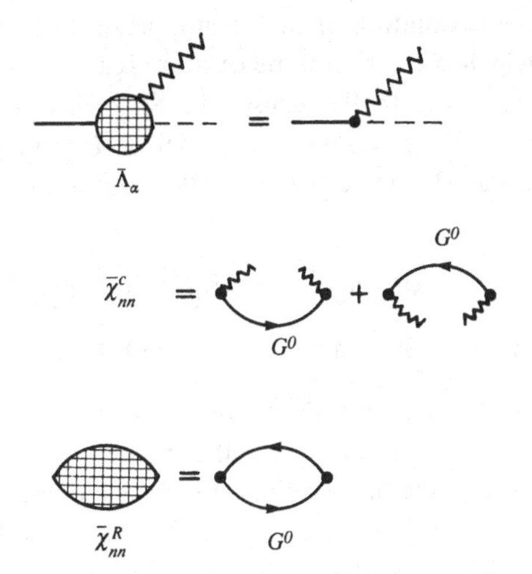

Fig. 5.7. Diagrammatic approximation for $\bar{\Lambda}_\alpha$, $\bar{\chi}^c_{nn}$ and $\bar{\chi}^R_{nn}$ used in the SPA.

It may be easily verified that the results (5.39)–(5.43) are equivalent to those obtained by our mean-field analysis in Section 3.4.

This "shielded potential approximation" was first discussed by Tserkovnikov (1965) using an equation-of-motion decoupling technique, and by Cheung and Griffin (1970). It was only later realized (Cheung, 1971; Griffin and Cheung, 1973; Szépfalusy and Kondor, 1974) that (5.41) and (5.42) define the *simplest* model approximation within the dielectric formalism which (a) is valid at finite temperatures, (b) satisfies the generalized Ward identities, (c) guarantees that both $G_{\alpha\beta}$ and χ_{nn} would exhibit the same excitations, albeit with different spectral weights. This approximation is also referred to as the RPA because of the familar form of the expressions (5.41) and (5.44). In Fig. 5.7, we show the diagrams for $\bar{\Lambda}_\alpha$, $\bar{\chi}^c_{nn}$ and $\bar{\chi}^R_{nn}$ used in the SPA.

In contrast to the standard Bogoliubov approximation in Section 3.2, the SPA includes the dynamical effects of the non-condensate atoms as well as their coupling to the condensate reservoir. Thus the SPA is not limited to $T \simeq 0$, where $n_0 \simeq n$. The finite-temperature behaviour of $G_{\alpha\beta}$ and χ_{nn} based on (5.41)–(5.43) has been discussed in some detail by Szépfalusy and Kondor (1974) and, more recently, by Payne and Griffin (1985). The SPA lets one see how $G^B_{\alpha\beta}$ in (3.36) and χ^B_{nn} in (3.42) are modified as the condensate fraction is depleted (as T increases towards T_λ). Within the SPA, the original Bogoliubov approximation (3.31) corresponds to setting $\bar{\chi}^{R0}_{nn}$ in (5.37) to zero.

The zero-loop approximation or SPA is important in that $G_{\alpha\beta}$ and χ_{nn} exhibit structure which is the prototype of more realistic approximations for the regular quantities. In this sense, the SPA plays a role at finite temperatures similar to the Bogoliubov model at $T = 0$. In particular, it is useful to distinguish between two limits (Szépfalusy and Kondor, 1974):

$$\left.\begin{array}{ll}\text{Weak-coupling (WC) limit,} & nV(\mathbf{p}=0) \ll k_B T \ , \\ \text{Strong-coupling (SC) limit,} & nV(\mathbf{p}=0) \gg k_B T \ . \end{array}\right\} \tag{5.45}$$

In the weak-coupling (WC) limit, the thermal (kinetic) energy dominates over the effect of the interactions. The self-consistent field of the excited atoms is *not* large enough to give rise to a well defined zero sound mode (zero of ϵ^R). In this WC limit, the only well defined pole of $G_{\alpha\beta}$ and χ_{nn} (given by the zeros of $C(\mathbf{Q},\omega)$ in (5.22)) may be viewed as a condensate-renormalized single-particle excitation ω_2 associated with $\bar{\omega}_2$, the zero of $\bar{D}$. Since one finds (Payne and Griffin, 1985) that $\epsilon^R \simeq 1$ in this WC limit, we conclude that the ω_2 mode has the dispersion relation

$$\omega_Q^2 = \varepsilon_Q^2 + 2n_0(T)V(Q)\varepsilon_Q \ . \tag{5.46}$$

At $T = 0$, this expression reduces to the Bogoliubov spectrum (3.35), but at finite T it involves the depleted Bose gas condensate fraction $n_0(T)$. The $T \neq 0$ Bogoliubov dispersion relation given by (5.46) was first derived by Popov (1965) using a direct analysis of the leading order diagrams in a dilute Bose gas (see also Chapter 6 of Popov, 1987).

It is important to emphasize that (5.46) has been derived in the intermediate-temperature region of a WIDBG by showing that, even though there are an appreciable number of atoms *outside* the condensate, they produce *no* appreciable self-consistent fields. It is the condensate effective field which has renormalized the particle-like excitation ε_Q (appropriate to the normal phase) to ω_Q as given in (5.46). As a consequence, the ω_Q mode exhibits the long-wavelength behaviour of a Goldstone mode (see Section 6.3) associated with oscillations of a broken-symmetry order parameter, i.e., it is phonon-like with a velocity which goes to zero with the Bose order parameter $\sqrt{n_0(T)}$.

In the literature, the Bogoliubov $T = 0$ calculation (where $n_0 \simeq n$) is often extrapolated to finite temperature by simply replacing n_0 by n. This leads to a temperature-independent phonon velocity, as is observed in superfluid ^{4}He. However, there is *no* theoretical basis for such a finite-temperature extension in the WC limit (see discussion at the end of this section). In a WIDBG, $S(\mathbf{Q},\omega)$ should exhibit a phonon peak at low Q whose frequency *and* weight vanish with the condensate order parameter (like a spin wave in a ferromagnet). Such a soft phonon mode would be expected, for example, in Bose-condensed spin-polarized Hydrogen gas (for a review, see Greytak and Kleppner, 1984). For further discussion, see Gay and Griffin (1985) and Payne and Griffin (1985).

We next turn to the strong-coupling SC limit defined in (5.45). In the SC limit of the SPA, the self-consistent field associated with the excited (non-condensate) atoms dominates over their thermal motion. In contrast to the weak-coupling SP limit, $\epsilon^R(\mathbf{Q},\omega)$ in (5.43) now leads to a zero sound density fluctuation in the non-condensate atoms. To examine this region in more detail, we use the high-frequency limit of $\bar{\chi}^{R0}_{nn}$ in (5.37), which gives

$$\epsilon^R(\mathbf{Q},\omega) \simeq 1 - V(Q)\frac{\tilde{n}Q^2}{m\omega^2} \,, \tag{5.47}$$

where $\tilde{n} = n - n_0$ is the density of atoms *outside* the condensate. In the language introduced at the end of Section 5.1, we are dealing with free-particle excitations ($\bar{\omega}_2 = \varepsilon_Q$) coupled through the condensate order parameter to a zero-sound density fluctuation ($\bar{\omega}_1 = \tilde{c}Q$, where $\tilde{c}^2 \equiv \tilde{n}V(Q)/m$). This leads to the following expressions:

$$\left.\begin{aligned} G_{11}(\mathbf{Q},\omega) &= \frac{(\omega+\varepsilon_Q)\epsilon^R + n_0V(Q)}{\epsilon^R\bar{D} - c_0^2Q^2} \,, \\ \chi_{nn}(\mathbf{Q},\omega) &= \frac{\bar{\chi}^{R0}_{nn}\bar{D} + n_0Q^2/m}{\epsilon^R\bar{D} - c_0^2Q^2} \,, \end{aligned}\right\} \tag{5.48}$$

where $c_0^2 \equiv n_0V(Q)/m$, $\bar{D} = \omega^2 - \varepsilon_Q^2$ and $\epsilon^R = 1 - \tilde{c}^2Q^2/\omega^2$. Both correlation functions in (5.48) exhibit poles given by the two solutions of $C(\mathbf{Q},\omega) = 0$ (see (5.22)), which in our simple SPA model reduces to

$$(\omega^2 - \tilde{c}^2Q^2)(\omega^2 - \varepsilon_Q^2) \; - \; c_0^2Q^2\omega^2 = 0 \,. \tag{5.49}$$

The usefulness of (5.48) and (5.49) lies in the fact that they illustrate the kind of behaviour one might expect in a Bose-condensed fluid. In this over-simplified model, the dielectric function is given by

$$\epsilon(\mathbf{Q},\omega) = 1 - V(Q)\left[\frac{n_0Q^2/m}{\omega^2 - \varepsilon_Q^2} + \frac{\tilde{n}Q^2/m}{\omega^2}\right] \,. \tag{5.50}$$

Two comments are relevant concerning this result. One could clearly use an improved single-particle energy ε_Q in place of $Q^2/2m$. Secondly, in our SPA gas model, the depletion $\tilde{n} = n - n_0$ is entirely due to thermal excitation of atoms out of the condensate. Using it as a guide to more realistic models suggests that we view $\bar{\chi}_{nn}^{R0}$ as describing the thermally excited quasiparticles, i.e., the "normal fluid". Thus $\tilde{n}$ in (5.50) is best interpreted as being proportional to the number of thermally excited quasiparticles (i.e., the normal fluid ρ_N/m.) Accordingly, n_0 in the first term in (5.50) would be best interpreted as the superfluid density ρ_S/m.

Examining the solutions of $\epsilon(\mathbf{Q}, \omega) = 0$ using (5.50), we obtain

$$\omega^2 = \frac{n_0 V(Q)}{m} Q^2 + \frac{\tilde{n} V(Q)}{m} Q^2 = \frac{n V(Q)}{m} Q^2 \tag{5.51}$$

if $\omega \gg \varepsilon_Q$. This result illustrates how we can end up with a temperature-independent zero sound phonon frequency using the SPA, even though its origin lies equally in the condensate and normal parts of $\bar{\chi}_{nn}$ which are individually very temperature-dependent. We also note that condensate self-energy in (5.42) reduces to $\Sigma^c(\mathbf{Q}, \omega = cQ) = nV(Q)$ using (5.47) and (5.51). This has the effect of ensuring that the strong-coupling single-particle spectrum is renormalized to cQ, with the velocity going as $\sqrt{n}$, instead of $\sqrt{n_0}$ as in the weak-coupling limit (5.46). In later chapters of this book, (5.50) in a suitably generalized version will be used to suggest how the phonon mode frequency is essentially temperature-independent in superfluid ^{4}He. As far as χ_{nn} is concerned, the derivation of (5.51) based on (5.50) goes back to a suggestion of Pines (1966). The analogous discussion of a temperature-independent phonon pole of $G_{\alpha\beta}$ is due to Tserkovnikov (1965).

To summarize, we emphasize that while the phonon spectrum exhibited by both $G_{\alpha\beta}$ and χ_{nn} is quite different in the WC and SC limits, in both limits it is given by the zeros of $\epsilon(\mathbf{Q}, \omega)$ in (5.2). In the limit of low Q, the phonon excitations are best viewed as zero sound arising from some combination of mean fields associated with the condensate ($\bar{\chi}_{nn}^c$ in (5.11*b*)) and the non-condensate atoms ($\bar{\chi}_{nn}^R$). This interpretation, which is exhibited in a simplified form by (5.50), was first emphasized by Ma, Gould and Wong (1971).

A very basic question is whether, and under what conditions, we are working in the SC or WC limit. One finds that at $T = 0$, the low-Q excitation spectrum is always in the SC region (in a Bose gas, this is clear from (5.45)). As many microscopic calculations at $T = 0$ have shown, when proper account is taken of the interaction-induced depletion of the condensate fraction, the Bogoliubov phonon velocity in a WIDBG involving n_0 is essentially renormalized to involve the *total* density of atoms (see, for example, Hohenberg and Martin, 1965; Ma, Gould and

Wong, 1971). It may be thought of as a compressional sound wave (involving a speed given by the usual thermodynamic derivative). Going from this SC domain at $T = 0$ to finite temperatures, one may either continue to be in the SC domain (as is probable in the case of liquids) or enter into the WC domain (as is probable in the case of gases). As we have seen, the dispersion relation of the phonon mode is quite different in the two domains, and this difference also shows up in the relative weight of the mode in the dynamic structure factor $S(\mathbf{Q}, \omega)$ and the single-particle spectral density $A(\mathbf{Q}, \omega)$.

We can summarize the SC and WC limits as follows:

SC: The collisionless phonon orginates as a pole of χ_{nn}, with a temperature-independent speed. It is the natural extension of first sound and is best viewed as a zero sound mode involving *all* the atoms. This phonon has a weight in $S(\mathbf{Q}, \omega)$ proportional to n but a weight in $G_{\alpha\beta}(\mathbf{Q}, \omega)$ proportional to $n_0(T)$. This description is most appropriate in a liquid but can also be formally considered in a gas (Szépfalusy and Kondor, 1974; Griffin, 1988). One expects that as the temperature increases, additional single-particle states must appear to take over the spectral weight in $A(\mathbf{Q}, \omega)$. This is nicely illustrated by the Bose gas model calculations of Szépfalusy and Kondor (1974), where overdamped high-energy excitations leave the imaginary frequency axis at some intermediate temperature and, above T_λ, ultimately become the free-particle states of energy $Q^2/2m$. The analogue of this in a Bose liquid has not been investigated.

WC: This limit naturally arises in a WIDBG at intermediate temperatures (SK, 1974; Payne and Griffin, 1985). In this domain, the phonon mode is completely associated with the condensate mean field even at finite temperatures. It thus has a speed which is very temperature-dependent, decreasing as $\sqrt{n_0(T)}$, as predicted by (5.46). It is a soft mode and is seen to be the natural extension of second sound in a Bose gas (Gay and Griffin, 1985) into the collisionless domain. It has a weight in $S(\mathbf{Q}, \omega)$ proportional to $n_0(T)$. In a WIDBG described by this WC limit, there is no well defined zero sound mode involving a non-condensate mean field.

An expanded discussion of the nature of the low-Q phonons in a Bose-condensate fluid is given in Section 6.3. The discussion there is quite general and not based on an analysis of dilute Bose gas models used in this section.

5.3 One-loop approximation for regular quantities

In Section 5.2, we discussed the simplest diagrammatic approximation for the regular functions, namely those of a (Hartree) Bose gas. We now briefly comment on the next level of approximation for the regular functions $\bar{\Lambda}^{\ell}_{\alpha}$, $\bar{\chi}^{\ell R}_{JJ}$ and $\bar{\Sigma}_{\alpha\beta}$, given by the "one-loop" diagrams. As with the SPA discussed in the preceding section, our main interest in the one-loop approximation is to develop further insight into superfluid ^{4}He, rather than using it to find perturbative corrections to the Bogoliubov approximation. In Section 6.1, we show that several non-trivial zero-frequency sum rules are satisfied by this one-loop approximation, which emphasizes that it captures the right physics.

The one-loop regular diagrams are shown in Fig. 5.8. Referring to Wong and Gould (1974) as well as Talbot and Griffin (1983) for details, the rules of diagrammatic perturbation give

$$
\begin{aligned}
\bar{\Sigma}_{11}(\mathbf{Q},\omega) &= \int \frac{d\mathbf{p}}{(2\pi)^3}\left[V(0)\tilde{n}_p + V(\mathbf{p}+\mathbf{Q})\tilde{n}_p\right] + n_0 V(0) \\
&\quad - n_0 \int \frac{d\mathbf{p}}{(2\pi)^3} V^2(\mathbf{p})(u_p - v_p)^2 \\
&\quad \times \left\{u^2_{p+Q}[R_1(\omega) - R_2(\omega)] + v^2_{p+Q}[R_1(-\omega) - R_2(-\omega)]\right\} \\
&\quad + n_0 \int \frac{d\mathbf{p}}{(2\pi)^3} V(\mathbf{p})V(\mathbf{p}+\mathbf{Q})(u_p - v_p)(u_{p+Q} - v_{p+Q}) \\
&\quad \times \left\{v_p u_{p+Q} R_1(\omega) + u_p u_{p+Q} R_2(\omega) + u_p v_{p+Q} R_1(-\omega) + v_p v_{p+Q} R_2(-\omega)\right\} ,
\end{aligned}
\tag{5.52}
$$

$$
\begin{aligned}
\bar{\Sigma}_{12}(\mathbf{Q},\omega) &= \int \frac{d\mathbf{p}}{(2\pi)^3} V(\mathbf{p}+\mathbf{Q})\tilde{m}_p + n_0 \int \frac{d\mathbf{p}}{(2\pi)^3} V^2(\mathbf{p})(u_p - v_p)^2 u_{p+Q} v_{p+Q} \\
&\quad \times \left\{R_1(\omega) - R_2(\omega) + R_1(-\omega) - R_2(-\omega)\right\} \\
&\quad - n_0 \int \frac{d\mathbf{p}}{(2\pi)^3} V(\mathbf{p})V(\mathbf{p}+\mathbf{Q})(u_p - v_p)(u_{p+Q} - v_{p+Q}) \\
&\quad \times \left\{(u_q u_{p+Q} + v_p v_{p+Q})R_1(\omega) + \frac{1}{2}(u_p v_{p+Q} + v_p u_{p+Q})[R_2(\omega) + R_2(-\omega)]\right\} ,
\end{aligned}
\tag{5.53}
$$

$$
\begin{aligned}
\bar{\chi}^{\ell R}_{JJ}(\mathbf{Q},\omega) &= -\int \frac{d\mathbf{p}}{(2\pi)^3}\left[\left(\mathbf{p} + \frac{\mathbf{Q}}{2}\right)\cdot\hat{\mathbf{Q}}\right]^2 \left\{(u_p u_{p+Q} - v_p v_{p+Q})^2 R_1(\omega)\right. \\
&\quad \left. - \frac{1}{2}(u_p v_{p+Q} - v_p u_{p+Q})^2[R_2(\omega) + R_2(-\omega)]\right\} ,
\end{aligned}
\tag{5.54}
$$

$$
\begin{aligned}
\bar{\Lambda}^{\ell}_1(\mathbf{Q},\omega) &= \frac{\sqrt{n_0}Q}{2} - \sqrt{n_0}\int \frac{d\mathbf{p}}{(2\pi)^3} V(\mathbf{p})\left[\left(\mathbf{p} + \frac{1}{2}\mathbf{Q}\right)\cdot\hat{\mathbf{Q}}\right] \\
&\quad \times \left\{(u_p - v_p)(u_p u_{p+Q} - v_p v_{p+Q})[u_{p+Q}R_1(\omega) + v_{p+Q}R_1(-\omega)]\right.
\end{aligned}
$$

$$-(u_p - v_p)(u_p v_{p+Q} - v_p u_{p+Q})[u_{p+Q}R_2(\omega) + v_{p+Q}R_2(-\omega)]\} \ . \tag{5.55}$$

The single-particle distributions are (in the Bogoliubov approximation)

$$\tilde{n}_p = \langle a_p^+ a_p \rangle_B = -\frac{1}{\beta}\sum_{\omega_n} G_{11}^B(\mathbf{p}, i\omega_n) = v_p^2 + (u_p^2 + v_p^2)N(\omega_p) \ , \tag{5.56}$$

$$\tilde{m}_p = \langle a_p^+ a_{-p}^+ \rangle_B = -\frac{1}{\beta}\sum_{\omega_n} G_{12}^B(\mathbf{p}, i\omega_n) = -u_p v_p[1 + 2N(\omega_p)] \ , \tag{5.57}$$

and the functions R_1 and R_2 are as defined in (3.46). While looking somewhat daunting, these one-loop expressions show several characteristic features of regular functions. Strictly speaking, we should calculate the one-loop diagrams in Fig. 5.8 using the full zero-loop or SPA single-particle Green's functions for the *internal* propagator lines. As we discussed in Section 5.2, however, the SPA $G_{\alpha\beta}$ is only simple in the single-particle (SP) limit, where it can be approximated by the Bogoliubov form given by (3.35)–(3.37). The one-loop results written down above have, in fact, been derived using this SP approximation to $G_{\alpha\beta}$. Thus the quasiparticle dispersion relation ω_p which enters these expressions (including the u_p and v_p amplitudes) is given by (5.46). We refer to this as the SPA–Bogoliubov approximation for $G_{\alpha\beta}$.

In the zero sound (ZS) limit for the SPA $G_{\alpha\beta}$, the structure is much more complicated than the single-particle SP limit. There is not much pedagogical value in working out the one-loop diagrams with such improved propagators. The basic physics is already clear using the zero-loop or SPA propagators. The one-loop regular functions based on the SPA–Bogoliubov propagators provide a well defined microscopic model which is still simple enough to understand the structure of a non-trivial approximation past the Hartree Bose gas expressions in (5.34).

Terms to first order in the interaction V in $\bar{\Sigma}_{\alpha\beta}$ ((5.52) and (5.53)) correspond to the *complete* Hartree–Fock proper self-energies (in contrast with the Bogoliubov approximation discussed at the end of Section 3.2 and shown in Fig. 3.3). In the normal phase ($T > T_\lambda$), the one-loop approximation consists of the Hartree–Fock self-energies.

The new physics involved in the one-loop approximation comes from the terms which are second-order in the two-body interaction. Referring to (3.46), R_1 clearly describes scattering processes involving thermally excited quasiparticles ($\mathbf{p} \to \mathbf{p} + \mathbf{Q}$), which are only present at finite temperatures. In contrast, R_2 describes the creation (or destruction) of two quasiparticles, processes which occur even at $T = 0$. The regular functions given by (5.52)–(5.55) lead to expressions for both $G_{\alpha\beta}(\mathbf{Q}, \omega)$

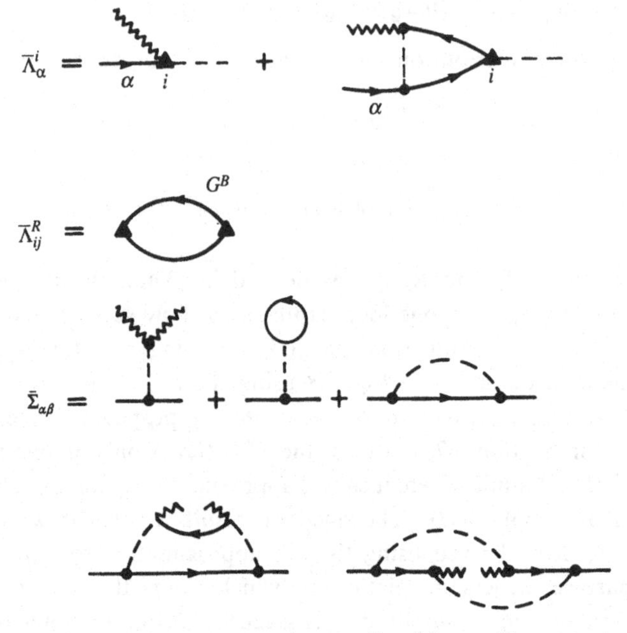

Fig. 5.8. The proper, irreducible diagrams for $\bar{\Sigma}_{\alpha\beta}$, $\bar{\Lambda}^\ell_\alpha$ and $\bar{\chi}^{\ell R}_{JJ}$ used in the one-loop approximation.

and $\chi_{nn}(\mathbf{Q}, \omega)$ (using the procedure given at the end of Section 5.1) which will involve two-quasiparticle processes. In contrast, the Bogoliubov approximation in Section 3.2 includes the creation or destruction of only a *single* quasiparticle.

In Fig. 5.9, we show an example of a second-order reducible self-energy diagram which is included in $\Sigma^c_{\alpha\beta}$ in (5.12) when we work within the one-loop approximation. It may be viewed as a broken-symmetry vertex correction to the simplest reducible self-energy exchange diagram shown in Fig. 3.3. It arises from using a $\bar{\Lambda}_\alpha$ in (5.11) which is associated with the $\bar{\Lambda}^\ell_\alpha$ given in Fig. 5.8. We note, however, that the one-loop approximation does *not* include the normal second-order contributions to $\Sigma_{\alpha\beta}$ shown in Fig. 5.10. Since the one-loop approximation does not include the effect of such real two-body collisions, it cannot be used to discuss hydrodynamic modes at *finite* temperatures.

While (3.44) and (3.45) of Section 3.3 illustrate the kind of structure expected in $\bar{\chi}^R_{nn}$, we note that calculating $\bar{\chi}_{nn}$ from the related current–current response function using (5.16*b*) is a superior approach. This automatically includes "backflow" effects and thus gives a density-response

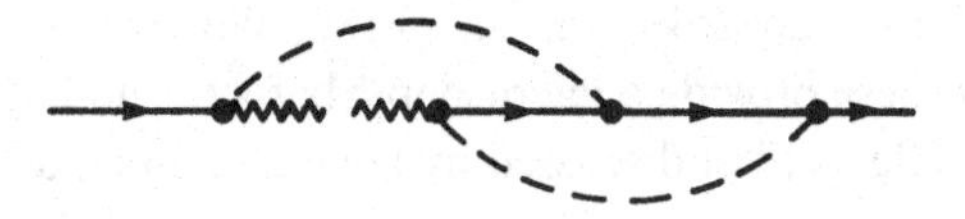

Fig. 5.9. A second-order reducible self-energy diagram included in the one-loop approximation for $\Sigma^{c}_{\alpha\beta}$.

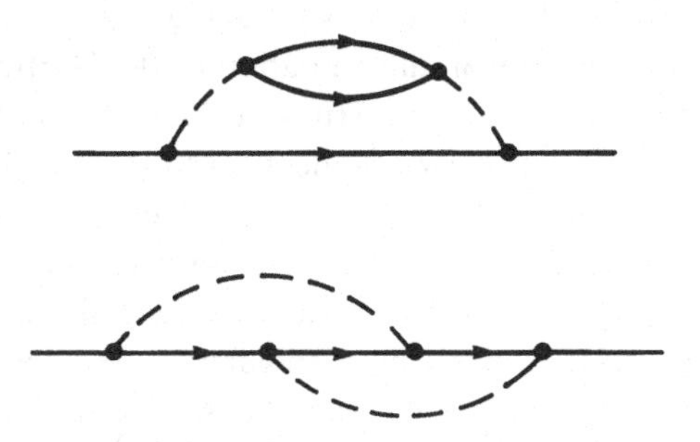

Fig. 5.10. Second-order irreducible self-energies not included in the one-loop approximation for $\tilde{\Sigma}_{\alpha\beta}$.

function χ_{nn} which is consistent with the f-sum rule (2.24), in contrast with (3.45). This approach was first used by Miller, Pines and Nozières (1962) as well as in Section 7.3 of Nozières and Pines (1964, 1990), and was later emphasized by Wong and Gould (1974) in the context of the dielectric formalism. Using the one-loop expression for $\bar{\chi}^{\prime R}_{JJ}$ given in (5.54), the second and third lines in (3.45) are then replaced by Eqs. (A7) and (A8), respectively, of Griffin and Talbot (1981).

Wong and Gould (1974, 1976) have used the one-loop approximation in a WIDBG at low temperatures as a systematic way of calculating the temperature-dependent corrections to the lowest-order Bogoliubov results discussed in Section 3.3. Since they assumed that there is *no* well defined zero sound mode in the non-condensate atoms, they were justified in using the expansion

$$\frac{1}{\epsilon^R} \simeq 1 + V(Q)\bar{\chi}^R_{nn} + \ldots \tag{5.58}$$

in both (5.24) and (5.12). That is, χ_{nn} in (5.24) was expanded around the Bogoliubov SPA approximation, whose spectrum is given by (5.46) in the SP limit. Such an expansion gives a result for $S(\mathbf{Q}, \omega)$ which has a "one-phonon" resonance (pole of $G_{\alpha\beta}$) as well as a "two-phonon" continuum from the second term in (5.24) and the Bose vertex functions $\bar{\Lambda}_\alpha/\epsilon^R$. The latter are often called "interference" contributions and correspond to the $\tilde{\rho}\hat{A}$ contributions in (3.18). Wong and Gould (1974) have shown how these provide a microscopic basis for understanding how such "backflow" effects (first discussed by Feynman and Cohen, 1956, in a different context) modify the elementary excitations in a Bose-condensed fluid.

In connection with the approximation (5.58), the work of Cheung and Griffin (1971b) is of interest. They compute χ_{nn} in a conserving approximation using functional differentiation of the Hartree–Fock self-energies (as given by Girardeau and Arnowitt, 1959). Cheung and Griffin then prove that the pole of this χ_{nn} is identical to that of $G_{\alpha\beta}$ computed in the second-order Beliaev (1958b) approximation, for all T and Q. To show this equivalence, however, it was necessary to assume that there were no zeros of $\epsilon^R(\mathbf{Q}, \omega)$ so that one could use an expansion analogous to (5.58). It is no accident that such a "regularity assumption" leads to a single identical mode in both $G_{\alpha\beta}$ and χ_{nn}. A similar assumption was made by Beliaev (1958b) in working out the explicit form of the poles of $G_{\alpha\beta}$.

An expression like (3.45) is often *implicitly* assumed in discussing the observed spectrum of $S(\mathbf{Q}, \omega)$ in superfluid ^{4}He. In particular, it is the basis of the classic paper by Miller, Pines and Nozières (1962). However, this sort of approximation to (5.24) breaks down when the dynamics of the "normal fluid" excitations play a crucial role, in which case one cannot ignore the structure arising from the zeros of ϵ^R. If the ZS limit is appropriate, the decomposition in (5.24) (and simple approximations to it such as (3.45)) is not as useful as (5.2) as a basis for interpreting $S(\mathbf{Q}, \omega)$. The low-Q phonon region in liquid ^{4}He is a case in point (see Section 6.3).

5.4 Gavoret–Nozières analysis

In this section, we summarize the seminal work by Gavoret and Nozières (GN, 1964). Their analysis gave the first rigorous treatment of the role of the Bose condensate in coupling the single-particle spectrum with the density fluctuations. They obtained expressions of the type given in (3.47) within a well defined diagrammatic scheme. In fact, GN analyse the structure of an arbitrary two-particle Green's function G_2. While this section concerns itself mainly with the particle–hole correlation function (involved in the density and current response functions), GN's general expressions for G_2 will be useful in Chapter 10, where we discuss the two-particle (or pair) spectrum in a Bose-condensed fluid. The formal results of GN are valid at all Q and ω and are easily rewritten at finite temperatures (using imaginary Bose frequencies in the usual way). In our summary, we work at $T = 0$ and follow GN's notation closely.

A second, equally famous, aspect of the GN paper is their determination of the explicit forms of $G_{\alpha\beta}$ and χ_{nn} in the $Q, \omega \to 0$ limit. Using regularity arguments specifically restricted to $T = 0$, they argue that χ_{nn}^R in (3.47) and the single-particle self-energies $\Sigma_{\alpha\beta}$ are non-singular in the limit $Q, \omega \to 0$. This is shown to be consistent, in that a direct evaluation of the pole of $G_{\alpha\beta}(\mathbf{Q}, \omega)$ gives a phonon with the expected (compressional) sound velocity. We defer further discussion of these results (and their extension to finite T) to Section 6.3, after we have derived various rigorous results in the zero-frequency limit in Section 6.1.

Using the matrix notation for field operators given in (3.21), GN define a general two-particle Green's function ($\alpha, \beta, \gamma, \delta = +, -$)

$$K_{\alpha\beta}^{\gamma\delta}(\mathbf{p}_\delta t_\delta, \mathbf{p}_\gamma t_\gamma; \mathbf{p}_\beta t_\beta, \mathbf{p}_\alpha t_\alpha) \equiv \langle \hat{T} \hat{a}_{\mathbf{p}_\delta,\delta}(t_\delta) \hat{a}_{\mathbf{p}_\gamma,\gamma}(t_\gamma) \hat{a}^+_{\mathbf{p}_\beta,\beta}(t_\beta) \hat{a}^+_{\mathbf{p}_\alpha,\alpha}(t_\alpha) \rangle \ . \tag{5.59}$$

In the Bose-condensed phase, there are 2^4 or 16 different two-particle Green's functions. We recall that $\hat{a}^+_{q\alpha} = \hat{a}_{-q,-\alpha}$. The Fourier transform of (5.59) is proportional to $K_{\alpha\beta}^{\gamma\delta}(p_\delta, p_\gamma; p_\beta, p_\alpha)$, where it is convenient to use a 4-momentum notation $p_\alpha = (\mathbf{p}_\alpha, \omega_\alpha)$, etc. Using the fact that $K_{\alpha\beta}^{\gamma\delta}$ is invariant under translations in space and time results in the requirement that $p_\delta + p_\gamma = p_\beta + p_\alpha$, i.e., only three of the "4-momentum" vectors are independent variables. Following GN, we use this fact and work with (see Fig. 5.11)

$$K_{\alpha\beta}^{\gamma\delta}\left(-p + \frac{Q}{2}, p + \frac{Q}{2}; p' + \frac{Q}{2}, -p' + \frac{Q}{2}\right) \equiv K_{\alpha\beta}^{\gamma\delta}(p', p; Q) \ , \tag{5.60}$$

where $Q \equiv (\mathbf{Q}, \omega)$ is the total 4-momentum of the pair of excitations. As an illustration of this notation in a concrete case, we note that the density-response function is given by (at $T = 0$)

$$\begin{aligned} \chi_{nn}(\mathbf{Q}, \omega) &= -i \int_{-\infty}^{\infty} \frac{dt}{2\pi} e^{i\omega t} \langle \hat{T} \hat{\rho}(\mathbf{Q}, t) \hat{\rho}(-\mathbf{Q}, 0) \rangle \\ &= -i \sum_{p', p} K_{-+}^{+-}(p', p; Q) \ , \end{aligned} \tag{5.61}$$

where

$$\sum_p \equiv \int \frac{d\mathbf{p}}{(2\pi)^3} \int \frac{d\omega_p}{2\pi} \ . \tag{5.62}$$

For further details, we refer to §15 of Lifshitz and Pitaevskii (1980) and Chapter 6 of Nozières (1964).

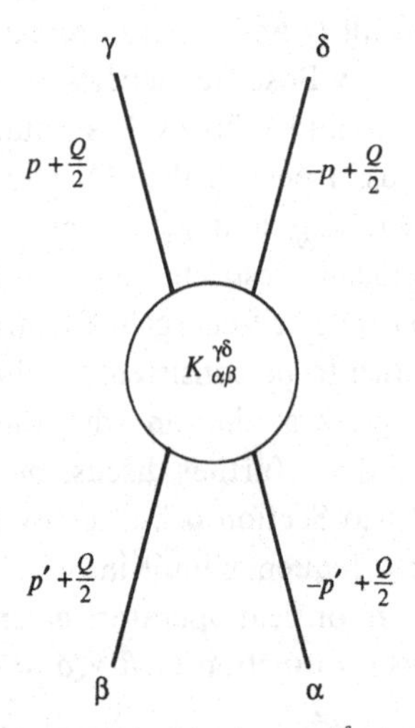

Fig. 5.11. Two-particle Green's functions $K^{\gamma\delta}_{\alpha\beta}(p', p; Q)$ defined in (5.60).

As discussed in Section 5.1, the diagrammatic analysis of $K^{\gamma\delta}_{\alpha\beta}(p', p; Q)$ in (5.60) leads naturally to improper contributions which carry a single (renormalized) propagator. In terms of three-point kernels $P^{\eta}_{\xi\varepsilon}(p', Q)$ and $P^{\eta\xi}_{\zeta}(p, Q)$, these improper contributions are shown in Fig. 5.12 and form what GN call the singular or condensate contribution to $K^{\gamma\delta}_{\alpha\beta}$:

$$^{c}K^{\gamma\delta}_{\alpha\beta}(p', p; Q) = Q^{\rho}_{\alpha\beta}(p', Q) G_{\rho\sigma}(Q) Q^{\gamma\delta}_{\sigma}(p, Q) \ . \tag{5.63}$$

As usual, the *repeated* Greek indices (ρ, σ) are summed over. The Bose vertex functions in (5.63) are clearly related to the $\Lambda_{\alpha}(Q)$ vertex functions introduced in Section 5.1, with, for example,

$$\begin{aligned} Q^{\gamma\delta}_{\sigma}(p, Q) &= \sqrt{n_0}\left[\delta_{\sigma,\gamma}(2\pi)^4\delta^{(4)}\left(p - \frac{Q}{2}\right) + \delta_{\sigma,\delta}(2\pi)^4\delta^{(4)}\left(p + \frac{Q}{2}\right)\right] \\ &\quad + P^{\eta\xi}_{\sigma}(p, Q) G_{\eta\gamma}\left(p + \frac{Q}{2}\right) G_{\xi\delta}\left(-p + \frac{Q}{2}\right) \ . \end{aligned} \tag{5.64}$$

In addition to (5.63), one has "regular" diagrammatic contributions to $K^{\gamma\delta}_{\alpha\beta}$ which arise in a normal system, proper diagrams which cannot be split into two by cutting a single propagator line. These are discussed at great length in microscopic derivations of Fermi liquid theory (see §15 of Lifshitz and Pitaevskii, 1980). Besides the two diagrams involving independent single-particle propagation, the others are usefully described

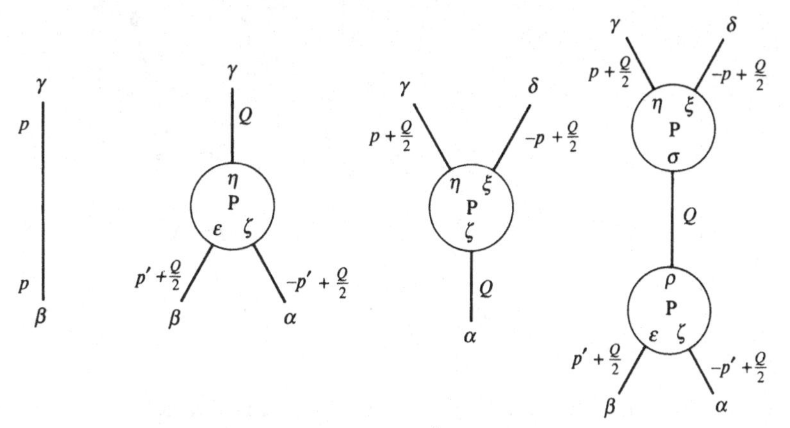

Fig. 5.12. The singular contributions to $K^{\gamma\delta}_{\alpha\beta}$ (see (5.63) and (5.64)) involving a single propagator as an intermediate state.

by introducing an interaction vertex function $\Gamma(p', p; Q)$ (see Fig. 5.13)

$$\begin{aligned} {}^{R}K^{\gamma\delta}_{\alpha\beta}(p', p; Q) &= G_{\alpha\delta}\left(-p + \frac{Q}{2}\right) G_{\beta\gamma}\left(p + \frac{Q}{2}\right) \delta^{(4)}(p - p')(2\pi)^4 \\ &+ G_{\beta\delta}\left(-p + \frac{Q}{2}\right) G_{\alpha\gamma}\left(p + \frac{Q}{2}\right) \delta^{(4)}(p + p')(2\pi)^4 \\ &+ G_{\alpha\zeta}\left(-p' + \frac{Q}{2}\right) G_{\beta\varepsilon}\left(p' + \frac{Q}{2}\right) \Gamma^{\eta\xi}_{\zeta\varepsilon}(p', p, Q) \\ &\times G_{\eta\gamma}\left(p + \frac{Q}{2}\right) G_{\xi\delta}\left(-p + \frac{Q}{2}\right) . \end{aligned} \tag{5.65}$$

Clearly the regular contribution in (3.47) is given by (see (5.61))

$$\chi^{R}_{nn}(\mathbf{Q}, \omega) = -i \sum_{p,p'} {}^{R}K^{+-}_{-+}(p', p; Q) . \tag{5.66}$$

The free propagation contribution to (5.66) from the first two terms in (5.65) is easily seen to reduce to

$$\begin{aligned} \chi^{R0}_{nn}(\mathbf{Q}, \omega) &= -i \sum_{p} \left[G_{--}\left(-p + \frac{Q}{2}\right) G_{++}\left(p + \frac{Q}{2}\right) \right. \\ &\left. + G_{+-}\left(-p + \frac{Q}{2}\right) G_{-+}\left(p + \frac{Q}{2}\right) \right] . \end{aligned} \tag{5.67}$$

This result is identical to the expression given in (3.44) at $T = 0$. It corresponds to the "bubble" skeleton diagram for χ_{nn} using the matrix renormalized single-particle propagators.

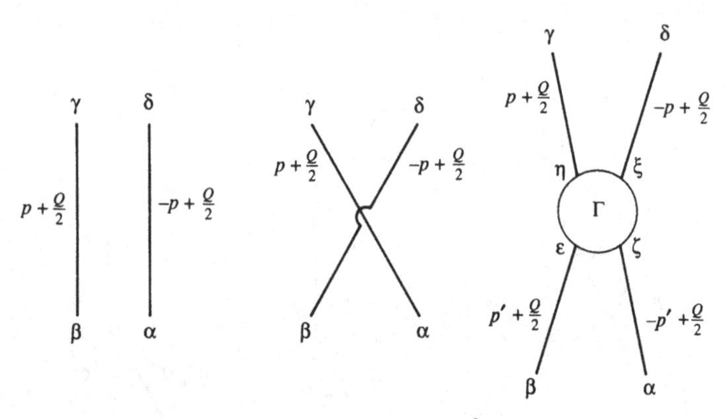

Fig. 5.13. Regular contributions to $K^{\gamma\delta}_{\alpha\beta}$, as given by (5.65).

The results of this section so far are exact. To proceed further, one must calculate the single-particle self-energies $\Sigma_{\alpha\beta}$ which determine $G_{\alpha\beta}$, the three-point kernels $P^{\eta\xi}_{\sigma}$ which determine the Bose vertex functions $Q^{\gamma\delta}_{\sigma}$ in (5.63), and finally the interaction vertex $\Gamma^{\eta\xi}_{\zeta\varepsilon}$ in (5.65). GN limit themselves to contributions involving intermediate states with *two* single-particle propagators, arguing these will give rise to the most important singularities in the $\mathbf{Q}, \omega \to 0$ limit, over and above those that arise from the single propagator in ${}^{c}K$ as given by (5.63). For the interaction vertex, this corresponds to the usual Bethe–Salpeter integral equation, except that it is now a matrix equation:

$$\begin{aligned}\Gamma^{\gamma\delta}_{\alpha\beta}(p', p; Q) &= I^{\gamma\delta}_{\alpha\beta}(p', p; Q) \\ &\quad + \frac{1}{2}\sum_{p''} I^{\eta\xi}_{\alpha\beta}(p', p''; Q) G_{\eta\varepsilon}\left(p'' + \frac{Q}{2}\right) G_{\xi\zeta}\left(-p'' + \frac{Q}{2}\right) \Gamma^{\gamma\delta}_{\zeta\varepsilon}(p'', p; Q) \\ &\equiv I + \frac{1}{2} IGG\,\Gamma \,. \end{aligned} \tag{5.68}$$

This equation is shown in Fig. 5.14. Here I is the "irreducible" contribution, defined by GN as all diagrams which do not involve two propagators in any intermediate state. Similarly, the three-point kernel P can be expressed in terms of I and the irreducible part of P, schematically given by

$$P = J + \frac{1}{2} IGGP \,. \tag{5.69}$$

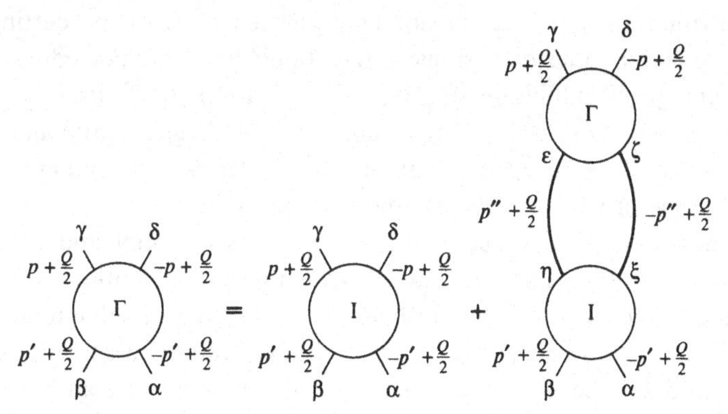

Fig. 5.14. Matrix Bethe–Salpeter equation for the interaction vertex Γ in (5.68), in terms of contributions I which do not involve two propagators as intermediate states.

Finally the self-energy is given by (schematically)

$$\Sigma = \tilde{\Sigma} + JGGP \ , \tag{5.70}$$

where $\tilde{\Sigma}$ is the "irreducible" contribution as defined by GN. As discussed in detail by GN, the solutions of (5.68)–(5.70) will be strongly dependent on the $Q \to 0$ singularities arising from the product $G(p+\frac{Q}{2})G(-p+\frac{Q}{2})$, which occurs in all the integrands of the preceding integral equations. However, we emphasize that (5.68)–(5.70) are not necessarily restricted to the study of the $Q \to 0$ limit. Moreover, they can easily be formally generalized to finite T.

For further discussion, it is convenient to write the solutions (5.68)–(5.70) in the schematic form

$$\Gamma = \frac{I}{1-\frac{1}{2}IGG} \quad \text{and hence} \quad {}^{R}K^{+-}_{-+} = \frac{2GG}{1-\frac{1}{2}IGG} \tag{5.71}$$

and

$$P = \frac{J}{1-\frac{1}{2}IGG} \ , \quad \Sigma = \tilde{\Sigma} + \frac{JGGJ}{1-\frac{1}{2}IGG} \ . \tag{5.72}$$

It is clear that any singularity of Γ will show up in both the three-point kernel P and the self-energy Σ, with a weight determined by the "irreducible" three-point kernel J.

The distinction between these results and the formally exact expression of the dielectric formalism of Section 5.1 should be kept clearly in mind. In the dielectric formalism, the key step was a partial re-summation of diagrams through the introduction of regular diagrams, i.e., by separating out contributions which could not be separated into two by cutting a single propagator line *or* a single interaction line. This procedure has the advantage of exhibiting the role of $\epsilon^R \equiv 1 - V(Q)\overline{\chi}^R_{nn}$ in a natural way, as discussed in Section 5.1. In contrast, the diagrammatic analysis of GN is based on separating out the role of terms with one and two single-particle propagators as intermediate states.

In order to make contact between the analysis of GN and that of the dielectric formalism, one must isolate those contributions of GN which include a *single* interaction line. The GN "irreducible" interaction function I in (5.68) and (5.69) contains "bubble" diagram contributions to Γ, P and Σ. The dielectric formalism sums up explicitly all bubbles, to infinite order. In contrast, the ladder diagrams (in a generalized sense) are still buried in the regular functions of the dielectric formalism. One should not confuse $\tilde{\Sigma}$ in (5.70) with the irreducible proper self-energy defined in (5.12) in Section 5.1. In the GN analysis, there is no equivalent of $\bar{G}_{\alpha\beta}$, not to mention $\bar{\Lambda}_\alpha$ and $\overline{\chi}^R_{nn}$. One could use an approach analogous to GN to isolate all contributions to *regular* quantities involving *two* isolated propagator lines. Further work along this line would be very useful, especially in the small Q, ω limit. The explicit consideration of ladder diagrams to all orders is also necessary if one wants to introduce (in a controlled fashion) a t-matrix in place of the bare interaction in the dielectric formalism.

It is clear that GN are going well past the simple physics of the one-loop approximation described in Section 5.3. This can already be seen in the normal phase, where the one-loop diagrams reduce to free Bose gas approximations for $\overline{\chi}^R_{nn}$ and $\overline{\chi}'^R_{JJ}$. In contrast, with proper attention given to the irreducible interaction function I and the use of Ward identities, a Bethe–Salpeter integral equation like (5.68) can be used to derive kinetic equations and collective modes in both the hydrodynamic and collisionless limits. This analogous derivation has been given in great detail for normal Fermi systems (see, in particular, Chapter 6 of Nozières, 1964) as well as for superfluid Fermi liquids (see Vollhardt and Wölfle, 1990).

Starting from the formulation summarized above, Gavoret and Nozières have evaluated the poles of $G_{\alpha\beta}$ and χ_{nn} in the limit of small Q and ω at $T = 0$. We defer further discussion of this *tour de force* calculation to Section 6.3.

5.5 The collective vs. single-particle scenarios

In this concluding section, we collect together the main formulas of the dielectric formalism and discuss the physical picture which these results imply concerning the excitations of a Bose fluid. Sections 5.1–5.4 give the microscopic calculations which lie behind the discussion of the present section. However, these earlier sections are fairly technical and lengthy, hiding the essential simplicity of the final results. The present section summarizes the main ideas which will be needed in Chapter 7 in interpreting experimental data on $S(\mathbf{Q},\omega)$ in liquid ^{4}He.

We recall from Chapter 3 that in the presence of a Bose condensate order parameter, the Fourier component of the density operator $\hat{\rho}(\mathbf{Q})$ naturally separates into two parts, given in (3.12) and (3.13). One immediately sees that the density-response function χ_{nn} will contain terms involving the creation (and destruction) of single-particle excitations (via the condensate reservoir). Thus it is not unexpected that χ_{nn} is given by an expression of the kind in (5.24), namely

$$\chi_{nn} = \Lambda_\alpha G_{\alpha\beta}\Lambda_\beta + \frac{\bar{\chi}^R_{nn}}{\epsilon^R} ,$$

where the Bose broken-symmetry vertex function Λ_α vanishes if $n_0 = 0$. The second term in the above expression describes the density fluctuations in the non-condensate atoms described by $\tilde{\rho}_Q$ in (3.13), with the "normal" dielectric function defined by (5.10), namely

$$\epsilon^R = 1 - V(Q)\bar{\chi}^R_{nn} .$$

Above T_λ, only the second or "normal" term on the r.h.s. of χ_{nn} in (5.24) is present. Since it only appears below T_λ, the first term $\Lambda G \Lambda$ in (5.24) may be viewed as the "superfluid" or condensate-related part of χ_{nn}.

The preceding discussion naturally leads to (5.24) (given in the preceding paragraph) being interpreted as the sum of "superfluid" and "normal" response functions (Nozières and Pines, 1964; 1990)

$$\chi_{nn} = \chi^c_{nn} + \chi^R_{nn} . \tag{5.73}$$

However this two-component form hides the fact that both contributions involve ϵ^R. This is made explicit by the fact that (see (5.9))

$$\Lambda_\alpha = \frac{\bar{\Lambda}_\alpha}{\epsilon^R} \tag{5.74}$$

and, in addition, the presence of reducible single-particle self-energies of the kind given in (5.12), i.e.,

$$\bar{\Lambda}_\alpha \frac{V(Q)}{\epsilon^R} \bar{\Lambda}_\beta \ . \tag{5.75}$$

As we discussed in Section 5.1, (5.24) is completely equivalent to the expression in (5.2),

$$\chi_{nn} = \frac{\bar{\chi}^c_{nn} + \bar{\chi}^R_{nn}}{1 - V(Q)(\bar{\chi}^c_{nn} + \bar{\chi}^R_{nn})} \ , \tag{5.76}$$

where $\bar{\chi}_{nn}$ is given by the sum of two terms given in (5.11*b*), with the "condensate" part

$$\bar{\chi}^c_{nn} = \bar{\Lambda}_\alpha \bar{G}_{\alpha\beta} \bar{\Lambda}_\beta \ . \tag{5.77}$$

Here $\bar{G}_{\alpha\beta}$ is the matrix single-particle Green's function as determined by the "normal" self-energy contributions but not those in (5.75), which are only present below T_λ. The density fluctuation spectrum is given by the zeros of the dielectric function

$$\epsilon(\mathbf{Q}, \omega) = 1 - V(Q)\bar{\chi}_{nn} = \epsilon^R - V(Q)\bar{\Lambda}_\alpha \frac{\bar{N}_{\alpha\beta}}{\bar{D}} \bar{\Lambda}_\beta \ , \tag{5.78}$$

where $\bar{D}$ is the denominator of the "normal" Green's function $\bar{G}_{\alpha\beta} = \bar{N}_{\alpha\beta}/\bar{D}$. One sees that because of the broken-symmetry vertex functions $\bar{\Lambda}_\alpha$, $\epsilon(\mathbf{Q}, \omega)$ in (5.78) will exhibit the hybridized zeros of both ϵ^R and $\bar{D}$. In physical terms, the single-particle and density excitations, which are uncoupled *above* T_λ, are hybridized *below* T_λ.

The two-component nature of $\bar{\chi}_{nn}$ in (5.11*b*) is the basis of the two-component form of χ_{nn} in (5.24). The existence and interplay of the two distinct contributions to $\bar{\chi}_{nn}$ lie at the root of the varying behaviour of $S(\mathbf{Q}, \omega)$ at different values of Q, as we discuss in Section 7.2. Consequently, in the search for parameterized expressions for $S(\mathbf{Q}, \omega)$, it seems best to work directly with the two components of $\bar{\chi}_{nn}$ rather than with the components of χ_{nn}. We also note that

$$S(\mathbf{Q}, \omega) = -\frac{[N(\omega) + 1]}{\pi n} \frac{1}{V(Q)} \operatorname{Im} \frac{1}{\epsilon(\mathbf{Q}, \omega + i\eta)} \ , \tag{5.79}$$

as may be easily verified using (5.2) in (3.40). Thus only $\epsilon(\mathbf{Q}, \omega)$ is needed to find $S(\mathbf{Q}, \omega)$. However, we note that (5.79) is equivalent to

$$\operatorname{Im} \chi_{nn}(\mathbf{Q}, \omega) = \frac{\operatorname{Im} \bar{\chi}_{nn}}{[1 - V(Q)\operatorname{Re} \bar{\chi}_{nn}]^2 + [V(Q)\operatorname{Im} \bar{\chi}_{nn}]^2} \ . \tag{5.80}$$

This form emphasizes that, in general, structure in Im χ_{nn} comes from the zeros of $\epsilon(\mathbf{Q}, \omega)$. On the other hand, if $V(Q)$ is very small (as it is for

$Q \gtrsim 2$ Å^{-1}), the structure in (5.80) is best thought of as arising directly from Im $\bar{\chi}_{nn}$, rather than a zero of $\epsilon(\mathbf{Q}, \omega)$.

The well known analysis of Gavoret and Nozières (1964) is based on isolating diagrammatic contributions involving intermediate states with one and two particle lines (see Section 5.4). By not separating out diagrams with a single interaction line, GN do not obtain expressions of the kind (5.76), but work directly with the less transparent two-component forms such as (5.73). The dielectric formalism is based on expressing various correlation functions in terms of regular functions, diagrammatic contributions which do not contain isolated single-particle *or* interaction lines. Specific calculations are then concerned with approximations to these regular functions.

As Wong and Gould (1974) have emphasized, working with regular functions involving the momentum current (rather than the density) has the advantage of building in the local conservation laws related to the equation of continuity. Perturbative calculations are then automatically "conserving" to the order one works at. Thus, in computing the density-response function χ_{nn}, one first expresses $\bar{\chi}_{nn}$ in terms of $\bar{\chi}^{\ell}_{JJ}$ using (5.16*b*), which in turn is given by (5.13*b*) in terms of $\bar{\chi}^{\ell R}_{JJ}$, $\bar{\Lambda}^{\ell}_{\alpha}$ and $\bar{G}_{\alpha\beta}$. The complete excitation spectrum of both $G_{\alpha\beta}$ and χ_{nn} is then determined by the zeros of

$$\epsilon(\mathbf{Q}, \omega) = 1 - \frac{nV(Q)}{m}\frac{Q^2}{\omega^2}\left[1 + \frac{\bar{\chi}^{\ell}_{JJ}(\mathbf{Q}, \omega)}{\rho}\right]. \tag{5.81}$$

This form clearly exhibits the zero sound mode (involving both the condensate and non-condensate contributions to $\bar{\chi}^{\ell}_{JJ}$) in an explicit manner. It is particularly useful in discussing the low-Q phonon region, as Ma, Gould and Wong (1971) have emphasized. Calculations based on (5.81) show clearly that a phonon in a Bose fluid in the $Q \to 0$ limit parallels the discussion of zero sound in a Fermi gas. Working with $\bar{\chi}^{\ell}_{JJ}$ in (5.81) has a technical advantage in that the integrand has two extra factors of momentum, which makes it better behaved than $\bar{\chi}_{nn}$ in the low-Q limit.

After the work of Gavoret and Nozières (1964) and Hohenberg and Martin (1965), it was implicitly accepted in much of the theoretical literature that two-component expressions such as in (3.47) were the key to understanding superfluid ^{4}He. The scenario was that in superfluid ^{4}He, $G_{\alpha\beta}$ had a sharp single-particle excitation and that this appeared as a resonance in χ_{nn} with a strength related to the Bose vertex function Λ_{α}. This scenario seems to ignore the condensate-induced hybridization of the single-particle and density fluctuations, whose possibility the dielectric formalism of Section 5.1 exposes so clearly. It ignores the fact that both terms in (3.47) or (5.24) are intimately connected and can not always be viewed as two separate terms. This is shown most dramatically by the fact that (5.24) is formally equivalent to (5.76). We shall argue that the answer to the question of which expression is most appropriate as a starting point in discussing $\chi_{nn}(\mathbf{Q},\omega)$ depends very much on the wavevector region one is dealing with.

Broadly speaking, one can distinguish the resulting spectra of χ_{nn} and $G_{\alpha\beta}$ by whether or not there is any significant structure above T_λ arising from $\bar{\chi}^R_{nn}$ in (5.10) or (3.47). A zero of ϵ^R corresponds to a zero sound pole. If there is *no* such structure, the two terms in (5.24) have a distinct physical significance and (5.24) then gives a better representation of the physics than (5.2). The χ_{nn} spectrum is modified below T_λ by the appearance of the single-particle excitation peaks (appropriately renormalized by the Bose condensate) which are associated with the first term in (5.24) or (3.47). Besides the scattering from thermally excited particle–hole excitations, which have their analogue in the normal phase, contributions from pair excitations appear with an intensity dependent on the condensate. We call this the single-particle or SP limit. The simplest illustration of this SP scenario is given by (3.45).

In the opposite limit, if there is a zero sound density fluctuation in χ_{nn} for temperatures above T_λ, there are important cancellations between the two terms in (5.24). In this case, the expressions in (5.76) and (5.81) are then more useful than (3.47) or (5.24). As illustrated by (5.50) and (5.51), it is possible that the dominant pole of *both* χ_{nn} and $G_{\alpha\beta}$ below T_λ will be a renormalized zero sound mode involving both the condensate and non-condensate mean fields. For this reason, we call this the collective zero sound or ZS limit.

Which scenario is more relevant in superfluid ^{4}He is not at all obvious, and here experimental data on $S(\mathbf{Q},\omega)$ play a crucial role. In the context of the Bose gas models discussed in Section 5.2, the strong-coupling limit corresponds to the ZS scenario defined above and the weak-coupling limit corresponds to the SP scenario. However, the SP and ZS scenarios are not limited to small Q. Moreover, we assumed in the dilute Bose gas models that SP structure arising from $\bar{G}_{\alpha\beta}$ was at low energy compared to the phonon frequency. In liquid ^{4}He, in contrast, we shall argue that the maxon–roton excitations are SP modes of the normal phase and, due to their high energy, continue to exist in the superfluid phase largely unchanged. At high $Q \gtrsim 1$ Å^{-1}, these SP modes will also be the dominant poles of χ_{nn} as a result of condensate-induced hybridization. We refer to Section 7.2 for further discussion of this scenario.

6

Response functions in the low-frequency, long-wavelength limit

In this chapter, we use the formalism developed in Sections 3.2, 5.1 and 5.4 to discuss various correlation functions in the long-wavelength, low-frequency limit. It is important that the microscopic theory based on a Bose broken symmetry used to describe the high-frequency excitations probed by neutrons *also* explains the low-frequency behaviour which characterizes superfluidity. In Section 6.1, we show how the generalized Ward identities given in Section 5.1 lead in a simple way to several rigorous zero-frequency sum rules. We discuss the structure of the low-frequency, long-wavelength response functions and make contact with the two-fluid description of Landau. In Section 6.2, we discuss the structure of the correlation functions in the hydrodynamic region, as given by the two-fluid equations of Landau (see Khalatnikov, 1965). While the hydrodynamic region of $S(\mathbf{Q},\omega)$ is difficult to probe by thermal neutron scattering, it can be studied by inelastic Brillouin light scattering (for excellent reviews, see Stephen, 1976; Greytak, 1978).

In Section 6.3, starting from the Gavoret–Nozières formalism summarized in Section 5.4, we review GN's explicit calculation of the phonon spectrum of $G_{\alpha\beta}$ and χ_{nn} (at T=0). We comment on the significance of the infrared divergences in the $Q,\omega \rightarrow 0$ limit first noted by Gavoret and Nozières (1964) and clarified in later work, by Nepomnyashchii and Nepomnyashchii (1978), Popov and Serendniakov (1979), and Nepomnyashchii (1983). We also discuss the relation between first and second sound which occurs in the hydrodynamic region and the phonons which arise in the collisionless region.

6.1 Zero-frequency sum rules and the normal fluid density

A microscopic derivation of the two-fluid model was accomplished in the early 1960's, especially by the work of Martin and Hohenberg (1965), Gavoret and Nozières (1964) and Pines (1965). The crucial step is to formulate a general definition of the normal fluid density ρ_N which is valid for *all* temperatures $0 \leq T \leq T_\lambda$ and which is not dependent on the existence of long-lived quasiparticles. The appropriate definition of ρ_N originates in Landau's argument (see Section 1.1) that in a slowly rotating bucket of superfluid ^{4}He, only the normal fluid component rotates. Thus ρ_N is simply related to the moment of inertia of the rotating fluid. By considering the momentum current produced by a constant (transverse) Coriolis force, standard linear response theory gives

$$\rho_N(T) = -\lim_{Q\to 0} \chi^t_{JJ}(\mathbf{Q}, \omega = 0) \ . \tag{6.1}$$

The normal fluid density is then formally determined by the long-wavelength limit of the zero-frequency transverse current–current correlation function, as defined in (5.4) and (5.5). Taking the $\omega = 0$ limit, the continuity equations in (5.16*a*) and (5.16*b*) reduce to the requirement

$$\rho = -\chi^{\ell}_{JJ}(\mathbf{Q}, \omega = 0) = -\bar{\chi}^{\ell}_{JJ}(\mathbf{Q}, \omega = 0) \ . \tag{6.2}$$

This is equivalent to the f-sum rule for $S(\mathbf{Q}, \omega)$ given by (2.24). Combining (6.1) and (6.2) gives a general definition of the superfluid density $\rho_S \equiv \rho - \rho_N$, namely

$$\rho_S(T) = -\lim_{Q\to 0} \left[\chi^{\ell}_{JJ}(\mathbf{Q}, \omega = 0) - \chi^t_{JJ}(\mathbf{Q}, \omega = 0)\right] . \tag{6.3}$$

Thus ρ_S is given by the *difference* between the zero-frequency longitudinal and transverse momentum current-response functions in the long-wavelength limit. In a normal Bose fluid, these are identical and thus $\rho_S = 0$.

For a detailed account of the derivation of (6.1) and the long-range spatial correlations which are implied by the fact that $\chi^{\ell}_{JJ} \neq \chi^t_{JJ}$, we refer to the excellent review article by Baym (1969). Chapters 4 and 6 of Nozières and Pines (1964, 1990) also give a lucid description of the physics implied by the relations (6.1)–(6.3). As we reviewed in Section 1.2, Pollock and Ceperley (1987) have successfully evaluated $\rho_S(T)$ using a Monte Carlo path-integral technique (see Fig. 1.9). Their work starts with the equivalent of (6.3) written as an integral over real-space current response functions. This paper by Pollock and Ceperley is highly recommended for the physical picture it gives concerning the long-range spatial correlations characteristic of a Bose superfluid.

Making use of the dielectric formalism of Section 5.1, we note that for arbitrary $(\mathbf{Q}, \omega)$, all diagrams contributing to $\chi^t_{JJ}(\mathbf{Q}, \omega)$ are regular (proper and irreducible), and hence $\chi^t_{JJ} = \bar{\chi}^t_{JJ} = \bar{\chi}^{tR}_{JJ}$ (see (5.13*b*)). Moreover, the diagrams contributing to $\bar{\chi}^{tR}_{JJ}$ and $\bar{\chi}^{\ell R}_{JJ}$ are similar except for their external points and, in the long-wavelength limit, these two functions are identical. Thus we can rewrite (6.1) in the equivalent form

$$\rho_N(T) = -\lim_{Q \to 0} \bar{\chi}^{\ell R}_{JJ}(\mathbf{Q}, \omega = 0) \ . \tag{6.4}$$

Combining this result with the first equation in (5.13*b*) and (6.2), we obtain

$$\rho_S(T) = -\lim_{Q \to 0} \sum_{\alpha,\beta} \bar{\Lambda}^{\ell}_{\alpha}(\mathbf{Q}, \omega = 0)\, \bar{G}_{\alpha\beta}(\mathbf{Q}, \omega = 0)\, \bar{\Lambda}^{\ell}_{\beta}(\mathbf{Q}, \omega = 0) \ . \tag{6.5}$$

This shows very explicitly how the superfluid density is related to the anomalous Bose current–field correlation functions $\bar{\Lambda}^{\ell}_{\beta}$ which describe the Bose broken symmetry. More technically, ρ_S is given by the sum of the *improper* diagrams contributing to the long-wavelength limit of $\bar{\chi}^{\ell}_{JJ}(\mathbf{Q}, \omega = 0)$, i.e., $\rho_S = -(\bar{\chi}^{l}_{JJ} - \bar{\chi}^{lR}_{JJ})$.

We now turn to the implications of the generalized Ward identities given by (5.17)–(5.19) in the $\omega = 0$ limit. One can show that (5.17) reduces to (Talbot and Griffin, 1983)

$$\bar{\Lambda}^{\ell}_{\alpha}(\mathbf{Q}, \omega = 0) = \frac{\sqrt{n_0}m}{Q} \alpha[\varepsilon_Q + B(Q)] \ , \tag{6.6}$$

where we have defined

$$B(Q) \equiv \bar{\Sigma}_{11}(\mathbf{Q}, \omega = 0) - \bar{\Sigma}_{12}(\mathbf{Q}, \omega = 0) - \mu \tag{6.7}$$

and have made use of the equivalence $\Sigma_{21}(\mathbf{Q}, \omega = 0) = \Sigma_{12}(\mathbf{Q}, \omega = 0)$ (see (3.28)). At $\omega = 0$, (5.19) reduces to

$$\bar{\chi}^{\ell R}_{JJ}(\mathbf{Q}, \omega = 0) = -\rho + \frac{\sqrt{n_0}m}{Q} \left[\bar{\Lambda}^{\ell}_{1}(\mathbf{Q}, \omega = 0) - \bar{\Lambda}^{\ell}_{2}(\mathbf{Q}, \omega = 0)\right] . \tag{6.8}$$

Combining (6.7) and (6.8) with (6.6), we obtain two very important long-wavelength results (valid at arbitrary temperatures):

$$\lim_{Q \to 0} \bar{\Lambda}^{\ell}_{\alpha}(\mathbf{Q}, \omega = 0) = \frac{(\operatorname{sgn} \alpha)\rho_S Q}{2m\sqrt{n_0}} \ , \tag{6.9}$$

$$\lim_{Q \to 0} B(Q) = \frac{Q^2}{2m}\left[\frac{\rho_S}{m n_0} - 1\right] . \tag{6.10}$$

We note that (5.17) does not allow us to derive any exact result for $\bar{\Lambda}_\alpha(\mathbf{Q}, \omega = 0)$.

Using (5.12) and the relation $\bar{\Lambda}_1(Q, \omega = 0) = \bar{\Lambda}_2(\mathbf{Q}, \omega = 0)$, one obtains in the $Q \to 0$ limit

$$\begin{aligned} \Sigma_{11}(\mathbf{Q}, \omega = 0) - \Sigma_{12}(\mathbf{Q}, \omega = 0) &= \bar{\Sigma}_{11}(\mathbf{Q}, \omega = 0) - \bar{\Sigma}_{12}(\mathbf{Q}, \omega = 0) \\ &= \mu + \frac{Q^2}{2m}\left(\frac{\rho_S}{mn_0} - 1\right), \end{aligned} \tag{6.11}$$

where the second line follows from (6.10). The results (6.9)–(6.11) were first derived at $T = 0$ (where $\rho_S = nm$) by Gavoret and Nozières (1964) using a direct analysis of Feynman diagrams. It may be viewed as a generalized version of the Hugenholtz–Pines theorem quoted in (3.32). The relations (6.6)–(6.8) emphasize the role which $\bar{\Sigma}_{\alpha\beta}$, $\bar{\chi}^{\ell R}_{JJ}$ and $\bar{\Lambda}^{\ell}_{\alpha}$ play as the basic building blocks of the theory (see discussion at the end of Section 5.1).

A direct consequence of (6.11) is that the regular self-energies $\bar{\Sigma}_{\alpha\beta}$ satisfy the Hugenholtz–Pines theorem (3.32). This means that the poles of $\bar{G}_{\alpha\beta}$ have no energy gap in the $Q \to 0$ limit. Moreover, if one can expand $\bar{\Sigma}_{\alpha\beta}(\mathbf{Q}, \omega)$ around $Q = 0$, $\omega = 0$, the modes will be phonon-like (see, however, Payne and Griffin, 1985).

The preceding derivation emphasizes that (6.11) is a direct consequence of the generalized zero-frequency Ward identities and hence of the continuity equations (5.16*a*) and (5.16*b*). This close relationship between the Hugenholtz–Pines relation (3.32) and the continuity equation was first demonstrated by Hohenberg and Martin (1965) and Huang and Klein (1964). Since $\bar{G}_{\alpha\beta}$ is given by the same equations (3.29) as $G_{\alpha\beta}$ (with $\Sigma_{\alpha\beta}$ replaced by $\bar{\Sigma}_{\alpha\beta}$), one finds (in the limit $Q \to 0$)

$$\begin{aligned} G_{11}(\mathbf{Q}, \omega = 0) - G_{12}(\mathbf{Q}, \omega = 0) &= \bar{G}_{11}(\mathbf{Q}, \omega = 0) - \bar{G}_{12}(\mathbf{Q}, \omega = 0) \\ &= -\frac{1}{\varepsilon_Q + B(Q)} \\ &= -\frac{2m}{Q^2}\frac{mn_0}{\rho_S}, \end{aligned} \tag{6.12}$$

where we have used (3.29), (3.30) and (6.11). This last result was first obtained by Bogoliubov (1963, 1970) and Hohenberg and Martin (1965). The ^{4}He atom momentum distribution n_Q for small Q can be directly related to (6.12), as discussed in Section 4.1.

We call attention to the fact that ρ_N and ρ_S arise naturally when we work with correlation functions involving the momentum current density (rather than the number density). It will be especially important when working in the two-fluid frequency region to base the discussion on $\bar{\chi}^{\ell R}_{JJ}$ and $\bar{\Lambda}^{\ell}_{\alpha}$, rather than $\bar{\chi}^{R}_{nn}$ and $\bar{\Lambda}_{\alpha}$. We also note that the high-frequency limit, $\bar{\Lambda}^{\ell}_{\alpha}(\mathbf{Q}, \omega \to \infty)$ is identical to the zero-frequency result (6.9) but with the superfluid density ρ_S replaced by the condensate density mn_0.

As Talbot and Griffin (1983) have emphasized, the above systematic derivation shows that the exact zero-frequency results (6.12), (6.11), and (6.9) are all really equivalent to (6.4). This follows when one takes into account that the generalized Ward identities (see Section 5.1) are direct consequences of the equation of continuity in a Bose-condensed system. This shows the power the dielectric formalism has in exposing how the Bose order parameter determines the structure of various correlation functions. The relation in (6.12) is of especial interest since it relates the zero-frequency, long-wavelength limit of the single-particle Green's functions to the ratio of the condensate density $n_0(T)$ *and* the superfluid density $\rho_S(T)$. In connection with the rigorous result (6.9), Talbot and Griffin (1984b) have used (6.12) to prove that the *total* mass current density associated with a moving condensate (velocity $\mathbf{v}_S = \mathbf{Q}/m$) is indeed equal to the two-fluid result,

$$\langle \mathbf{J} \rangle_Q = \rho_S \mathbf{v}_S \ . \tag{6.13}$$

Talbot and Griffin (1983) have calculated the zero-frequency regular functions using the one-loop diagrams given in Section 5.3. Starting from the expression for ρ_N in (6.4), one can verify that only the thermal scattering part of (5.54) remains in the $Q, \omega \to 0$ limit (Fetter, 1970)

$$\lim_{Q \to 0} \bar{\chi}^{\ell R}_{JJ}(\mathbf{Q}, \omega = 0) = -\int \frac{d\mathbf{p}}{(2\pi)^3} (\hat{\mathbf{Q}} \cdot \mathbf{p})^2 (u_p^2 - v_p^2) \left[-\frac{\partial N(\omega_p)}{\partial \omega_p} \right] \ . \tag{6.14}$$

Recalling that $u_p^2 - v_p^2 = 1$, this reduces to Landau's well known quasi-particle formula for $\rho_N(T)$. In addition, one can show that $\bar{\Lambda}^{\ell}_{\alpha}$ in (5.55) gives (6.9) while the self-energies in (5.52) and (5.53) reproduce the Gavoret–Nozières sum rule (6.11) with a superfluid density $\rho_S = \rho - \rho_N$, with ρ_N again being given by (1.3). The (somewhat lengthy) calculations leading to these results (see Talbot and Griffin, 1983) show explicitly how the one-loop corrections in (5.52)–(5.55) provide additional terms which result in the natural appearance of a superfluid density $\rho_S(T)$ which is different from $mn_0(T)$. Taken together, these results give further weight to the argument made in Section 5.3 that the one-loop diagrams for the regular functions (in the dielectric formalism) give a consistent, non-trivial description of an interacting Bose-condensed fluid at finite temperatures.

In this section, we have emphasized the close relation between a Bose broken symmetry and superfluidity. As is well known (for a review of Kosterlitz–Thouless ideas, see Nelson, 1983), there is no long-range order (LRO) in a two-dimensional (2D) Bose fluid (at $T \neq 0$), even though it exhibits superfluidity as described by ρ_S and $\mathbf{v}_S$. However, quasi-LRO and superfluidity in 2D Bose fluids can still be understood as the consequence of a macroscopic occupation of some state. The difference is that because of strong *phase* fluctuations, the "condensate" in 2D is no longer simply associated with the fraction of atoms in the zero-momentum state but rather with long-range spatial correlations which decay with a characteristic power law. For further discussion, see p. 1372 of Griffin (1987) and Popov (1983, 1987).

6.2 Hydrodynamic (two-fluid) limit

In discussing the excitations of quantum liquids, a very basic distinction arises between the low-frequency "hydrodynamic" region and the high-frequency "collisionless" region. These two domains are most easily distinguished by introducing the average lifetime $\bar{\tau}$ of the elementary excitations which make the dominant contribution to the thermodynamic and transport properties of the fluid in question. Roughly speaking, $\bar{\tau}$ is the lifetime of the elementary excitations with thermal energies of order $k_B T$ (measured with respect to the chemical potential μ). The two frequency domains can be defined as

$$\omega \ll 1/\bar{\tau} \ : \text{hydrodynamic,} \tag{6.15a}$$

$$\omega \gg 1/\bar{\tau} \ : \text{collisionless.} \tag{6.15b}$$

Clearly the cross-over frequency $\bar{\omega} \equiv 1/\bar{\tau}$ is very dependent on the temperature, since $1/\bar{\tau}$ usually increases as some power of the temperature. In particular, the distinction between the two regions is lost at $T = 0$. In superfluid ^{4}He in the temperature region ~ 1 K, a typical value of $\bar{\omega}$ might be of order 10^9 Hz. (For a more detailed discussion of quasi-particle lifetimes, see Khalatnikov, 1965.) We conclude that inelastic (Brillouin) light scattering probes the hydrodynamic modes (described by the two-fluid equations), while inelastic neutron scattering probes the collisionless modes.

The distinction between the collisionless and hydrodynamic domains has been extensively discussed in the case of liquid ^{3}He using the well known Landau kinetic equations that describe the quasiparticle dynamics of a Fermi liquid (see, for example, Chapter 1 of Pines and Nozières, 1966). In a Bose-condensed fluid, the equivalent kinetic equations governing the elementary excitations are available for a WIDBG only in the weak-coupling limit (Kirkpatrick and Dorfman, 1985). However, the basic concepts involved in distinguishing the high- and low-frequency regions are the same in all quantum liquids and solids (solid ^{4}He is discussed in Chapter 11). The various kinds of collisionless and hydrodynamic modes have been worked out in great detail in superfluid ^{3}He. The mode structure in this case is very rich due to the p-wave Cooper pair condensate (for further discussion, see Vollhardt and Wölfle, 1990).

In this section, we examine the structure of $G_{\alpha\beta}$ and χ_{nn} for a Bose-condensed fluid in the hydrodynamic region. The discussion uses the two-fluid equations of motion (Khalatnikov, 1965) to obtain explicit expressions for the correlation functions in the region of low Q and ω, where this description is correct. Only in Section 6.3 do we turn to the question of deriving the low Q, ω correlation functions directly from a diagrammatic analysis of the structure of a Bose-condensed fluid, which shows the role of the Bose condensate more clearly.

The hydrodynamic domain is also referred to as the "collision-dominated" region. Assuming that we are dealing with an acoustic dispersion relation, one sees from (6.15a) that for hydrodynamic modes there are many collisions between the elementary excitations in one wavelength $\lambda = 2\pi/Q$ of the collective mode. For this reason, the hydrodynamic limit is often difficult to derive in a liquid, when one starts from the microscopic or atomic level. From another point of view, it is simple, in that this limit can be described by equations of motion for "coarse-grained" local variables like pressure, temperature and velocity. These equations are basically conservation laws plus constitutive equations involving transport coefficients (such as thermal conductivity and the various viscosities). This description can be written down from general considerations, a detailed microscopic theory being needed only to evaluate the various thermodynamic derivatives and transport coefficients which occur in the hydrodynamic equations. In the case of a Bose-condensed fluid, these are the well known two-fluid equations as developed by Landau and Khalatnikov (Khalatnikov, 1965). They predict two kinds of wave-like phenomena to occur (first and second sound). Second sound has been studied in great detail in the context of superfluid ^{4}He, but it also exists in a WIDBG, as discussed by Ma (1971), Popov (1983), Kirkpatrick and Dorfman (1985) and Gay and Griffin (1985).

Starting with the linearized equations of motion describing the hydrodynamic domain of any system, one can evaluate the various correlation functions in the low Q and ω region (Kadanoff and Martin, 1963). The calculation is straightforward, but algebraically rather lengthy. The correlation functions predicted by the two-fluid equations were first worked out by Hohenberg and Martin (1964, 1965) as well as by Bogoliubov (1963, 1970). We refer the reader to Chapter 10 of Forster (1975) for more details, as well as the classic review article by Martin (1968). In our subsequent analysis, we omit the transport coefficients (dissipation) in order to expose the structure of the correlation functions in as simple a manner as possible and, moreover, we neglect vorticity in the superfluid velocity field ($\nabla \times \mathbf{v}_S = 0$).

The dynamic structure factor $S(\mathbf{Q}, \omega)$ for $Q \lesssim 10^{-3}$ Å^{-1} can be studied by using inelastic (Brillouin) light scattering as well as ultrasonic measurements, as discussed in Section 4 of the review article by Woods and Cowley (1973). The theory of Brillouin light scattering in liquid ^{4}He is reviewed by Stephen (1976). One is usually probing the hydrodynamic region (6.15a) when one uses these experimental techniques. For $T > T_\lambda$, $S(\mathbf{Q}, \omega)$ as measured by light scattering will exhibit a Rayleigh central peak at $\omega \simeq 0$ due to scattering from (diffusive) temperature fluctuations, in addition to two Brillouin sound-wave peaks at $\omega = \pm cQ$, where c is the first sound velocity (see Pike, Vaughan and Vinen, 1970). Light-scattering experiments on liquid ^{4}He are intrinsically difficult since the electronic polarizability of the closed-shell He atom is very small and thus the scattering is weak. Moreover, the intensity of the diffusive central component relative to the sound-wave peaks is proportional to the Landau–Placzek ratio $\gamma - 1$ (where $\gamma = C_P/C_V$), which rapidly decreases as the temperature is lowered below the gas–liquid critical point (see Fig. 6.1).

In spite of the kinematical difficulties, neutron scattering has also been successfully used (Woods, Svensson and Martel, 1975, 1978) to measure $S(\mathbf{Q}, \omega)$ at $T = 4.2$ K and SVP, for momentum transfers as small as $Q \simeq 0.1$ Å^{-1}. The results are consistent with the expected hydrodynamic structure discussed above. Neutron scattering can be used to show the cross-over from the hydrodynamic to the slightly higher collisionless sound velocity (Woods, Svensson and Martel, 1976). At $T = 2.3$ K, the velocity increases from 220 ms^{-1} to 255 ms^{-1} at around $Q \sim 0.25$ Å^{-1} (see Fig. 6.2).

In the superfluid phase below T_λ, the main difference in the hydrodynamic region is the appearance of propagating second sound modes in

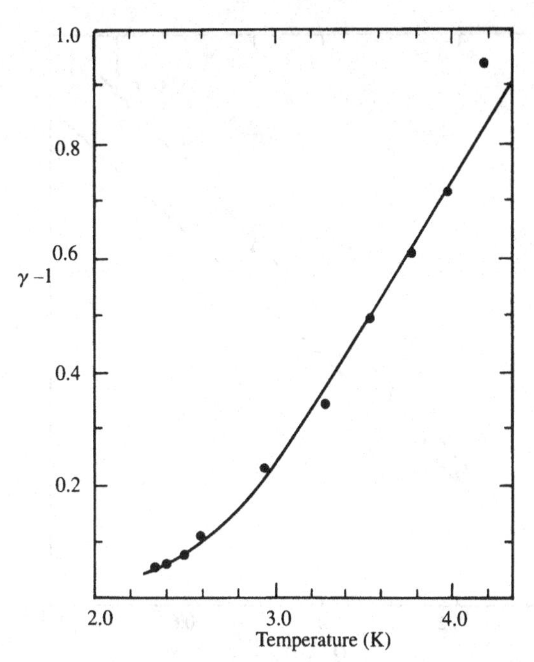

Fig. 6.1. The temperature dependence of the Landau–Placzek ratio in normal liquid ^{4}He as given by thermodynamic data up to the boiling point at 4.2 K (1 bar) [Source: Pike, 1972].

place of the diffusive central peak. The dynamic structure factor implied by the two-fluid equations is given by (Ginzburg, 1943; Hohenberg and Martin, 1964, 1965)

$$S(\mathbf{Q}, \omega) = [N(\omega)+1]\frac{Q^2}{m}[Z_\mathrm{I}\delta(\omega^2 - u_\mathrm{I}^2 Q^2) + (1 - Z_\mathrm{I})\delta(\omega^2 - u_\mathrm{II}^2 q^2)] \ , \quad (6.16)$$

where the weight of the first sound mode is

$$Z_\mathrm{I} = \frac{u_\mathrm{I}^2 - v^2}{u_\mathrm{I}^2 - u_\mathrm{II}^2} \ , \quad (6.17a)$$

with (s is the entropy and C_V the specific heat)

$$v^2 \equiv T\frac{s^2}{C_V}\frac{\rho_S}{\rho_N} \ . \quad (6.17b)$$

The first and second sound velocities in (6.16) satisfy the exact relations

$$u_\mathrm{I}^2 + u_\mathrm{II}^2 = v^2 + \left.\frac{\partial p}{\partial \rho}\right|_S \ , \quad u_\mathrm{I}^2 u_\mathrm{II}^2 = v^2 \left.\frac{\partial p}{\partial \rho}\right|_T \ . \quad (6.18)$$

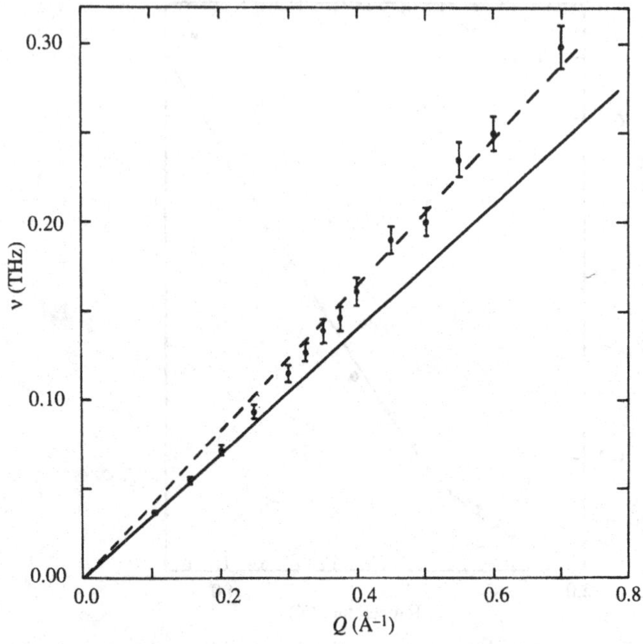

Fig. 6.2. The phonon dispersion relation at $T = 2.3$ K showing the transition from first sound to zero sound. The dashed line is drawn as a guide to the eye. The full line has a slope equal to thermodynamic sound velocity [Source: Woods, Svensson and Martel, 1976].

In the case of a dilute Bose *gas*, where $(C_P/C_V) - 1$ is appreciable, one finds that *both* first and second sound have appreciable weight in the density fluctuation spectrum in (6.16) (see, for example, Gay and Griffin, 1985). In superfluid ^{4}He at SVP, in contrast, we have $C_P \simeq C_V$ to a very good approximation, whence it follows that $u_{\rm II} \simeq v$ and $u_{\rm I} \simeq (\partial p/\partial \rho)^{\frac{1}{2}}$. Consequently we have $Z_{\rm I} \simeq 1$ in (6.16) and thus the second sound mode has negligible weight in $S(\mathbf{Q}, \omega)$ in superfluid ^{4}He. In physical terms, this means that second sound at SVP is mainly a temperature wave, which is only weakly coupled into the density fluctuation spectrum (Khalatnikov, 1965).

However, the thermal expansion coefficient and hence the Landau–Placzek ratio $\gamma - 1$ diverges at T_λ; moreover, this expansion coefficient is considerably increased as we approach the superfluid transition by working under high pressure (Ferrell *et al.*, 1968). This feature has been used by several groups to study the behaviour of the second sound component in $S(\mathbf{Q}, \omega)$ near T_λ in great detail (for reviews, see Greytak, 1978;

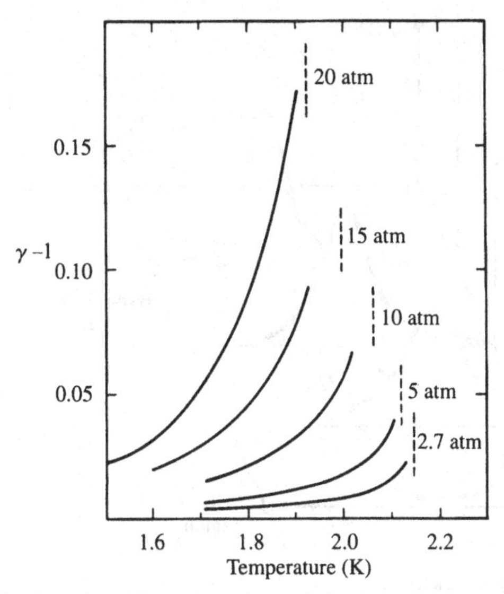

Fig. 6.3. The Landau–Placzek ratio $(C_P/C_V) - 1$ vs. temperature for various pressures in superfluid ^{4}He. The vertical dashed lines indicate the value of the transition temperature $T_\lambda(p)$ at various pressures [Source: Vinen, 1971].

Stephen, 1976). The ratio of the second sound to first sound scattering intensity can be approximated by the Landau–Placzek ratio $\gamma - 1$ within a few per cent if we are outside the critical region ($\Delta T \gtrsim 1\,\mathrm{mK}$), as discussed by Hohenberg (1973) and O'Connor, Palin and Vinen (1975). In Fig. 6.3, we plot $\gamma - 1$ as a function of the temperature for different pressures, as given by thermodynamic data (Vinen, 1971). These results may also be relevant to neutron-scattering studies of $S(\mathbf{Q}, \omega)$ in superfluid ^{4}He at very small momentum transfers, which are usually analysed (see Section 7.1) without considering any residual contribution from second sound. We recall that the second sound component is the critical mode associated with the Bose broken symmetry near T_λ, where its behaviour is somewhat complicated (see Fig. 6.4). For further details and references, we refer to Tarvin, Vidal and Greytak (1977).

Green's functions in the two-fluid region

A unique aspect of a Bose-condensed fluid is that the single-particle Green's function $G_{\alpha\beta}$ can be *directly* related to correlation functions such as χ_{nn} and χ^{ℓ}_{JJ} in the hydrodynamic domain. This is a consequence of

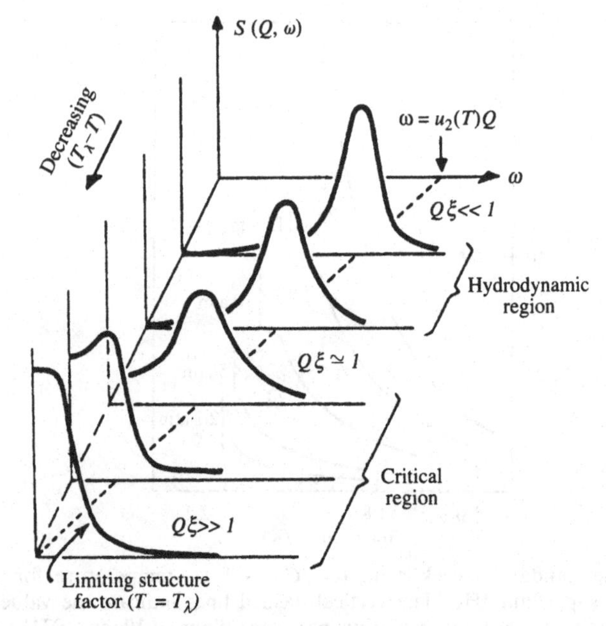

Fig. 6.4. Schematic plot of the observed second sound contribution to $S(\mathbf{Q}, \omega)$ in the superfluid phase just below T_λ. The fluctuations in the Bose order parameter are characterized by a correlation length $\xi(T)$, which diverges as T_λ is approached. For wavevectors probed in light scattering, $Q\xi(T)$ becomes of order unity for temperatures within about 1 mK of T_λ [Source: Tarvin, Vidal and Greytak, 1977].

the fact that the local superfluid velocity (a hydrodynamic variable) is related to the gradient of the phase fluctuations. The following heuristic discussion captures the essential physics (see Appendix B of Griffin, 1981; p. 107 of Lifshitz and Pitaevskii, 1980; Chapter 10 of Forster, 1975).

Quantum field operators can be expressed in terms of amplitude and phase operators. A systematic study shows that the slow, long-wavelength *phase* fluctuations dominate over the *amplitude* fluctuations of an interacting Bose-condensed system; hence one can use

$$\hat{\psi}(\mathbf{r}) = \sqrt{n_0} e^{i\hat{\phi}(\mathbf{r})} , \ \hat{\psi}^+(\mathbf{r}) = \sqrt{n_0} e^{-i\hat{\phi}(\mathbf{r})} \tag{6.19}$$

in discussing the dynamics in the limit of small Q and ω. Using (6.19) in the usual quantum mechanical expression for the current operator, one can identify the part of the current directly associated with the motion of the condensate as

$$\hat{\mathbf{J}}_S(\mathbf{r}) = mn_0 \hat{\mathbf{v}}_S(\mathbf{r}) , \tag{6.20}$$

where the superfluid velocity operator is given by the gradient of the phase

$$\hat{\mathbf{v}}_S(\mathbf{r}) = \frac{1}{m}\nabla\hat{\phi}(\mathbf{r}) . \tag{6.21}$$

(As Talbot and Griffin (1984b) have discussed, the additional current associated with the non-condensate atoms which are dragged along by the condensate atoms combines with the current in (6.20) to give (6.13).)

With (6.21), the longitudinal part of the superfluid velocity correlation function (as defined in (5.4) and (5.5)) is directly related to the phase fluctuation spectrum

$$\chi^{\ell}_{v_S v_S}(\mathbf{Q},\omega) = \frac{Q^2}{m^2}\chi_{\phi\phi}(\mathbf{Q},\omega) . \tag{6.22}$$

Moreover, using (6.19) in (3.10) and expanding the exponentials for small phase fluctuations, we find

$$\left.\begin{aligned} \tilde{\psi}(\mathbf{r}) &= \hat{\psi}(\mathbf{r}) - \sqrt{n_0} \simeq i\sqrt{n_0}\hat{\phi}(\mathbf{r}) , \\ \tilde{\psi}^{+}(\mathbf{r}) &\simeq -i\sqrt{n_0}\hat{\phi}(\mathbf{r}) . \end{aligned}\right\} \tag{6.23}$$

Thus the single-particle Green's functions reduce to

$$\begin{aligned} G_{11}(\mathbf{Q},\omega) &= -G_{12}(\mathbf{Q},\omega) \\ &= n_0\,\chi_{\phi\phi}(\mathbf{Q},\omega) = n_0\frac{m^2}{Q^2}\chi^{\ell}_{v_S v_S}(\mathbf{Q},\omega) , \end{aligned} \tag{6.24}$$

where we have used (6.22) in the second line. The expression (6.24) gives a *direct* relation in the small Q,ω limit between $\chi^{\ell}_{v_S v_S}$ (as determined by the two-fluid equations) and $G_{\alpha\beta}$. This important relation arises only because of the Bose broken symmetry. A systematic way of separating out the slow, long-wavelength fluctuations from the fast, short-wavelength motions in field-theoretical calculations is discussed in Sections 18 and 19 of Popov (1983). His procedure gives a more precise basis to the preceding heuristic proof of (6.24).

$G_{11}(\mathbf{Q},i\omega_n)$ can be expressed in terms of the single-particle spectral density (see pp. 150ff of Mahan, 1990)

$$G_{11}(\mathbf{Q},i\omega_n) = \int_{-\infty}^{\infty}\frac{d\omega}{2\pi}\frac{A(\mathbf{Q},\omega)}{i\omega_n - \omega} . \tag{6.25}$$

This spectral density has been discussed at length in Section 4.1. The single-particle spectral density $A(\mathbf{Q},\omega)$ can be obtained from the two-fluid hydrodynamic equations once we have the key relation given in (6.24). From our general analysis in Sections 5.1 and 5.5, $G_{\alpha\beta}$ can be expected to exhibit the same resonances as χ_{nn}, corresponding to first and second sound (albeit with different weights). The expression for $A(\mathbf{Q},\omega)$ valid in the two-fluid domain has been discussed with some generality by Hohenberg and Martin (1964). In superfluid ^{4}He away from T_λ, one can ignore terms of order $\gamma-1$ (i.e., $C_P \simeq C_V$), in which case the general expression simplifies to (Cheung and Griffin, 1971a)

$$A(\mathbf{Q},\omega) = 2\pi\frac{n_0 m^2}{\rho_S}\mathrm{sgn}\ \omega\left[\frac{\rho_S}{\rho}u_{\mathrm{I}}^2\delta(\omega^2-u_{\mathrm{I}}^2Q^2)+\frac{\rho_N}{\rho}u_{\mathrm{II}}^2\delta(\omega^2-u_{\mathrm{II}}^2Q^2)\right], \tag{6.26}$$

where u_{I} and u_{II} are defined in (6.18). Inserting (6.26) into (6.25), one may verify that at zero frequency

$$G_{11}(\mathbf{Q},\omega=0) = -\frac{n_0 m^2}{\rho_S Q^2}. \tag{6.27}$$

This is, as expected, in perfect agreement with the Bogoliubov–Hohenberg–Martin sum rule (6.12).

In contrast to the expression for $S(\mathbf{Q},\omega)$ given in (6.16), second sound has appreciable weight in the single-particle spectral density $A(\mathbf{Q},\omega)$ given in (6.26) within the hydrodynamic domain of superfluid ^{4}He, as long as $T \neq 0$. (In the limit where (6.26) is valid, u_{II} reduces to v in (6.17*b*)). The strong amplitude of second sound in (6.26) emphasizes that this acoustic mode is associated with the Bose order-parameter fluctuations, which are in turn closely related to the field fluctuations described by the single-particle Green's functions. The second sound pole dominates $A(\mathbf{Q},\omega)$ in the region near T_λ (see, for example, Tarvin, Vidal and Greytak, 1977).

In comparing the spectrum exhibited by $G_{\alpha\beta}$ and χ_{nn} in superfluid ^{4}He in the two-fluid region, we can do no better than quote the classic paper by Hohenberg and Martin (1964):

(1) "When $T=0$, the second sound velocity u_{II} approaches a finite value but its contribution to all correlation functions (i.e. the residue at $\omega = u_{\mathrm{II}}Q$) vanishes."
(2) "At $T=0$, both χ_{nn} and $G_{\alpha\beta}$ are both sharply peaked about $\omega = u_{\mathrm{I}}Q$, where $u_I^2 = \partial p/\partial\rho$. This pole exhausts the f-sum rule for χ_{nn} but not the corresponding sum rule for $G_{\alpha\beta}$."
(3) "With increasing temperature, χ_{nn} is not greatly altered; the oscillation at $\omega = u_{\mathrm{I}}Q$ continues to dominate it, merging smoothly at $T = T_\lambda$ with the dominant contribution from ordinary sound in the normal phase. There is also a contribution to χ_{nn} (with relative weight $(C_P - C_V)/C_V \ll 1$) from second sound (the mode of tempera-

ture transport for $T < T_\lambda$). This contribution to χ_{nn} merges with the corresponding one for $T < T_\lambda$ (the thermal conduction mode $\omega = iD_T Q^2$)."

(4) "As the temperature increases, the correlation function $G_{\alpha\beta}$ is substantially altered. In fact, at $T = T_\lambda$, the oscillation at $\omega = u_\mathrm{I} Q$ has vanishing weight in $G_{\alpha\beta}$. Likewise, the contribution at $\omega = u_\mathrm{II} Q$ vanishes as it 'joins' the oscillation $O(Q^2)$ describing a single-particle energy of a non-superfluid system."

6.3 The nature of phonons in Bose fluids

As we mentioned in our overview in Section 1.1, the fact that the quasiparticle spectrum at low Q is phonon-like in superfluid ^{4}He was usually viewed as obvious in the older literature. This phonon mode was identified as a compressional sound wave familiar in all liquids and gases. In the microscopic field-theoretic literature, however, this phonon spectrum has always presented a challenge. The key requirement is to prove that because of the Bose condensate, the phonon pole exhibited by the density-response function $\chi_{nn}(\mathbf{Q}, \omega)$ is indeed the *only* low-energy pole of the single-particle Green's function $G_{\alpha\beta}(\mathbf{Q}, \omega)$; that is to say, the elementary excitations, which completely determine the thermodynamic and transport properties at low temperatures, are exhausted by the compressional density fluctuations. Moreover, one must also show that this phonon frequency is essentially temperature-independent in superfluid ^{4}He (see Section 7.1 for discussion of neutron data concerning this), and smoothly transforms into the zero sound pole of χ_{nn} above T_λ.

We have seen in Chapter 5 that a Bose broken symmetry naturally leads to a mixing of the spectra exhibited by $G_{\alpha\beta}$ and χ_{nn}, as shown by (3.47) or, more explicitly, (5.24). This sharing of poles is also present in the hydrodynamic region described by the two-fluid equations, as shown by the results in Section 6.2.

As we mentioned in Section 3.2, Gavoret and Nozières (1964) gave the first convincing proof that both $G_{\alpha\beta}$ and χ_{nn} exhibited the same phonon mode in the Bose-condensed phase. Their argument proceeds as follows. As indicated in our review of GN in Section 5.4, a key role is played by the singularities arising from the product of two fully renormalized Beliaev propagators. These occur in the intermediate state of Feynman diagrams contributing to Γ, P and $\Sigma_{\alpha\beta}$ (see (5.68)–(5.70)). In Fermi liquid theory, similar products of two propagators give rise to singularities related to the low-energy particle–hole excitations. The well known Lindhard function is given by (the variables represent a 4-momentum, as in Section 5.4)

$$-i\sum_p G_0\left(p+\frac{Q}{2}\right)G_0\left(-p+\frac{Q}{2}\right) = -i\sum_p G_0(p+Q)G_0(p)\ . \quad (6.28)$$

The analogue of this expression in a Bose-condensed system is given by (5.67). At $T = 0$ (to which GN restrict themselves), there is no contribution to (5.67) from thermally excited quasiparticles (particle–hole excitations) but only from the creation of two quasiparticles. This is illustrated explicitly by the model expression in (3.45).

Carrying out the 4-momentum integration in expressions like (5.67), GN find a logarithmic divergence

$$\sum_p G_{\eta\varepsilon}\left(p+\frac{Q}{2}\right)G_{\xi\zeta}\left(-p+\frac{Q}{2}\right) \sim \ln|\omega^2 - c^2Q^2|\ . \quad (6.29)$$

This infrared divergence is a direct consequence of assuming that $G_{\alpha\beta}(\mathbf{p},\omega)$ has a phonon pole

$$G_{\alpha\beta}(\mathbf{p},\omega) \sim \frac{1}{\omega^2 - c^2p^2}\ . \quad (6.30)$$

It comes from the fact that the pair energy $\omega_{p+\frac{1}{2}Q} + \omega_{p-\frac{1}{2}Q}$, which occurs in the denominator of the integrand of (6.29), becomes vanishingly small in the limit of small Q and p. However GN note that such divergences cancel out in the final results for $G_{\alpha\beta}$ and χ_{nn} in the $Q,\omega \to 0$ limit and hence conclude it is safe to eliminate these divergences by introducing a fictitious energy gap Δ into the excitation spectrum of $G_{\alpha\beta}$. This "regularity assumption" allows perturbative calculations about $Q = 0$ and $\omega = 0$, a procedure which makes sense since the final results are well defined when the gap $\Delta \to 0$ is set to zero at the end.

Working within the above scheme, GN argue that it is sufficient to evaluate χ^R_{nn} at $\mathbf{Q} = 0$, $\omega = 0$. Moreover they compute the three-point kernel $P^{\eta\xi}_\sigma$ given by (5.69) and the Beliaev self-energies $\Sigma_{\alpha\beta}$ using Taylor expansions about $\mathbf{Q} = 0, \omega = 0$. Making a careful diagrammatic analysis, the terms at $\mathbf{Q} = 0$, $\omega = 0$ as well as the Taylor expansion coefficients all reduce to various thermodynamic quantities and associated derivatives. This whole procedure makes use of exact Ward identities and is, technically, very similar to the Green's function analysis of the response functions of a Fermi liquid in the $\mathbf{Q},\omega \to 0$ limit (see Chapter 6 of Nozières, 1964).

The final results of this very lengthy procedure are deceptively simple (at least at $T = 0$):

$$G_{11}(\mathbf{Q},\omega) = G_{22}(\mathbf{Q},\omega) = -G_{12}(\mathbf{Q},\omega) = \frac{n_0}{n}\frac{mc^2}{\omega^2 - c^2Q^2}, \tag{6.31}$$

$$\chi^c_{nn}(\mathbf{Q},\omega) \equiv \sum_{\alpha,\beta} \Lambda_\alpha G_{\alpha\beta}\Lambda_\beta = \frac{n}{m}\frac{Q^2}{\omega^2 - c^2Q^2} + \frac{\partial\tilde{n}}{\partial\mu}, \tag{6.32}$$

$$\chi^R_{nn}(\mathbf{Q} = 0, \omega = 0) = -\frac{\partial\tilde{n}}{\partial\mu} = \frac{n}{mc^2}, \tag{6.33}$$

where c^2 is the sound velocity given by compressibility zero-frequency sum rule (2.28) and $\tilde{n} = n - n_0$ is the depletion. We note that (6.32) and (6.33) are equivalent to

$$\left.\begin{aligned} \chi^c_{nn}(\mathbf{Q},\omega) &= \frac{n}{mc^2}\left(\frac{\omega^2}{\omega^2 - c^2Q^2}\right), \\ \chi_{nn}(\mathbf{Q},\omega) &= \chi^c_{nn} + \chi^R_{nn} = \frac{n}{m}\frac{Q^2}{\omega^2 - c^2Q^2}. \end{aligned}\right\} \tag{6.34}$$

The amazing feature of the GN calculation is their explicit proof that the phonon velocity which enters into the pole of $G_{\alpha\beta}$ is given precisely by the thermodynamic derivative which enters the compressibility sum rule. This is proven rigorously, within the calculational procedure previously outlined. In the process of obtaining the above results, GN also obtained the zero-frequency results given by (6.9), (6.10) and (6.11) using a direct diagrammatic analysis. In addition, they explicitly evaluated the vertex function $\Lambda_\alpha(\mathbf{Q},\omega)$ to obtain the rigorous result (at $T = 0$)

$$\Lambda_\alpha(\mathbf{Q} = 0, \omega) = \frac{1}{\sqrt{n_0}}\left[n_0\left(1 + \frac{\partial\tilde{n}}{\partial n_0}\right) + \operatorname{sgn}\alpha\frac{\partial\tilde{n}}{\partial\mu}\frac{\omega}{2}\right], \tag{6.35}$$

to lowest order in ω.

Nepomnyashchii and Nepomnyashchii (1978) later gave a careful analysis of the infrared divergences discovered by GN. They concluded that the final results summarized by (6.31)–(6.35) were correct but that $\Sigma_{12}(\mathbf{Q},\omega) \sim 1/\ln|\omega^2 - c^2Q^2|$ and hence

$$\left.\begin{aligned} \Sigma_{12}(\mathbf{Q} = 0, \omega = 0) &= 0, \\ \Lambda_\alpha(\mathbf{Q} = 0, \omega = 0) &= 0. \end{aligned}\right\} \tag{6.36}$$

These results required extensive changes in the original analysis of GN, which assumed that $\Sigma_{12}(\mathbf{Q} = 0, \omega = 0)$ was finite. The results in (6.36) can be related to the thermodynamic derivatives (at $T = 0$)

$$\frac{\partial \tilde{n}}{\partial \mu} = \frac{\partial n}{\partial \mu} \equiv \frac{n}{mc^2} \, , \quad \frac{\partial \tilde{n}}{\partial n_0} = -1 \, . \tag{6.37}$$

As noted by Nepomnyashchii and Nepomnyashchii (1978), (6.35)–(6.37) imply that $\Lambda_\alpha(\mathbf{Q} = 0, \omega) \propto \omega$ and $\chi^c_{nn}(\mathbf{Q}, \omega = 0) = 0$. This last result is consistent with the expression given in (6.34). It means that the condensate part of the density-response function makes no contribution to the compressibility sum rule (2.28) (see also Griffin, 1981). The physics behind (6.36) will be discussed shortly.

It is useful to recall briefly the heuristic approach used by Hohenberg and Martin (1965) to derive (6.31) and (6.34). They first give general arguments (see HM (6.38)) that the longitudinal current correlation function χ^ℓ_{JJ} has a two-component form similar to (3.47). They write it in the form (see HM (5.16) and also Griffin, 1979*a*)

$$\chi^\ell_{JJ}(\mathbf{Q}, \omega) = \rho_S(\mathbf{Q}, \omega)\chi^\ell_{v_S, v_S}(\mathbf{Q}, \omega)\rho_S(\mathbf{Q}, \omega) - \rho_N(\mathbf{Q}, \omega) \, , \tag{6.38}$$

where the superfluid velocity correlation function is related to the single-particle Green's function as in (6.24). Within the dielectric formalism, such an explicit two-component expression for χ^ℓ_{JJ} is given in Section 8.3. The functions $\rho_N(\mathbf{Q}, \omega)$ and $\rho_S(\mathbf{Q}, \omega)$ are so denoted because HM show that these reduce to the normal and superfluid densities, defined in Section 6.1,

$$\left.\begin{aligned} \lim_{Q \to 0} \rho_N(\mathbf{Q}, \omega = 0) &= \rho_N \, , \\ \lim_{Q \to 0} \rho_S(\mathbf{Q}, \omega = 0) &= \rho_S \, . \end{aligned}\right\} \tag{6.39}$$

Within the dielectric formalism, the results in (6.39) are equivalent to the rigorous identities (see Sections 5.1 and 6.1)

$$\left.\begin{aligned} \lim_{Q \to 0} \chi^{\ell R}_{JJ}(\mathbf{Q}, \omega = 0) &= -\rho_N \, , \\ \lim_{Q \to 0} \chi^{\ell c}_{JJ}(\mathbf{Q}, \omega = 0) &= -\rho_S \, . \end{aligned}\right\} \tag{6.40}$$

Basically, HM assume that the functions $\rho_N(\mathbf{Q}, \omega)$ and $\rho_S(\mathbf{Q}, \omega)$ are non-singular functions and hence can be approximated by (6.39) in the $Q, \omega \to 0$ limit, reducing (6.38) to

$$\chi^\ell_{JJ}(\mathbf{Q}, \omega) = \rho_S^2 \chi^\ell_{v_S v_S}(\mathbf{Q}, \omega) - \rho_N \, . \tag{6.41}$$

Restricting their discussion to $T = 0$ (where $\rho_N = 0$ and $\rho_S = \rho$), HM further assume that $\chi^{\ell}_{v_s v_s}$ has the form

$$\chi^{\ell}_{v_s v_s}(\mathbf{Q}, \omega) = \frac{AQ^2}{\omega^2 - a^2 Q^2} \ . \tag{6.42}$$

The zero-frequency, long-wavelength sum rules (see (6.24), (6.27) and (2.28))

$$\left.\begin{aligned} \chi^{\ell}_{v_s v_s}(\mathbf{Q}, \omega = 0) &= -\frac{1}{\rho_S} \ , \\ \chi_{nn}(\mathbf{Q}, \omega = 0) &= -\frac{n}{mc^2} \ , \end{aligned}\right\} \tag{6.43}$$

in conjunction with the continuity equation (5.16a), allow one to show that $A = a^2/\rho$ and $a = c$ in (6.42). Taking (6.24) into account, we thus arrive at the Gavoret–Nozières results (6.31) and (6.34). This derivation manifestly gives results consistent with two-fluid hydrodynamics, since it makes crucial use of the zero-frequency sum rules (6.43). It has the advantage of showing how the phonon mode is related to the order-parameter phase fluctuations (see (6.22)–(6.24)), but it also suggests that the results are tied to the two-fluid region of Q–ω space.

Physical meaning of the infrared divergence

We now turn to a discussion of the significance of the infrared divergences which play such a crucial role (albeit indirect) in the derivation of (6.31)–(6.33). We have alluded to the fact that at $T = 0$, while infrared divergent terms such as (6.29) appear in many terms, they cancel out in the final expressions for the correlation functions in the limit $Q, \omega \to 0$. Popov (1983, 1987) has developed a general analysis which does not have such low-momentum divergent terms. He accomplished this by first doing perturbation theory with the field operator components involving wavevectors greater than some characteristic k_0 (such terms are well behaved). The remaining slowly varying parts of $\hat{\psi}(\mathbf{r}), \hat{\psi}^+(\mathbf{r})$ are written in terms of phase and amplitude operators. The equations of motion for these slowly varying parts are then shown to be essentially equivalent to the "quantum hydrodynamic" Hamiltonian introduced by Landau (1941). Moreover, Popov shows that the correlation functions involving the phase and amplitude operators exhibit damped first and second sound modes at finite temperatures. This procedure is the most physical way of deriving, from first principles, the two-fluid equations describing superfluid behaviour in the $Q, \omega \to 0$ limit (see Section 6.2).

On the other hand, the infrared divergences are not unphysical since they can be shown to be associated with phase fluctuations of the Bose order parameter. As with an isotropic Heisenberg ferromagnet, the appearance of the static Bose order parameter breaks the symmetry which is restored by dynamic fluctuations of this order parameter. This means that the static and dynamic aspects of the order parameter must be treated in a consistent fashion if one is to ensure that the excitations have no energy gap in the $Q \to 0$ limit. In order to properly consider the dynamics of the condensate, one has to treat $\hat{a}_0$ and $\hat{a}_0^+$ as quantum mechanical operators rather than *c*-number averages as in the Bogoliubov prescription. As discussed very lucidly by Weichman (1988), the most convenient way of proceeding is to work with new field operators (see also (5.33))

$$\left.\begin{aligned} \hat{b}_k &\equiv \hat{a}_k - \sqrt{N_0}\delta_{k,0} \ , \\ \hat{b}_k^+ &\equiv \hat{a}_k^+ - \sqrt{N_0}\delta_{k,0} \ . \end{aligned}\right\} \tag{6.44}$$

As we noted at the end of Section 5.1, Bose condensation is defined by the condition $\langle \hat{b}_0 \rangle = \langle \hat{b}_0^+ \rangle = 0$ in this representation. One may express $G_{\alpha\beta}(\mathbf{Q}, \omega)$ in terms of the $\hat{b}$ operators instead of the $\hat{a}$ operators. All the formulas (3.22)–(3.30) in Section 3.2 are still valid, except that $G_{\alpha\beta}(\mathbf{Q} = 0, i\omega_n)$ now describes the dynamics of the condensate atoms (Talbot, 1983; Appendix A of Talbot and Griffin, 1983).

In Section 3.4, we used linear response theory to discuss the expectation value of $\hat{A}_Q = \hat{a}_Q + \hat{a}_{-Q}^+$ induced by a symmetry-breaking field which couples into $\hat{A}_Q$ with a strength $\delta\eta^0$. One finds (see (3.49*b*))

$$\delta A(\mathbf{Q}, \omega) = \chi_{AA}(\mathbf{Q}, \omega)\delta\eta^0(\mathbf{Q}, \omega) \ , \tag{6.45}$$

where the response to the symmetry-breaking field in (3.48) is given by

$$\chi_{AA}(\mathbf{Q}, \omega) = \sum_{\alpha,\beta} G_{\alpha\beta}(\mathbf{Q}, \omega) \ . \tag{6.46}$$

Here $G_{\alpha\beta}(\mathbf{Q}, \omega)$ is defined as in (3.22) but in terms of the $\hat{b}^+$, $\hat{b}$ operators given in (6.44). Clearly $\chi_{AA}(\mathbf{Q} = 0, \omega = 0)$ gives the value of $\langle \hat{A}_0 \rangle = \langle \hat{a}_0 \rangle + \langle \hat{a}_0^+ \rangle$ induced by a homogeneous, static symmetry-breaking field $\delta\eta^0(\mathbf{Q} = 0, \omega = 0)$. Since the spontaneous Bose broken symmetry is described by finite values of $\langle \hat{a}_0 \rangle$ and $\langle \hat{a}_0^+ \rangle$ in the limit of an infinitesimal symmetry-breaking field $\delta\eta^0 \to 0$, we conclude from (6.45) that below T_λ, we must have

$$\chi_{AA}(\mathbf{Q} = 0, \omega = 0) = \infty \tag{6.47}$$

The fact that $\chi_{AA}(\mathbf{Q} = 0, \omega = 0)$ is singular is easily understood physically. No matter how small $\delta\eta^0$ is, it fixes the phase of $\langle \hat{a}_0 \rangle$ and thus breaks the phase degeneracy. The fluctuations of the order parameter restore the symmetry, with (6.47) guaranteeing that it costs zero energy to excite the $\mathbf{Q} = 0$ mode. This requirement naturally leads one to expect gapless excitations in the long-wavelength limit (Goldstone theorem).

Nepomnyashchii (1983) and Weichman (1988) have given a lucid discussion of the physics behind the requirement (6.47) that $\chi_{AA}(\mathbf{Q}, \omega)$ be singular in the limit $\mathbf{Q}, \omega \to 0$. More generally, $\chi_{AA}(\mathbf{Q}, \omega)$ is proportional to the so-called longitudinal susceptibility. One finds that longitudinal (or amplitude) fluctuations are inevitably coupled into the transverse (or phase) fluctuations of an order parameter with two components. This is nicely illustrated in Section V.A of Weichman (1988). In particular one finds that in the $\mathbf{Q}, \omega \to 0$ limit

$$\chi_{AA}(\mathbf{Q}, \omega) \sim \frac{1}{\ln |\,\omega^2 - c^2 Q^2|} \quad , \qquad T = 0 \ , \tag{6.48}$$

$$\chi_{AA}(\mathbf{Q}, i\omega_n = 0) \sim \frac{1}{Q} \quad , \qquad cQ \ll k_B T \ . \tag{6.49}$$

(At finite T, $\chi_{AA}(\mathbf{Q}, i\omega_n \neq 0)$ is non-singular because of the finite spacing of the Matsubara Bose frequencies.) We now see the role of the infrared divergences which were discussed earlier. Inserting the general Beliaev expressions for $G_{\alpha\beta}$ into (6.46) and using the Hugenholtz–Pines theorem (3.32), one can show that (Nepomnyashchii and Nepomnyashchii, 1978)

$$\lim_{Q \to 0} \chi_{AA}(\mathbf{Q}, \omega = 0) = \frac{-1}{\Sigma_{12}(\mathbf{Q} = 0, \omega = 0)} \ . \tag{6.50}$$

This result exposes the close relation between the requirement that $\Sigma_{12}(\mathbf{Q}, \omega)$ vanish (see (6.36)) and that the longitudinal susceptibility $\chi_{AA}(\mathbf{Q}, \omega)$ diverge in the long-wavelength, zero-frequency limit. This relationship was not explicitly brought out in the original calculations of Gavoret and Nozières (1964), although the infrared singularities were handled in a way that led to the correct final results in (6.31)–(6.34).

To lowest order in Q and ω, the infrared divergent terms do not appear explicitly in the correlation functions (6.31)–(6.34). We have seen, however, that their existence plays a crucial role in the derivation of these results, as first noted by GN at $T = 0$. These divergent terms do show up explicitly in the higher-order corrections. Expanding the exact Beliaev expressions in (3.29) and using the zero-frequency results derived in Section 6.1, one finds (Nepomnyashchii and Nepomnyashchii, 1978;

Griffin, 1984)

$$\left.\begin{aligned} G_{11}(\mathbf{Q},\omega=0) &= -\frac{n_0 m^2}{\rho_s Q^2} - \frac{1}{4\Sigma_{12}(\mathbf{Q},\omega=0)}\,, \\ G_{12}(\mathbf{Q},\omega=0) &= \frac{n_0 m^2}{\rho_s Q^2} - \frac{1}{4\Sigma_{12}(\mathbf{Q},\omega=0)} \end{aligned}\right\} \tag{6.51}$$

in the long-wavelength limit. One can use this result for G_{11} in conjunction with (6.48)–(6.50) to obtain the leading-order correction to the atomic momentum distribution $\tilde{n}_p$ discussed in Section 4.1. At finite temperatures, one finds that (4.10) is replaced by

$$\tilde{n}_p = -k_B T G_{11}(\mathbf{p},\omega=0) = \frac{n_0}{n}\frac{mk_B T}{p^2} + O\left(\frac{1}{p}\right) \tag{6.52}$$

for $cp \ll k_B T$. The coefficient of the correction in (6.52) has been calculated by Giorgini, Pitaevskii and Stringari (1992) in a very simple way terms of the order-parameter phase fluctuations. At $T = 0$, however, these authors show that the logarithmic divergences do not show up in the momentum distribution, with the result that in the long-wavelength limit

$$\tilde{n}_p = \frac{n_0}{n}\frac{mc}{2p} + \text{const}\,. \tag{6.53}$$

This should be of considerable interest in the analysis of high-momentum neutron-scattering data (see Sections 4.3 and 4.4).

Excitations outside the small Q, ω, T domain

As we noted at the end of Section 5.5, the analysis of Gavoret and Nozières (1964) and Hohenberg and Martin (1965) led to the general scenario that (a) $G_{\alpha\beta}$ has a sharp phonon excitation which is entirely due to the fluctuations of the condensate and (b) the only density fluctuation exhibited by χ_{nn} is due to the pole of $G_{\alpha\beta}$, which appears in χ_{nn} with a spectral weight also related to the Bose condensate. In this scenario, it is the first term in (3.47) which plays the crucial role. As we have seen, the explicit calculations of GN and HM (see also Nepomnyashchii and Nepomnyashchii, 1978) show the correctness of this picture in the $Q, \omega \to 0$ limit, at least at $T = 0$. We refer to this as the Gavoret–Nozières–Pines (GNP) scenario, since it was most explicitly formulated by these authors (see especially the review by Pines, 1965). A more recent formulation of this scenario within the language of the dielectric formalism has been given by Nepomnyashchy (1992).

This scenario became accepted as a general basis for understanding the excitations in superfluid ^{4}He at arbitrary temperatures and wavevectors. Remarks made by Hohenberg and Martin (1964) concerning the *collisionless* region illustrate this: "The well-defined oscillation frequency of density fluctuations (seen in χ_{nn} when $\omega\bar{\tau} \gg 1$) results entirely from their coupling to quantum field correlations as a result of Bose condensation. In an uncondensed Boson fluid this coupling is absent and χ_{nn} is not sharply peaked for $\omega\bar{\tau} \gg 1$." The analysis of the low-Q behaviour of $\chi_{nn}(\mathbf{Q}, \omega)$ given in Chapter 7 of Nozières and Pines (NP, 1964, 1990) is also based on the same scenario, as shown by the statement "in the long wavelength limit, the only density fluctuations of importance are those produced by exciting a single quasiparticle from the condensate" (quote from p. 97 of NP). Since the origin of the sharp phonon resonance in χ_{nn} was as a pole of $G_{\alpha\beta}$, NP introduced the term "quasi-particle" sound to distinguish it from ordinary sound. At finite T, it was argued that χ^R_{nn} in (3.47) would exhibit additional structure at low energies arising from the thermally excited quasiparticles. Above $T \gtrsim 1$ K, the dominant contribution would be from rotons, while for $T \lesssim 0.6$ K, it would be from phonons. This thermal scattering was assumed to produce a broad background, such as given by (3.45), without any particular structure. In the analysis of NP (see also Griffin, 1979a), the two terms in (3.47) were identified with the condensate (or "superfluid") and normal (or "normal fluid") contributions. (For further discussion of this identification, see Section 8.3.)

In the GN analysis, the role of any density fluctuations associated with the "normal" term χ^R_{nn} was effectively suppressed in χ_{nn} as a direct consequence of the infrared singularities. As a result, the resonances in χ_{nn} as given by (3.47) are associated completely with the first term containing $G_{\alpha\beta}$. Moreover, in the $Q, \omega \to 0$ limit, the poles of $G_{\alpha\beta}$ are tied in closely with the fluctuations of the underlying Bose order parameter. In this limit, the long-wavelength density response of a Bose-condensed liquid at $T = 0$ is very similar to that in a dilute Bose gas as described by the Bogoliubov approximation (see Section 3.3) but for different reasons. In the latter case, there are essentially no non-condensate atoms while in a Bose liquid, the normal fluid dynamics is "suppressed" due to the infrared anomaly (as emphasized by Nepomnyashchy, 1992).

We now turn to the question of the extent to which the GNP scenario described above is valid outside the small Q, ω region at low temperatures.

We have emphasized that (6.31) and (6.34) are rigorous results but have been only proved at $T = 0$. These expressions are in precise agreement with the zero-temperature limit of the two-fluid formulas (6.16) and (6.25).

In this limit, the weight of the second sound mode vanishes in both $G_{\alpha\beta}$ and χ_{nn}. Only the first sound pole remains, with a velocity u_{I} which is equivalent (at $T = 0$) to the compressional sound velocity c in (6.34). The fact that GN exhibited a phonon mode with the compressional sound velocity suggests that their results are, strictly speaking, valid only in the low-frequency domain. As shown in Fig. 6.2, there is a slight difference in magnitude between the first sound velocity c and the collisionless zero sound velocity u. Rather than c, the quasiparticle phonon velocity is given by u in the collisionless domain. We also note that the phonon dispersion curve ω_Q at low temperatures exhibits anomalous dispersion, with the slope being larger than the compressional sound velocity c up to $Q = 0.55$ Å^{-1} (see Fig. 7.2). This means that the expression in (6.29) would not be divergent at ω_Q for $Q \gtrsim 0.1$ Å^{-1}.

What the GNP scenario leaves out is that, in a Bose liquid, both $G_{\alpha\beta}$ and χ_{nn} may exhibit excitation branches that have nothing to do with fluctuations of the condensate (for an early criticism of this kind, see Straley, 1972). This is certainly the case in the normal phase. We recall that, to a certain degree, collisionless density fluctuations are present in any liquid. In addition, the single-particle excitations $Q^2/2m$ in a normal Bose gas are expected to be strongly renormalized but still present in a normal Bose liquid. A new scenario was formulated by Szépfalusy and Kondor (SK, 1974) in terms of understanding how the coupled spectra of $G_{\alpha\beta}$ and χ_{nn} in the superfluid phase uncoupled and smoothly merged with the spectra of the normal phase as the condensate vanished. The dielectric formalism exposes this structure in the most direct way. The interpretation of Glyde and Griffin (1990) is developed within this SK scenario (see Section 7.2).

The GNP scenario given above is based on (3.47) or (5.24), while the SK scenario is based on (5.76). While these expressions are formally equivalent, they can lead to quite different physical pictures. In particular, the dielectric formalism results based on (5.76) show how χ_{nn} and $G_{\alpha\beta}$ can both be dominated by a phonon whose velocity is temperature-independent. This is because the condensate and normal parts of $\overline{\chi}_{nn}$ add up coherently, as illustrated by (5.50). In the GNP analysis starting from two-component expressions like (3.47), in contrast, inclusion of the effect of the thermally excited quasiparticles (the "normal fluid") can lead to a strongly temperature-dependent phonon velocity (a difficulty which is illustrated by the results in Section 7.3 of NP and Appendix C of Hohenberg and Martin, 1965).

The dielectric formalism of Chapter 5 naturally leads to a way of interpreting the phonon excitation which is valid for both normal and superfluid Bose fluids (gas or liquid). In this interpretation, the phonon is viewed as a zero sound particle–hole excitation associated with two distinct kinds of effective fields. One is produced by the condensate atoms and is present in both gas and liquids. The other effective field is associated with the dynamics of the non-condensate atoms and may be thought of as the "normal fluid" effective field associated with thermally excited quasiparticles. This normal fluid mean field is well defined in a liquid but not in a gas. At $T = 0$, only the condensate mean field is important. However, as the temperature increases, the condensate field decreases in strength while the non-condensate mean field becomes increasingly important (in a liquid). Thus there is a smooth transition to the normal phase above T_λ in the case of a Bose liquid. This zero sound interpretation of phonons in Bose-condensed fluids has its roots in the work of Pines (1963) and Ma, Gould and Wong (1971), although these authors only considered the low-temperature limit.

The analysis of GN, like other analyses which we have been reviewing in this section, is strictly concerned with the excitation spectrum in the double limit of small Q and small ω. We have seen that the Hugenholtz–Pines theorem (3.32) requires that, in the $Q \to 0$ limit, there must be a zero-energy Goldstone mode associated with the breaking of gauge symmetry. The low-energy phonon spectrum exhibited by $G_{\alpha\beta}$ and χ_{nn}, however, does not *a priori* exclude additional high-energy modes from appearing in the long-wavelength limit. Indeed, we know that in superfluid ^{4}He, the condensate-induced hybridization results in the high-energy two-roton spectrum having finite weight in $G_{\alpha\beta}$ and χ_{nn} even at small Q (see Chapter 10).

GN note that their final result for χ_{nn} given by (6.34) in the long-wavelength limit is consistent with the compressibility sum rule as well as the f-sum rule (2.24). The fact that (6.34) exhausts the f-sum rule can be easily over-interpreted. It simply means that the phonon is the only *low*-energy mode at long wavelengths. A calculation which is limited to the small-ω region can clearly say nothing about the existence of high-energy modes. One has the example of quantum crystals, where $S(\mathbf{Q}, \omega)$ can exhibit a lot of structure which makes no contribution to the f-sum rule (see Section 11.2). A useful sum-rule analysis of the single-phonon, multiphonon and interference contributions to $S(\mathbf{Q}, \omega)$ in the small-Q limit is given by Wong and Gould (1974) within the one-loop approximation (see pp. 292ff of WG).

In the dielectric formalism of Chapter 5, a crucial role is played by the regular single-particle Green's function $\bar{G}_{\alpha\beta}$. As noted in Section 6.1, the regular self-energies $\bar{\Sigma}_{\alpha\beta}$ obey the Hugenholtz–Pines theorem and thus $\bar{G}_{\alpha\beta}$ may also exhibit a low-frequency branch (probably acoustic) in the long-wavelength limit. As with $G_{\alpha\beta}$, such a low-energy pole can be directly associated with the fluctuations of the Bose order parameter. However, $\bar{G}_{\alpha\beta}$ may also exhibit additional high-energy modes, corresponding to intrinsic single-particle excitations characteristic of the normal phase. These would not be expected to be modified by the appearance of a Bose condensate if they existed at relatively large energy. (In contrast, the two-roton multiparticle spectrum will only appear in $\bar{G}_{\alpha\beta}$ with finite weight in the condensed phase, as can be seen from the one-loop self-energies in (5.52) and (5.53)). Thus in parameterizing the spectrum for $\bar{G}_{\alpha\beta}$ for superfluid ^{4}He, one should allow for high-energy modes as well as a gapless low-energy excitation. These remarks will be the basis for the model spectrum we postulate in Section 7.2 in order to understand the $S(\mathbf{Q},\omega)$ line-shape data (see Fig. 7.22).

In the collisionless region, we have argued that the dielectric formalism naturally leads to viewing the phonon as a zero sound mode but one which is associated with the mean fields of both the condensate and non-condensate atoms. This generalized zero sound picture allows one to understand why the collisionless phonon velocity in superfluid ^{4}He is essentially unchanged as we go from $T = 0$ to above T_λ. Different scenarios are possible within the dielectric formalism framework and, as noted above, Nepomnyashchy (1992) interprets the phonon mode as being essentially a single-particle (SP) excitation at all T below T_λ, as in the GNP scenario described above. Further discussion of this alternative scenario is deferred to Section 12.1.

In summary, we have seen in this section that the nature of the phonon excitations in Bose-condensed fluids is quite subtle. We have tried to distinguish between the phonons in the two-fluid domain (where they are intimately tied to the oscillations of the condensate) and those in the collisionless domain. We have argued that the only available rigorous microscopic calculations of the structure of χ_{nn} and $G_{\alpha\beta}$ are, in fact, mainly concerned with the long-wavelength, low-frequency two-fluid domain. It is really only in this "macroscopic" domain that phonons can be directly related to the dynamics of the condensate (or superfluid), a point also emphasized by Nozières and Pines (1964, 1990). In contrast, the phonons which play the role of elementary excitations in thermodynamic and transport properties are in the collisionless region. These are most directly studied by inelastic neutron scattering, as we discuss in Chapter 7.

7

Phonons, maxons and rotons

In this chapter, we review the high-resolution neutron-scattering data for the dynamic structure factor $S(\mathbf{Q}, \omega)$ and suggest an interpretation within a unified picture of the excitations in liquid ^{4}He consistent with the ideas of Chapter 5. We argue that the phonons ($0.1 \lesssim Q \lesssim 0.7$ Å^{-1}) in the collisionless region and rotons ($Q \sim 1.9$ Å^{-1}) are really two *separate* branches of the density fluctuation spectrum in the superfluid phase which are hybridized by the condensate. The low-wavevector phonon is interpreted as a zero sound collective density fluctuation while the large-Q maxon–roton is interpreted as a strongly renormalized single-particle excitation. In the intermediate-wavevector region $0.8 \lesssim Q \lesssim 1.2$ Å^{-1}, we argue that there is evidence that *both* excitation branches, a sharp single-particle (or atomic-like) maxon excitation and a broad high-energy zero sound phonon, are observed in $S(\mathbf{Q}, \omega)$. Within this scenario, the appearance of the sharp maxon–roton resonance ($0.8 \lesssim Q \lesssim 2.4$ Å^{-1}) in $S(\mathbf{Q}, \omega)$ below the superfluid transition temperature T_λ is direct *dynamical* evidence for the Bose broken symmetry and the associated Bose condensate in superfluid ^{4}He.

In Section 7.1, we review the neutron-scattering intensity data for small, intermediate and large wavevectors. In Section 7.2, these results are interpreted starting from the assumption that superfluid ^{4}He is a Bose-condensed liquid. The condensate inevitably leads to a mixing of the single-particle spectrum described by $G_{\alpha\beta}(\mathbf{Q}, \omega)$ and the density fluctuation spectrum described by $S(\mathbf{Q}, \omega)$. In the scenario we have summarized in Section 5.5, the crucial question is: for a given wavevector Q, are we in the zero sound (ZS) or the single-particle (SP) regime?

While our specific microscopic picture is new (Glyde and Griffin, 1990; Griffin, 1991), it is a natural development within the field-theoretic approach which we use in this book. We attempt to give a consistent picture of the entire density fluctuation spectrum over a wide range of Q, ω and T. This scenario is still at an early stage of development and many specific aspects are still tentative. We hope our preliminary analysis will, however, be a stimulus and guide to future investigations.

While this chapter contains many plots of experimental data, it should be emphasized that this book is primarily concerned with theory. Thus, in our selection of neutron data, our main interest is in deciding which of several possible theoretical scenarios is being realized in superfluid ^{4}He. We do not give any critical discussion of the experimental data as such. The recent time-of-flight data taken at IN6 at ILL (Andersen, Stirling *et al.*, 1991) is especially useful since it covers such a wide range of $\mathbf{Q}, \omega$ and T. Earlier high-resolution data taken using triple-axis spectrometers is less extensive but has the advantage of giving $S(\mathbf{Q}, \omega)$ in a more direct fashion than time-of-flight methods. The triple-axis data is nicely summarized by Svensson (1989, 1991).

7.1 $S(\mathbf{Q}, \omega)$: neutron-scattering data

In Section 2.2, we have given a brief summary of the experimental results for $S(\mathbf{Q}, \omega)$ obtained from neutron-scattering studies over a wide range of Q and ω, and described how the characteristic features change as we go from low temperatures (1 K) to T_λ and above. Section 2.2 should be reviewed before reading the more detailed discussion of data given in the present section.

At low temperatures, $S(\mathbf{Q}, \omega)$ exhibits an extremely sharp quasiparticle peak up to about 2.4 Å^{-1}. For example, the intrinsic phonon half-width at half-maximum Γ_Q is of the order of 0.025 K at 1.2 K (Mezei and Stirling, 1983). Most plots of $S(\mathbf{Q}, \omega)$ in the literature have background scattering (empty-cell) removed but still include *instrumental* broadening, which depends on the particular wavevector Q being studied and the neutron spectrometer used. Since the intrinsic width of the quasiparticle peak is so small at low temperatures ($T \lesssim 1$ K), it can be ignored in view of the much larger instrumental resolution width. Thus the observed low-temperature quasiparticle peak in $S(\mathbf{Q}, \omega)$ can be fit to a (Gaussian) resolution function to find the *instrumental* half-width Γ_G. The measured $S_{\text{exp}}(\mathbf{Q}, \omega)$ is then viewed as the intrinsic density fluctuation spectrum $S(\mathbf{Q}, \omega)$ convoluted with this Gaussian resolution-function. In the recent study by Stirling and Glyde (1990) at $Q = 0.4$ Å^{-1}, for example, the Gaussian resolution half-width at half maximum is estimated to be $\Gamma_G = 0.022$ THz (1.1 K) assuming that the width of the phonon peak at

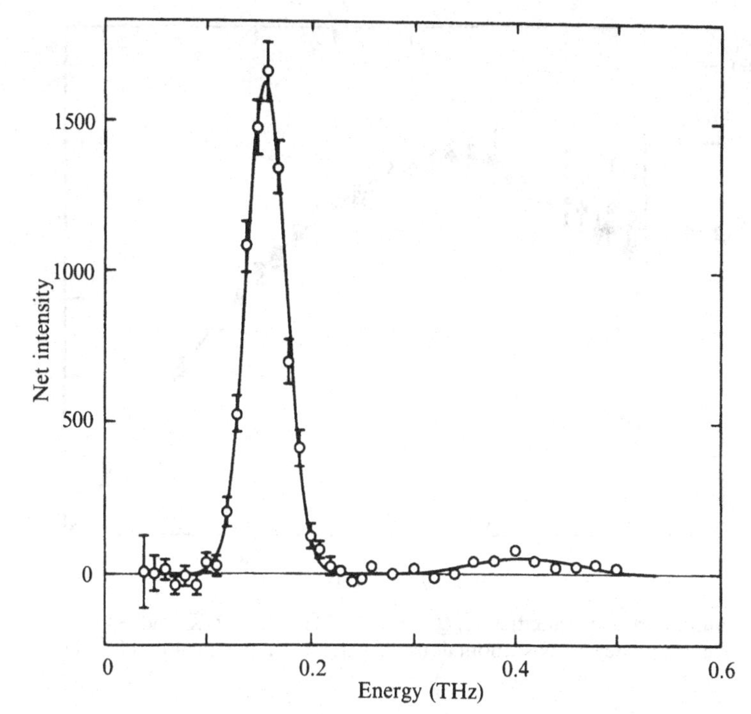

Fig. 7.1. The inelastic neutron-scattering intensity vs. frequency for $Q = 0.4$ Å^{-1}and $T = 1.35$ K. As usual, the background (empty-cell) scattering has been extracted. The line is a least-square fit to a convolution of (7.3) and (7.4) with a Gaussian resolution function, as described in the text [Source: Stirling and Glyde, 1990].

$T = 1.35$ K shown in Fig. 7.1 is entirely instrumental in origin. Following usual practice, both the intrinsic $S(\mathbf{Q},\omega)$ and the resolution-broadened $S_{\text{exp}}(\mathbf{Q},\omega)$ are referred to as $S(\mathbf{Q},\omega)$.

Low momentum: phonons

In this section, we will focus our attention on neutron data showing how the phonon line shape in $S(\mathbf{Q},\omega)$ changes with temperature for $T \gtrsim 1.3$ K. However, we first make a few remarks about the phonon line width at lower temperatures. Phonon lifetimes have been measured with great accuracy by ultrasonic techniques in the temperature region $T \lesssim 0.6$ K, where one may ignore phonon interactions with thermally excited rotons.

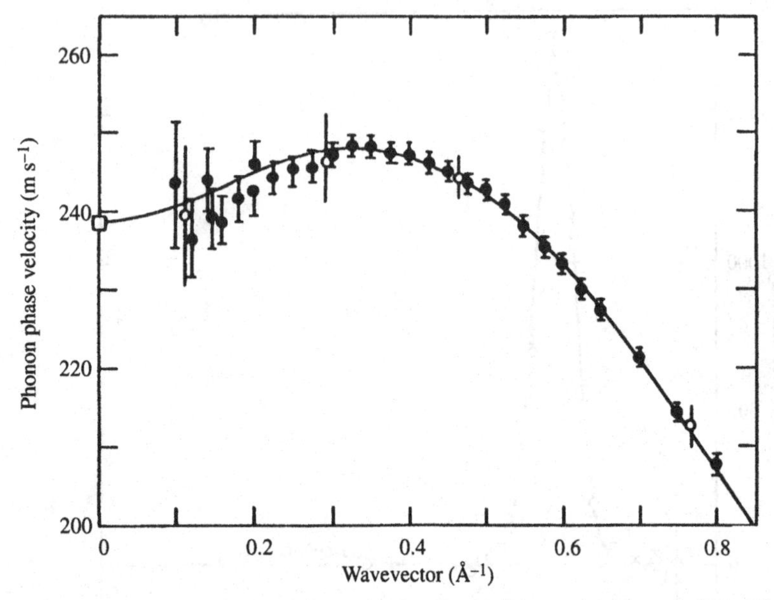

Fig. 7.2. Phonon phase velocities ω/Q for liquid ^{4}He at 1.2 K and SVP. The quasiparticle frequency shows anomalous dispersion in the region $0.1 < Q < 0.55$ Å^{-1} [Source: Stirling, 1983].

This is a well studied, although complicated, subject (see Section 4 of Woods and Cowley, 1973; Maris, 1977).

At sufficiently low temperatures, the main source of phonon damping is spontaneous decay via three-phonon processes (using the Landau–Khalatnikov picture reviewed in Section 1.1). Unless one includes the finite width of the phonons, this three-phonon process is controlled by kinematics and only occurs because of "anomalous" dispersion. Anomalous dispersion refers to the fact that, as first suggested by Maris and Massey (1970), the phonon dispersion curve in Fig. 1.3 curves upward slightly before bending over, i.e.,

$$\omega_Q = cQ(1 - \gamma Q^2) \ , \tag{7.1}$$

where $\gamma(T)$ is *negative*. This curvature has been studied by several groups using neutron scattering (for references, see p. 338 of Glyde and Svensson, 1987). The high-resolution data of Stirling (1983) at $T = 1.2$ K are shown in Fig. 7.2. As a result of this anomalous dispersion, a phonon can decay via the three-phonon process as long as its phase velocity is *greater* than c, i.e., as long as Q is *less* than some threshold wavevector Q_c.

Stirling (1983) obtained the SVP value $Q_c = 0.55$ Å^{-1}; Q_c decreases with pressure (Svensson, Martel and Woods, 1975). Evidence for the resulting three-phonon decay has been reported by Mezei and Stirling (1983), who observed the expected abrupt decrease in the phonon width (at $T = 0.95$ K) in the region $Q \sim 0.5$–0.6 Å^{-1}. At such low temperatures, it is argued that phonons with $Q > Q_c$ can decay only by the much weaker four-phonon process involving thermally excited phonons.

For temperatures above 1 K, in contrast, the main damping mechanism of phonons is scattering from thermally excited rotons; thus one expects that Γ_Q will be approximately proportional to the number of *rotons* present

$$N_R \sim \sqrt{T} e^{-\Delta/T} \ . \tag{7.2}$$

This prediction is borne out by finite-temperature studies on the width of phonons, as discussed in the remainder of this section (Cowley and Woods, 1971; Mezei and Stirling, 1983; Stirling and Glyde, 1990).

From our perspective, however, the most striking feature of the phonon peak in the collisionless region of $S(\mathbf{Q},\omega)$ is its persistence while the temperature increases, right through T_λ. This somewhat surprising result was first observed in a classic experiment by Woods (1965*b*) for $Q = 0.38$ Å^{-1}. More recent studies that confirm these original results include the work of Mezei and Stirling (1983) for $0.3 < Q < 0.7$ Å^{-1} at temperatures up to 1.6 K, Stirling and Glyde (1990) for $Q = 0.4$ Å^{-1} over the temperature range $1.35 < T < 3.94$ K, and the very complete ILL data of Andersen, Stirling *et al.* (1991). We concentrate on the data of Stirling and Glyde (SG), since it is the most extensive set of high-resolution data which has been analysed and published on the full temperature dependence of $S(\mathbf{Q},\omega)$ for Q in the phonon region (see Fig. 1.4).

As the temperature rises from 1.35 to 2.96 K, the phonon line at $\omega_Q \simeq 0.16$ THz (7.7 K) broadens and its peak intensity slowly decreases, but the peak *position* remains essentially constant (Fig. 1.4). In particular, there is no evidence of any *qualitative* change in the region around the superfluid transition at 2.17 K. There is also clear evidence (see Fig. 7.1) of a weak, high-frequency structure with an energy slightly larger than twice the roton energy 2Δ (i.e., it peaks at 0.40 THz (19.7 K) at 1.35 K). Evidence for this high-energy peak is also shown in Fig. 10.4. We defer analysis of such "multiparticle" contributions to Chapter 10.

The original study by Woods (1965*b*) did not have sufficient resolution to measure the intrinsic phonon line width, but later studies by Cowley

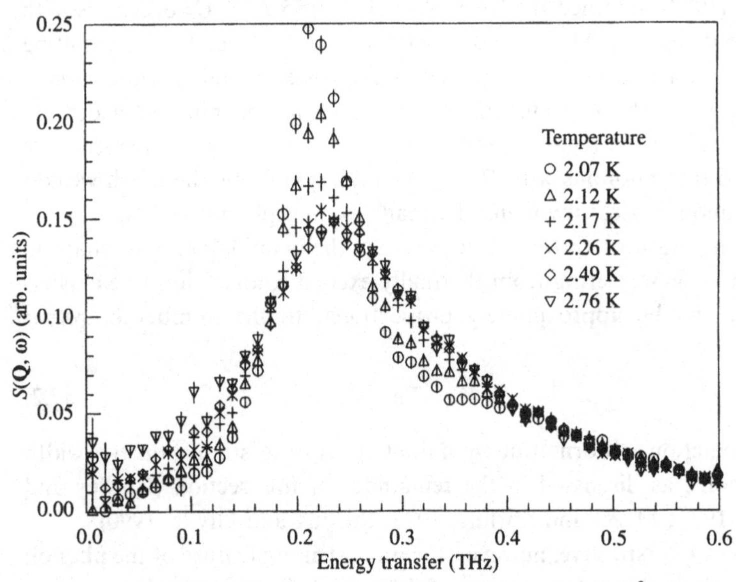

Fig. 7.3. Neutron-scattering intensity vs. frequency for $Q = 0.6$ Å^{-1} in the region around T_λ [Source: Andersen, Stirling *et al.*, 1991].

and Woods (1971) for small Q observed that the widths increased fairly smoothly through the superfluid transition. The high-resolution neutron data at 0.4 Å^{-1} obtained by SG have confirmed these early results, with a more complete examination of the region around T_λ. The only significant change in the peak position and line shape appears in the region from $T = 2.96$ K to 3.94 K, which may be connected to the fact that the Landau–Placzek ratio is steadily increasing with temperature (see Fig. 6.1) and becomes of order unity as we approach the gas–liquid critical point at 5.2 K. As discussed in Section 6.2, this means that the entropy fluctuations are coupled into the density fluctuations in the low-Q hydrodynamic domain; some residual effect of this may remain even for wavevectors as large as $Q = 0.4$ Å^{-1}. The line shape changes which are apparent in the high-temperature region $2.56 < T < 3.94$ K certainly deserve more careful theoretical study.

On the other hand, the data in Fig. 1.4 also clearly show that there is a temperature region extending from 1 K right up to about 2.5 K (i.e., well *above* $T_\lambda = 2.17$ K) in which the low-Q phonon line shape in $S(\mathbf{Q}, \omega)$ does not show much qualitative change. The same sort of line shapes are obtained for wavevectors up to about $Q = 0.7$ Å^{-1}. As illustration,

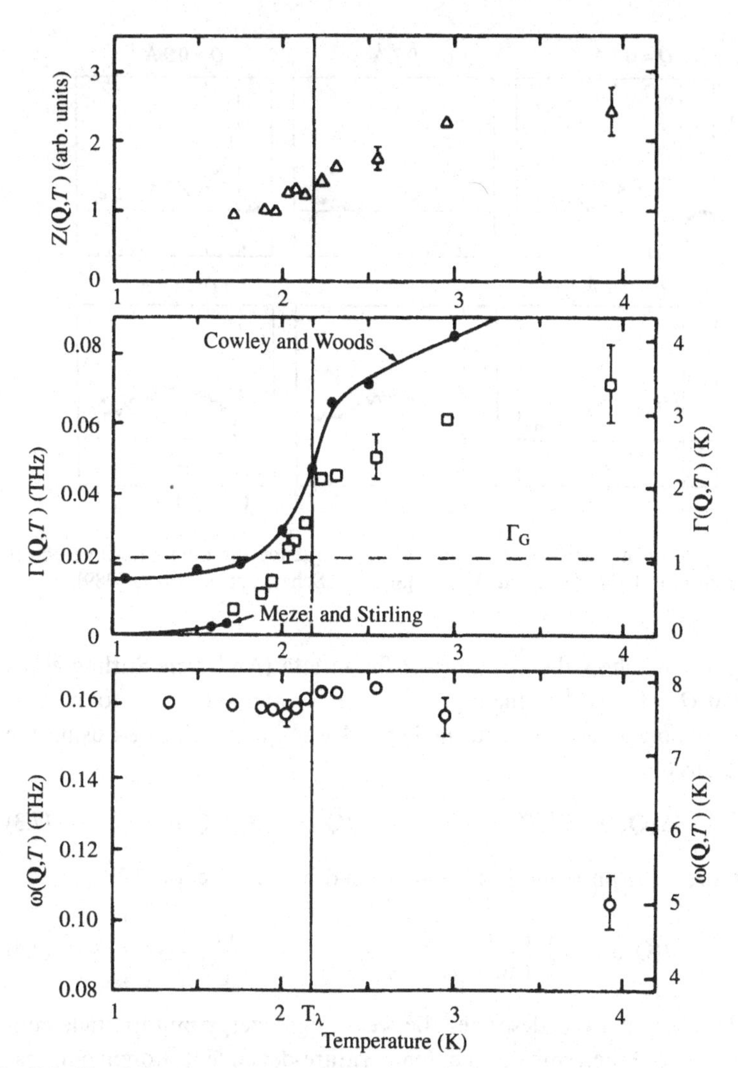

Fig. 7.4. The fitting parameters ω_Q, Γ_Q and $Z(Q)$ as a function of the temperature, for $Q = 0.4$ Å^{-1} at SVP. These results are based on using (7.3) and (7.4). Γ_G shows the Gaussian resolution width. Results are also shown for the total half-width obtained by Cowley and Woods (1971) and for the intrinsic half-width obtained by Mezei and Stirling (1983) at $T < 1.7$ K [Source: Stirling and Glyde, 1990].

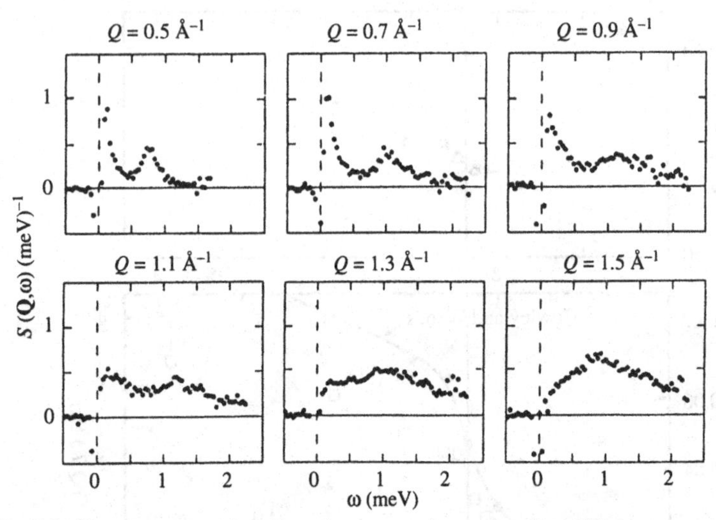

Fig. 7.5. Neutron-scattering line shapes vs. frequency for different wavevectors in normal liquid ^{3}He (SVP) at 0.12 K [Source: Scherm *et al.*, 1987, 1989].

in Fig. 7.3, we show the ILL time-of-flight data (Andersen, Stirling *et al.*, 1991) at $Q = 0.6$ Å^{-1} in the region around T_λ and above. SG found that the resolution-broadened data in Fig. 1.4 was quite well fitted using the simple ansatz

$$S(\mathbf{Q}, \omega) = [N(\omega) + 1]\{Z(Q)A(\mathbf{Q}, \omega) + G_M(\mathbf{Q}, \omega)\} , \qquad (7.3)$$

where the "one-phonon" peak is described by (see Section 2.2)

$$A(\mathbf{Q}, \omega) = \frac{1}{\pi}\left[\frac{\Gamma_Q}{(\omega - \omega_Q)^2 + \Gamma_Q^2} - \frac{\Gamma_Q}{(\omega + \omega_Q)^2 + \Gamma_Q^2}\right] \qquad (7.4)$$

and the Gaussian G_M describes the weak high-energy multiparticle contribution. SG determined the temperature-dependent Lorentzian parameters $\omega_Q(T)$ and $\Gamma_Q(T)$, as well as $Z(Q, T)$ which determines the peak height, by convoluting (7.3) with the Gaussian resolution function discussed above. The fit parameters so obtained are shown in Fig. 7.4.

We conclude that $S(\mathbf{Q}, \omega)$ in both superfluid and normal liquid ^{4}He exhibits a collisionless phonon mode and that its existence appears to be independent of the presence of superfluidity or a Bose condensate. Contrary to the view expressed in most of the literature, we argue that it is only at $T = 0$ that this mode is usefully related to the Bogoliubov phonon excitation in a dilute Bose gas which is entirely associated with

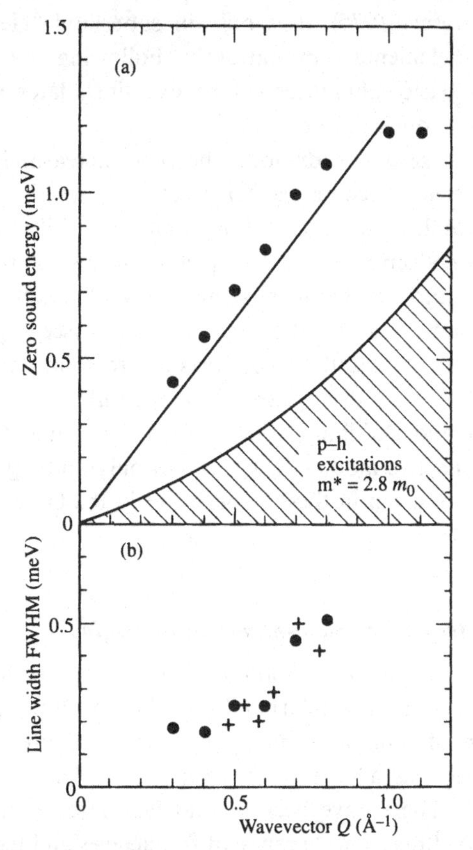

Fig. 7.6. Zero sound dispersion relation and width in normal liquid ^{3}He as a function of Q, at 0.12 K and SVP. The solid line gives the sound velocity. The crosses are the widths obtained by Hilton *et al.* (1980) [Source: Scherm *et al.*, 1987].

condensate fluctuations (see Sections 3.3, 3.4, 5.2 and 6.3). Pines (1963, 1966) first suggested that this phonon mode could be better interpreted as a collective zero sound mode, in analogy with zero sound propagation in normal liquid ^{3}He. In Fig. 7.5, we show the neutron-scattering intensity from liquid ^{3}He for various momentum transfers. The higher-energy peak position corresponds to zero sound, whose position and width are plotted in Fig. 7.6. Such collisionless modes arise from time-dependent self-consistent fields and are not very dependent on the quantum statistics obeyed by the atoms or the temperature. While such zero sound modes at low Q seem to be a general feature in many liquids (see the review by Copley and Lovesey, 1975), it is only in superfluid ^{4}He that they correspond to the "elementary excitations". Following our discussion in Chapter 5, this basic difference will be explained later as being a consequence of Bose condensation.

At higher Q, such zero sound modes become increasingly damped. The self-consistent fields become weaker as Q increases (i.e., at shorter wavelengths) as a result of stronger p–h fluctuations, until the scattering intensity from this collective mode disappears entirely. Zero sound in normal liquid ^{3}He starts to broaden significantly at about ~ 0.7 Å^{-1} and has largely disappeared by about $Q \sim 1.2$ Å^{-1} (see Figs. 7.5 and 7.6). As another example, liquid Ne supports a zero sound mode at low Q, but by $Q \sim 0.8$ Å^{-1} it disappears (Buyers *et al.*, 1975). A recent study by Dzugutov and Dahlborg (1989) shows that liquid Bi (a liquid metal) exhibits a collective mode in $S(\mathbf{Q}, \omega)$, but only up to $Q \sim 0.6$ Å^{-1}. Further references are given in these papers and in the book by Hansen and McDonald (1986).

Intermediate momentum: maxon region

At $Q \sim 1.1$ Å^{-1}, the dispersion relation ω_Q reaches its maximum energy ~ 0.30 THz (14.4 K). This is referred to as the "maxon", just as the $Q \sim 2$ Å^{-1} region of minimum energy is called the roton. We now turn our attention to this maxon wavevector region, defined roughly as $0.8 \lesssim Q \lesssim 1.5$ Å^{-1}. There have been several high-resolution studies of $S(\mathbf{Q}, \omega)$ for $Q \sim 1$ Å^{-1} over a wide range of frequencies and temperatures through T_λ. The original SVP work of Woods and Svensson (1978) made a detailed study at $Q = 0.8$, 1.13 (the maxon wavevector Q_M), 1.3 and 1.9 Å^{-1}. Talbot, Glyde, Stirling and Svensson (TGSS, 1988) have measured $S(\mathbf{Q}, \omega)$ for $Q = 1.13$ Å^{-1} at a pressure of 20 bar. Very recently, Andersen, Stirling and coworkers (1991) at ILL have carried out further high-resolution experiments at SVP. In Fig. 7.7(b), we show their recent data at $Q = 1.1$ Å^{-1} as a function of ω over a wide range of temperatures. In Fig. 7.7(a), we give an expanded view of the same data near T_λ.

The TGSS high-pressure data are shown in Fig. 7.8 at a series of temperatures below T_λ. At low T, the intensity is seen to consist of a very sharp peak at 0.30 THz (14.4 K) and a broad distribution (at this resolution) centred at about ~ 0.5 THz (24 K). The sharp peak broadens while its scattering intensity *decreases* dramatically as T approaches T_λ from below. Indeed, the sharp peak apparently disappears entirely at $T \simeq T_\lambda$, as the liquid passes from the superfluid to the normal phase

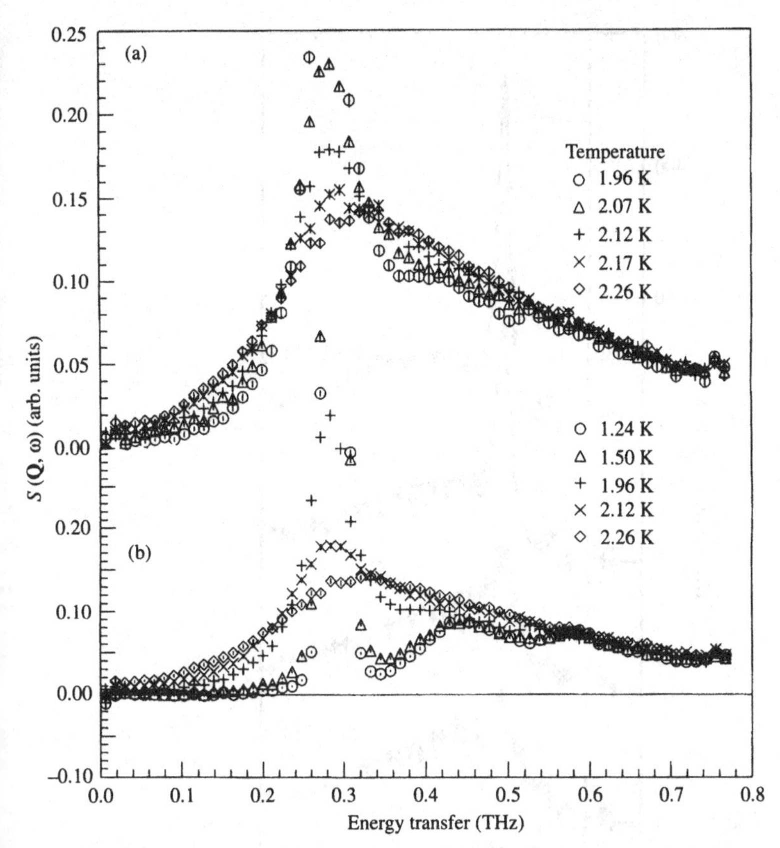

Fig. 7.7. $S(\mathbf{Q},\omega)$ vs. frequency at SVP for $Q = 1.1$ Å^{-1} at a series of temperatures above and below T_λ. The upper panel shows an expanded view of the region around T_λ [Source: Andersen, Stirling *et al.*, 1991; Stirling, 1991].

(see also Woods and Svensson, 1978). There is very little change in the remaining scattering intensity as T goes from 1.90 K to 3.94 K, as shown in the bottom panel of Fig. 2.4. In the high-pressure data in Fig. 7.8, this broad temperature-independent distribution, which is centred at an energy *much* higher than the sharp quasiparticle peak, appears to be present in the superfluid phase without much change. In particular, the multiparticle resonances so visible at low temperatures at SVP (see Fig. 7.7) are not so evident at higher pressures (see Fig. 7.8). The comparative data shown in Fig. 7.9 at 1.27 K also seem to be consistent with the "smearing" of the fine structure with pressure. However, recent high-resolution data at 1.25 K from ILL (see Fig. 7.10) does in fact show

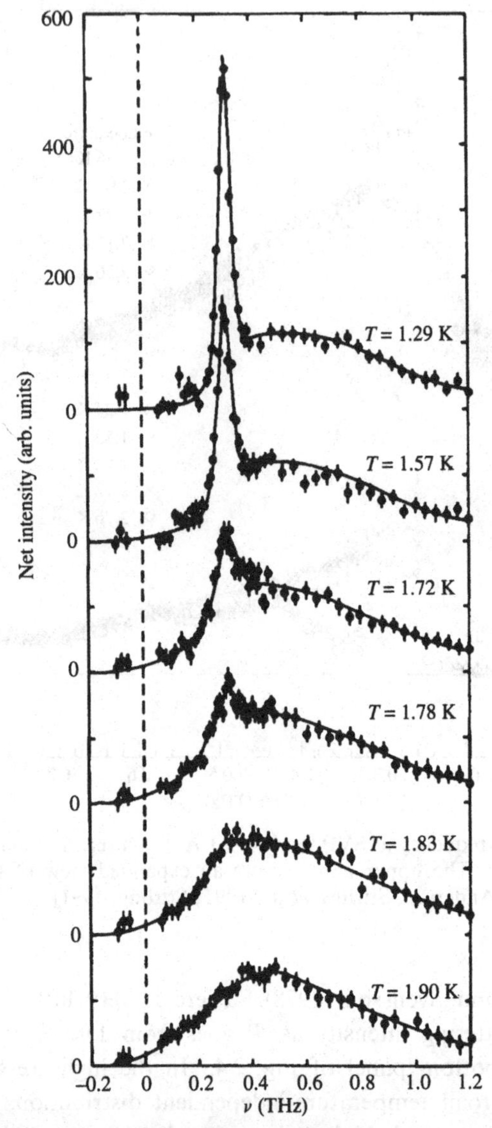

Fig. 7.8. The net scattering intensity vs. frequency for $Q = 1.13$ Å^{-1}, $T_\lambda = 1.928$ K, at $p = 20$ atm in the superfluid phase. Note the wide frequency range, up to 1.2 THz [Source: Talbot, Glyde, Stirling and Svensson, 1988].

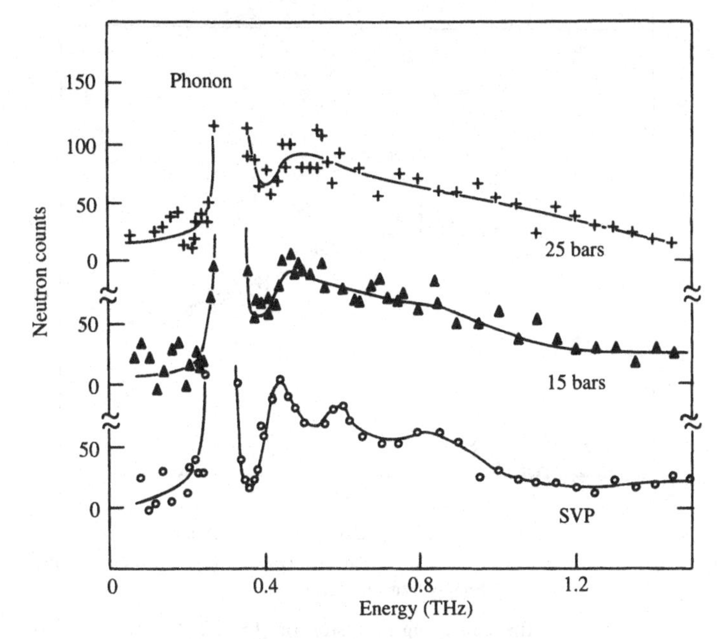

Fig. 7.9. The pressure dependence of the high-energy component of $S(\mathbf{Q},\omega)$ at $Q = 1.13$ Å^{-1} and $T = 1.27$ K [Source: Stirling, 1985].

a strong multiparticle resonance at $\sim$ 0.45 THz at $p = 15$ bar. This suggests that the temperature-independent high-energy spectrum shown in Fig. 7.8 would probably exhibit considerably more structure as the temperature decreases in higher-resolution data.

Above T_λ, it seems difficult not to interpret the broad high-energy distribution at $Q = 1.1$ Å^{-1} as a strongly damped zero sound mode. Thus the collisionless density mode which was well defined at $Q = 0.4$ Å^{-1} (Fig. 1.4) is now spread over a large energy range due to damping when we reach $Q \sim 1.1$ Å^{-1}, but it is still present. Strong evidence for this interpretation is that a simple extrapolation of the phonon dispersion relation cQ at 20 bar up to $Q = 1.13$ Å^{-1} predicts a peak centred at $\sim$ 0.6 THz (29 K). This prediction is in reasonable agreement with the observed peak at $\sim$ 0.5 THz (24 K). As in liquid ^{3}He, at such a large value of Q , the zero sound energy would be expected to have "softened" because of the decreasing strength of the self-consistent fields at short wavelengths.

In contrast to the high-pressure data, the SVP normal distribution (see Fig. 7.7) at $Q = 1.1$ Å^{-1} peaks at considerably *lower* frequencies

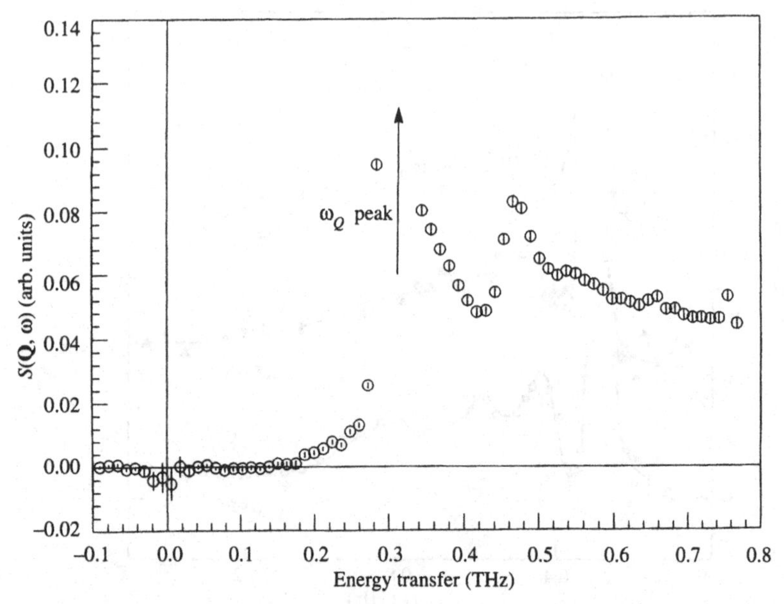

Fig. 7.10. Expanded view of the scattering intensity for $Q = 1.1$ Å^{-1}, at $T = 1.24$ K and $p = 15$ bar. The multiparticle peak at 0.45 THz is clearly visible [Source: Andersen, Stirling *et al.*, 1991].

$\sim$ 0.3 THz (14 K). This is again consistent with an extrapolation of the low-Q phonon dispersion relation to $Q = 1.1$ Å^{-1} using the reduced sound velocity at SVP. This sensitivity to pressure of the normal distribution peak frequency is, of course, exactly what one would expect when dealing with a (zero) sound mode. The fact that the normal distribution is peaked just above the emerging maxon quasiparticle peak (see Fig. 7.7(a)) makes it more difficult to separate out the zero sound mode scattering intensity as the temperature is lowered below T_λ at SVP than it is at high pressure.

What happens to this zero sound mode at $Q = 1.1$ Å^{-1} as we go below T_λ is somewhat complicated to follow because of the appearance of a multiparticle component (see Chapter 10) which overlaps the same frequency region. At low T ($\lesssim 1$ K), this high-energy multiparticle structure is increasingly visible while the scattering intensity from the broad normal zero sound peak appears to have largely vanished (see Figs. 1.6 and 7.7). Turning this around, we can say that as the temperature increases towards T_λ from below, scattering intensity associated with the zero sound mode steadily builds up, "submerging" the multiphonon resonances. In

Section 7.2, we argue that the zero sound intensity increases as a result of the decreasing strength of the condensate-induced hybridization, which is the origin of the weight of both the sharp quasiparticle peak and the pair-excitation spectrum in $S(\mathbf{Q},\omega)$ below T_λ.

The key point we are making is that the high-energy component to $S(\mathbf{Q},\omega)$ at intermediate values of $Q \sim 1.1$ Å^{-1} is some temperature-dependent mixture of multiphonon resonances associated with creating two quasiparticles (maxons and rotons) plus a broad zero sound component. The latter increasingly saturates the scattering intensity as T increases towards T_λ from below. We note that the zero sound frequency will be expected to show much more dispersion (i.e., Q-dependence) than the multiphonon resonances, which may allow their contributions to be separated.

The high-pressure maxon data in Fig. 7.8 appear to show the quasiparticle peak "growing out" of the normal distribution in a very clear fashion. As we have mentioned, the SVP data is less dramatic in this regard since the quasiparticle peak appears at a frequency just *slightly* below the normal distribution peak. However, careful study of the SVP data in Fig. 7.7(b) still shows a characteristic asymmetry of the rapidly growing quasiparticle peak which allows it to be distinguished from the normal distribution. This kind of asymmetry was first noted by Woods and Svensson (1978) in their pioneering study. It is also useful to compare the emerging line shape at 1.1 Å^{-1} in Fig. 7.7(a) with the line shape in the phonon region in Fig. 7.3. The qualitative difference between these two regions seems clear in such high-resolution data.

In Section 7.2, we interpret the sharp maxon peak observed *below* T_λ as a result of the neutron exciting a single atom out of the condensate, i.e., it is a pole of the single-particle Green's function $G(\mathbf{Q},\omega)$. We recall from (3.47) and (5.24) that $S(\mathbf{Q},\omega)$ can contain a term of the form $\Lambda(\mathbf{Q},\omega)G(\mathbf{Q},\omega)\Lambda(\mathbf{Q},\omega)$, where the Bose broken-symmetry vertex function $\Lambda(\mathbf{Q},\omega)$ depends on the condensate density n_0 (see Section 7.2 for more details). This immediately leads to the single-particle excitation peak at ω_Q having a finite weight in $S(\mathbf{Q},\omega)$ below T_λ, with a weight Λ^2 which *increases* with the value of n_0. We believe that these excitations exist *above* T_λ, but that they would not show up as a distinct peak in $S(\mathbf{Q},\omega)$ simply because the Bose vertex function $\Lambda(n_0)$ has vanished with n_0. The sharp maxon peak at ~ 0.3 THz corresponds to the same excitation branch as the roton peak at $Q \sim 2$ Å^{-1} discussed below, but we shall argue that it is quite different from the zero sound branch which dominates the spectrum at low Q.

Roton region

From the earliest days of inelastic neutron scattering, there has been special interest in studying the roton dispersion relation around the minimum shown in Fig. 1.3. At SVP, this minimum occurs at $Q_R = 1.925$ Å^{-1} and the formula in (1.2) is an excellent fit in the region ± 0.25 Å^{-1} around Q_R. The fact that the quasiparticle intensity $Z(Q)$ has a strong peak at Q (see Figs. 2.2 and 2.8) means that the roton scattering cross-section is especially large, allowing higher instrumental resolution to be used. The first high-resolution study of $S(\mathbf{Q}, \omega)$ for $Q \sim Q_R$ over a *wide* frequency range and at temperatures below and above T_λ was by Woods and Svensson (WS, 1978). New high-precision SVP data showing how the $S(\mathbf{Q}, \omega)$ line shape at Q_R changes near T_λ have been reported by Stirling and Glyde (1990). The most complete study is the time-of-flight data of Anderson, Stirling *et al.* (1991). In addition, there have been extensive studies of the effect of pressure on the roton spectrum (see Fig. 2.2). We mention, in particular, the older work of Dietrich, Graf, Huang and Passell (1972) and the more recent high-resolution study by Talbot, Glyde, Stirling and Svensson (TGSS, 1988) at $p = 20$ bar over the temperature range 1.29–2.97 K (at this pressure, $T_\lambda = 1.93$ K, $Q_R = 2.03$ Å^{-1} and the roton energy $\Delta(T = 1.3 \text{ K}) = 0.158$ THz (7.56 K)).

The SVP scattering intensity shown in Fig. 7.11 at 2.0 Å^{-1} clearly shows that there are *qualitative* changes in the line shape as one passes through T_λ. In Fig. 2.6, we show a contour plot of the scattering intensity in the roton wavevector region $Q = 2.0$ Å^{-1}. At low T, the observed intensity in Fig. 7.12 consists of a sharp peak at 0.16 THz (7.6 K). As T increases, the intensity in the sharp peak drops dramatically until it disappears at T_λ, just as in the maxon case $Q = 1.13$ Å^{-1}. As with the maxon, we interpret the sharp peak as scattering from a single-particle roton excitation ω_Q whose intensity in $S(\mathbf{Q}, \omega)$ is given by $\Lambda^2(Q_R, n_0)$. This single-particle excitation should also exist *above* T_λ but is no longer seen as a distinct peak in $S(\mathbf{Q}, \omega)$ since $n_0(T) = 0$ in the normal phase.

Above T_λ, we are left with a broad distribution, which is not very temperature-dependent (see Fig. 7.11). At very low temperatures ($T \sim 1.3$ K), we note that the high-pressure roton data of TGSS also exhibits a broad (at this level of resolution) multiparticle component peaked at ~ 0.5 THz (see Fig. 7.13).

In the data shown in Figs. 7.11–7.13, as well as in Fig. 2.6, there is no evidence that *normal* liquid ^{4}He supports even a remnant of a high-energy collective zero sound mode at $Q \sim 2$ Å^{-1}. Further evidence for

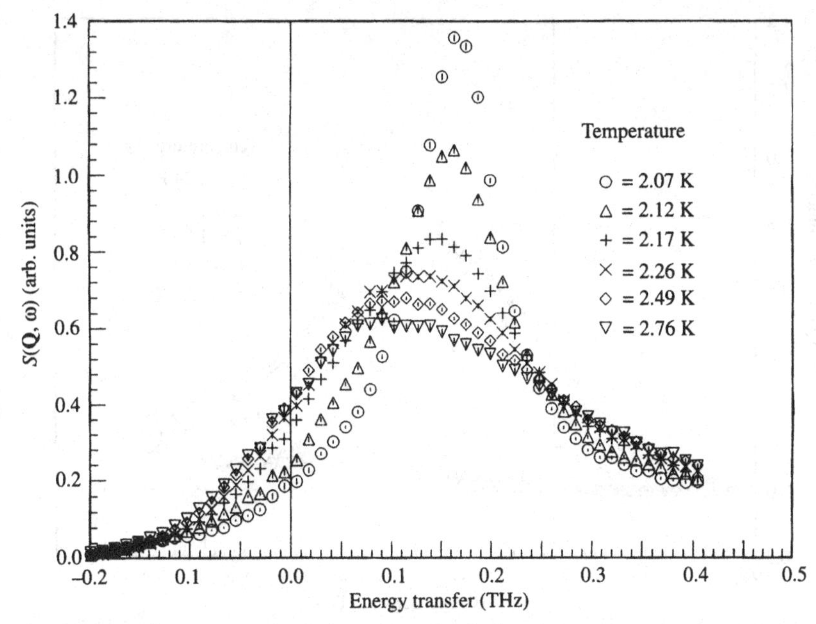

Fig. 7.11. $S(\mathbf{Q},\omega)$ vs. frequency at $Q = 2.0$ Å^{-1}and SVP for temperatures well above and just below T_λ [Source: Andersen, Stirling *et al.*, 1991].

the lack of zero sound at these wavevectors comes from a comparison with the $S(\mathbf{Q},\omega)$ spectrum in normal ^{3}He, the only other liquid which exists in the temperature range 1–3 K. Normal liquid ^{3}He exhibits a well defined zero sound mode only for $Q \lesssim 1$ Å^{-1} (see Fig. 7.6). At substantially larger Q, RPA calculations of $\chi_{nn}(\mathbf{Q},\omega)$ show that it can be well approximated by the incoherent particle–hole spectrum given by the Lindhard function $\chi^0_{nn}(\mathbf{Q},\omega)$ of a Fermi gas of quasiparticles.

The $S(\mathbf{Q},\omega)$ spectra of *normal* liquid ^{4}He and ^{3}He are compared in Fig. 7.14. The broad distribution in normal ^{4}He for $Q \sim 2$ Å^{-1} is seen to be very similar to that in normal ^{3}He, as to both width and peak position. We also remark that these spectra are not that different from classical liquids at values of $Q \gtrsim 1$ Å^{-1}, where there is no longer evidence for the collective behaviour which is apparent at lower wavevectors.

As a crude estimate, one expects that the zero sound frequency will be given by $\omega_Q \sim (nV(Q)/m)^{1/2}Q$. Since the effective potential $V(Q)$ starts to decrease when $Q \gtrsim 1$ Å^{-1} (see, for example, Section 9.2), so will the zero sound frequency. This phenomenon has been studied in detail in connection with zero sound in liquid ^{3}He (Pines, 1985). Such

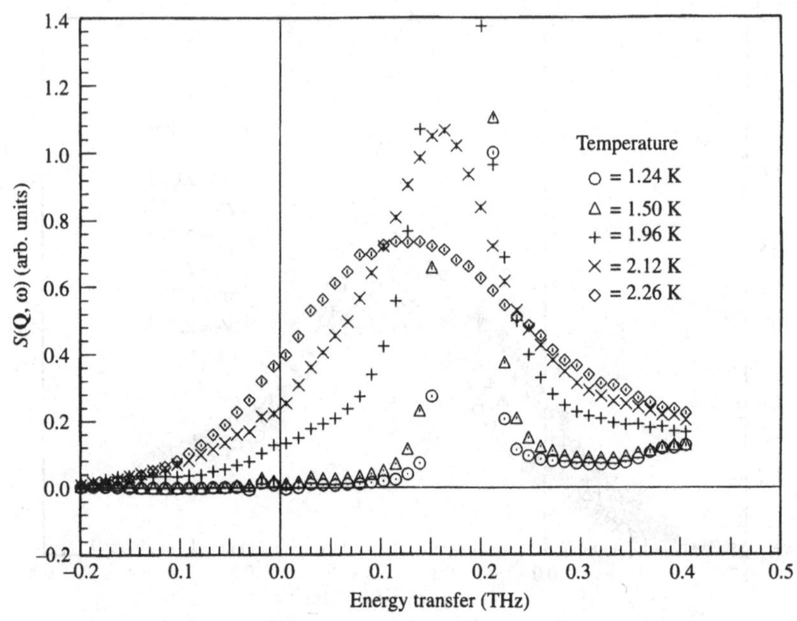

Fig. 7.12. $S(\mathbf{Q},\omega)$ vs. frequency in the roton wavevector region at $Q = 2.0$ Å^{-1} and SVP for temperatures well below and just above T_λ [Source: Andersen, Stirling *et al.*, 1991].

a decrease in the peak position of the normal distribution of liquid ^{4}He was observed (Fig. 7.15) at 4.2 K by Woods, Svensson and Martel (1975) but at such high temperatures, the line shape has significant distortions (recall that 4.2 K is the boiling temperature of liquid ^{4}He at a pressure of 1 bar). In Fig. 7.16, we estimate the peak position and width of the normal distribution *just* above T_λ using the recent ILL (Grenoble) results of Andersen, Stirling *et al.* (1991).

We note that the results in Figs. 7.15 and 7.16 have not had the effect of the detailed balance factor removed. We recall that at higher temperatures, the factor $[N(\omega)+1]$ in (2.20*b*) can lead to considerable difference between the observed $S(\mathbf{Q},\omega)$ and the more fundamental $\text{Im}\chi_{nn}(\mathbf{Q},\omega)$. If we could ignore instrumental broadening, the latter could be obtained by multiplying the $S(\mathbf{Q},\omega)$ data by $(1-e^{-\beta\omega})$.

Plots such as those in Figs. 7.15 and 7.16 are misleading, however. We believe there is a transition between the low $Q \lesssim 1$ Å^{-1} region (where zero sound arising from the near zero of the *denominator* of (5.80) dominates the scattering intensity) and the high-Q region, where the

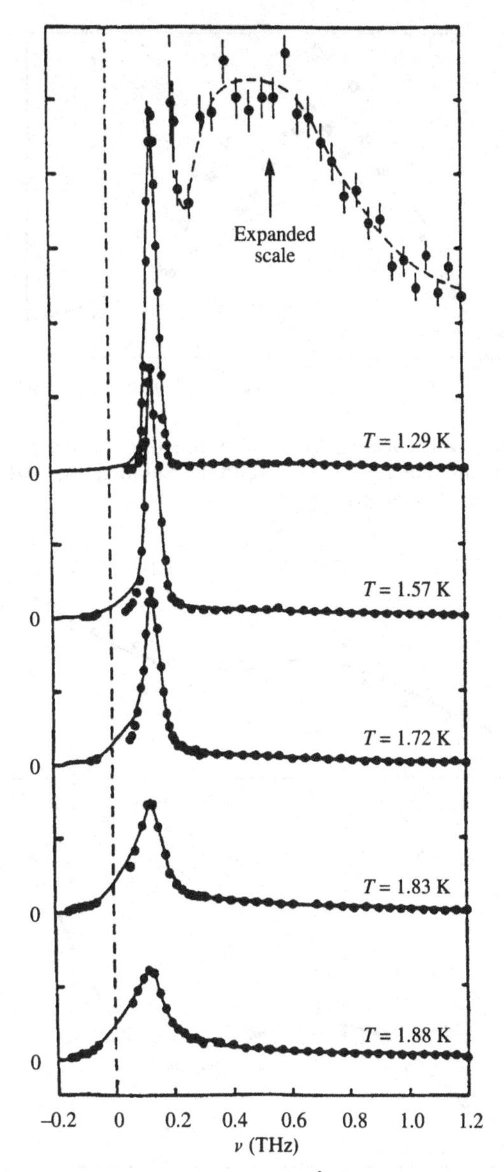

Fig. 7.13. $S(\mathbf{Q}, \omega)$ vs. frequency for $Q = 2.03$ Å^{-1}and $p = 20$ atm in the superfluid phase. The data at 1.29 K are shown on an expanded scale [Source: Talbot, Glyde, Stirling and Svensson, 1988].

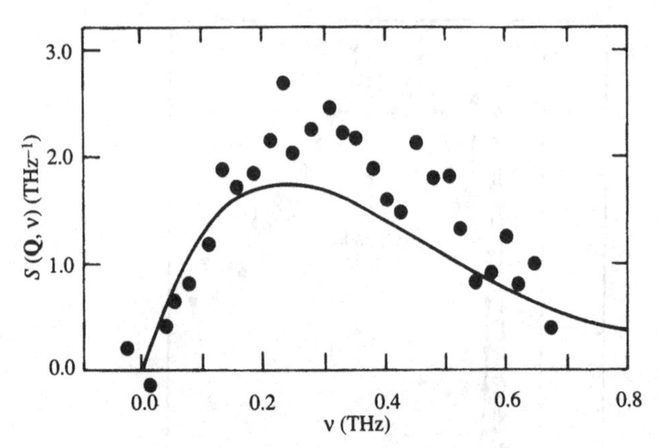

Fig. 7.14. Comparison between $S(\mathbf{Q}, \nu)$ as a function of the frequency ν at $Q = 2.0$ Å^{-1} for liquid ^{3}He at $T = 15$ mK (dots, Sköld *et al.*, 1976) and liquid ^{4}He at 4.2 K [Source: Woods, private communication].

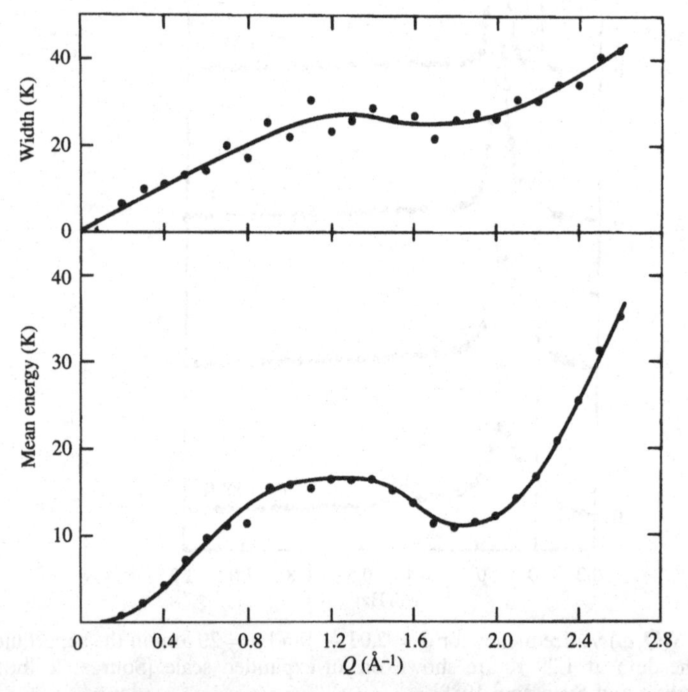

Fig. 7.15. The mean energy and the full width (at half maximum) of the peak in $S(\mathbf{Q}, \omega)$ at 4.2 K and SVP. It should be noted that, at this high temperature, the line shapes are quite asymmetrical [Source: Woods, Svensson and Martel, 1975].

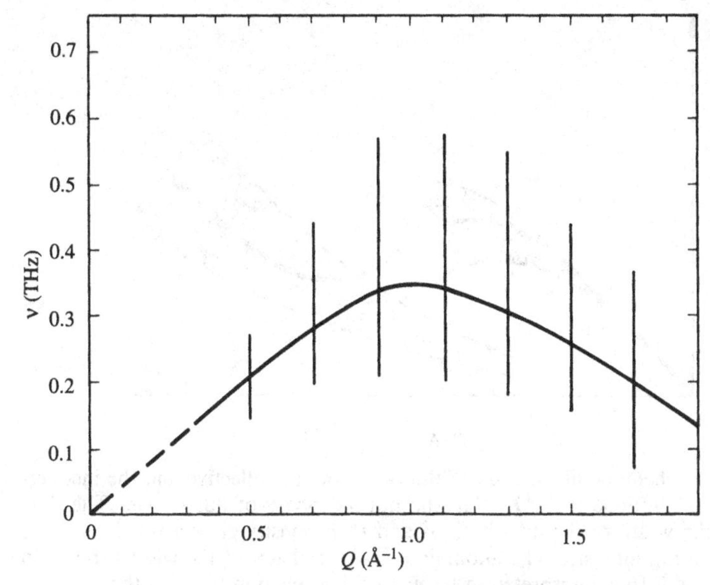

Fig. 7.16. Frequency of the peak position of the normal scattering intensity (T = 2.26 K, SVP) as a function of wavevector Q, just above the superfluid transition temperature. The total width at half maximum (including instrumental resolution) is also indicated. See also Fig. 7.15. These results are based on an analysis of unpublished ILL data [Source: Andersen, Stirling *et al.*, 1991].

contribution from the incoherent p–h spectrum (from the *numerator* of (5.80)) dominates. The fact that this transition is continuous hides the fact that the origin of the scattering intensity is completely different. Strong evidence for this transition is the result that, for $Q \gtrsim 1.3$ Å^{-1}, the width of the broad normal distribution *narrows* (see Figs. 7.15 and 7.16). This decrease in the width would be very puzzling if $S(\mathbf{Q}, \omega)$ was describing the same collective mode for both small and large Q. (We also note that in superfluid ^{4}He below T_λ, there is no such anomalous behaviour of the width of the maxon–roton excitation in the region around $Q \sim 1.5$ Å^{-1}. This is consistent with our interpretation of the maxon–roton mode as a single-particle-like excitation which is quite distinct from zero sound branch.) In Fig. 7.17, we attempt to illustrate the scattering intensity from the two different kinds of density fluctuations in the normal phase.

The same analysis is relevant to plots of the dispersion curve of the "peak" in $S(\mathbf{Q}, \omega)$ in normal liquid ^{3}He (see Fig. 7.18). The latter system has the advantage that one can clearly see how the scattering intensity from the p–h continuum crosses the remnant of the collective zero sound

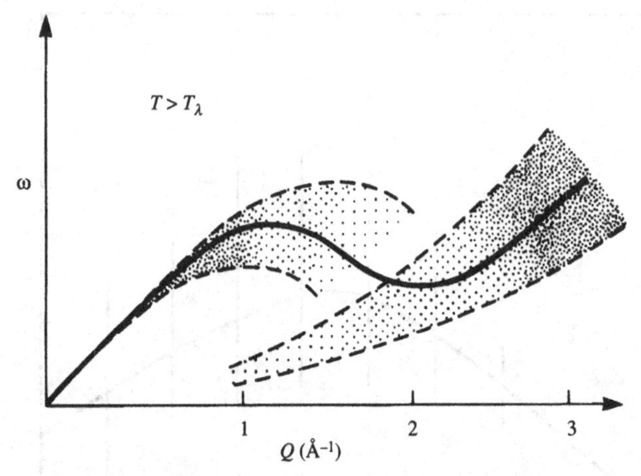

Fig. 7.17. A schematic illustration of the (zero sound) collective and the incoherent p–h contributions to $S(\mathbf{Q},\omega)$ in the normal phase of liquid ^{4}He. The dots represent the width of the distributions and their density gives a rough measure of the scattering intensity. The anomalous Q-dependence of the widths shown in Figs. 7.15 and 7.16 is interpreted in terms of this transition between the coherent zero sound region at lower Q and the incoherent p–h region at high Q.

mode, simulating a roton-like dip. Our conclusion is that a pseudo-"roton-like" depression (as in Fig. 7.15) in the dispersion relation of the peak position of $S(\mathbf{Q},\omega)$ may well appear in a non-Bose-condensed liquid, but its physical origin is completely different from the roton observed in superfluid ^{4}He.

The experimental data in the region just past the roton minimum, $2 \lesssim Q \lesssim 2.4$ Å^{-1}, are especially interesting. In Fig. 7.19, we show recent SVP data from ILL (Fåk and Andersen, 1991). As we have mentioned above, at $Q \simeq Q_R$, the normal distribution is peaked at a frequency just slightly *lower* than the roton peak which appears when we go below T_λ (see Fig. 7.11). However, we see that as Q approaches 2.4 Å^{-1}, the quasiparticle peak frequency rapidly increases towards the high-energy peak centred around 2 meV. We defer further analysis of this interesting intermediate-wavevector region to Section 7.2.

The neutron-scattering line shape changes very dramatically once we go past $Q \sim 2.4$ Å^{-1}. As we discuss in Chapter 10, this change is caused by the strong mixing of the one- and two-roton spectra in this region, where they overlap in energy. In the region $2.5 \lesssim Q \lesssim 3$ Å^{-1}, there is a low-intensity peak with an energy close to twice the roton energy (see Fig. 7.20). By $Q \sim 3$ Å^{-1}, however, most of the scattering intensity

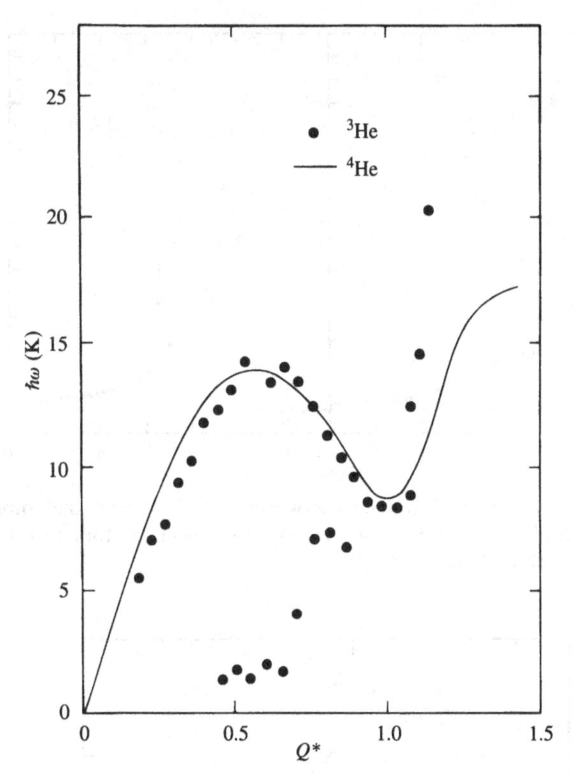

Fig. 7.18. A plot of the peak positions in $S(\mathbf{Q},\omega)$ vs. wavevector in liquid ^{3}He at 0.12 K and SVP (see also Figs. 7.5 and 7.6). The zero sound and incoherent particle–hole spectrum are seen to merge, giving rise to an apparent "roton-minimum". For comparison, the peak position in superfluid ^{4}He is shown. The wavevector Q^* is given in dimensionless units such that the static structure factor $S(\mathbf{Q})$ has its first maximum at $Q^* = 1$ [Source: Scherm *et al.*, 1989].

has gone into a broad temperature-independent distribution centred at much higher energy. This is approximately peaked at the recoil energy $\omega_R = Q^2/2m$, the free ^{4}He atom kinetic energy. In this high-momentum region, one is entering the region where the impulse approximation is increasingly valid (see Section 2.3 and Chapter 4).

Detailed studies of the broad peak in $S(\mathbf{Q},\omega)$ in the momentum region $3 \lesssim Q \lesssim 10$ Å^{-1} show clear evidence for oscillations in the position (relative to ω_R) and width as a function of Q (Martel *et al.*, 1976; Tanatar *et al.*, 1987; Stirling *et al.*, 1988). Martel *et al.* (1976) first suggested that these oscillations could be understood in terms of the impulse approximation, generalized to include collisional broadening. In

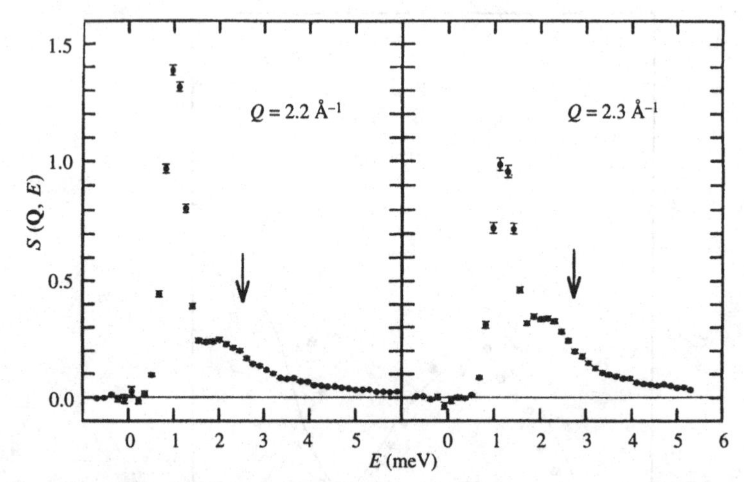

Fig. 7.19. $S(\mathbf{Q}, \omega)$ vs. energy transfer for wavevectors just past the roton minimum, at SVP and $T = 1.3$ K. The arrows show the free-atom recoil energy $Q^2/2m$ [Source: Fåk and Andersen, 1991].

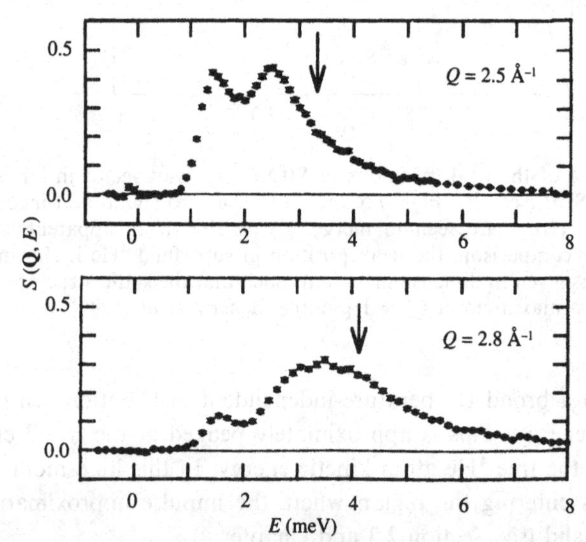

Fig. 7.20. $S(\mathbf{Q}, \omega)$ vs. energy transfer at SVP and $T = 1.3$ K, at wavevectors just above the cross-over with the two-roton continuum shown in Fig. 7.21 [Source: Fåk and Andersen, 1991].

their approximation, $S(\mathbf{Q},\omega)$ is given by (2.45), with $\Delta = 0$ and $\Gamma(\mathbf{Q})$ given by (2.46), corresponding to

$$\bar{\chi}_{nn}(\mathbf{Q},\omega) = \int \frac{d\mathbf{p}}{(2\pi)^3} \frac{n_p}{\omega - \varepsilon_Q - \frac{\mathbf{p}\cdot\mathbf{Q}}{m} + i\Gamma(\mathbf{Q})} . \tag{7.5}$$

The atomic collision cross-section $\sigma(Q)$ in (2.46) exhibits "glory oscillations" as a consequence of quantum interference effects associated with Bose statistics and these show up in $S(\mathbf{Q},\omega)$ through the broadening $\Gamma(\mathbf{Q})$ in (7.5). In terms of the general expression for the density response function χ_{nn} given by (5.2), this kind of analysis ignores the effect of the denominator (this is clear from the derivation given in Section 4.2). Stirling *et al.* (1988) and Tanatar *et al.* (1987) have pointed out that oscillations can also show up in the $S(\mathbf{Q},\omega)$ spectrum due to the effects of the interaction potential in the denominator of (5.2), in addition to those from collisional broadening as in (7.5). Whatever their point of origin, these oscillations are of considerable interest since they are a direct consequence of Bose statistics and deserve further study.

7.2 $S(\mathbf{Q},\omega)$: theoretical interpretation

We now turn to a more detailed interpretation of the $S(\mathbf{Q},\omega)$ data, using the microscopic theory of a Bose-condensed liquid. It is convenient to discuss the wavevector regions in *reverse* order, starting with large $Q \sim 2$ Å^{-1} (rotons). Our analysis will build on the ideas developed in Chapter 5 (summarized in Section 5.5) and in Section 6.3.

Roton region

In the presence of a condensate, the density fluctuation $\hat{\rho}(\mathbf{Q})$ naturally splits into two parts as in (3.13). The density-response function also naturally splits into two parts as given by (3.47) or (5.24),

$$\chi_{nn}(\mathbf{Q},\omega) = \sum_{\alpha,\beta} \Lambda_\alpha(\mathbf{Q},\omega) G_{\alpha\beta}(\mathbf{Q},\omega) \Lambda_\beta(\mathbf{Q},\omega) + \chi^R_{nn}(\mathbf{Q},\omega), \tag{7.6}$$

where Λ_α is the Bose broken-symmetry vertex function which vanishes with n_0. Here $\chi^R_{nn}(\mathbf{Q},\omega)$ is the density-response function associated with the non-condensate atoms and is defined by

$$\chi^R_{nn}(\mathbf{Q},\omega) = \frac{\bar{\chi}^R_{nn}(\mathbf{Q},\omega)}{1 - V(Q)\bar{\chi}^R_{nn}(\mathbf{Q},\omega)} \equiv \frac{\bar{\chi}^R_{nn}(\mathbf{Q},\omega)}{\epsilon^R(\mathbf{Q},\omega)} . \tag{7.7}$$

We also recall from (5.9) that $\Lambda_\alpha = \bar{\Lambda}_\alpha/\epsilon^R$. In simple physical terms, one may interpret the first term in (7.6) as a density fluctuation due to exciting a single atom out of the condensate, an interpretation going back to the work of Miller, Pines and Nozières (MPN, 1962) at $T = 0$. That the associated scattering intensity would vanish at T_λ was originally discussed on the basis of (7.6) by Griffin (1979a) as well as Griffin and Talbot (1981).

In (7.6), we have separated out the contribution ($\Lambda G \Lambda$) unique to a Bose-condensed liquid from the contribution expected to remain in the normal phase. This two-component form is especially useful at intermediate and high momentum, where there is no strong structure arising from a zero of ϵ^R in (7.7). This is ultimately a consequence of the fact that $V(Q)$ becomes very small at $Q \gtrsim 1.5$ Å^{-1}. This is the "weak-coupling" or single-particle (SP) region. As discussed in Section 7.1, for $Q \sim 2$ Å^{-1}, normal liquid ^{4}He does not exhibit any zero sound mode above T_λ. In this SP scenario, we can effectively set $\epsilon^R \simeq 1$ and approximate χ_{nn} in (7.6) by

$$\chi_{nn}(\mathbf{Q}, \omega) = \sum_{\alpha,\beta} \bar{\Lambda}_\alpha(\mathbf{Q}, \omega) G_{\alpha\beta}(\mathbf{Q}, \omega) \bar{\Lambda}_\beta(\mathbf{Q}, \omega) + \bar{\chi}^R_{nn}(\mathbf{Q}, \omega) \ . \tag{7.8}$$

In this wavevector and frequency region, it is implicit that $G_{\alpha\beta}$ will be quite similar to $\bar{G}_{\alpha\beta}$.

As suggested in Section 7.1, the first term in (7.8) can be naturally identified with the appearance of the sharp roton peak in $S(\mathbf{Q}, \omega)$ below T_λ, with the roton being associated with an intrinsic pole of the single-particle Green's function of liquid ^{4}He. Thus we are led to a very fundamental relation between the weight of the roton resonance and the value of the Bose vertex function $\bar{\Lambda}_\alpha(\mathbf{Q}, \omega)$, namely

$$Z(Q) \sim \bar{\Lambda}^2(Q, \omega_Q, n_0) \ . \tag{7.9}$$

This gives a microscopic basis for the disappearance of the roton peak as $T \to T_\lambda$, as first observed by Woods and Svensson (1978).

Strictly speaking, from the data, we are only justified in concluding that the rotons are poles of the single-particle Green's functions $G_{\alpha\beta}$ in the superfluid phase. However, the data clearly show that the roton peak position in $S(\mathbf{Q}, \omega)$ decreases only about 10–20% (depending on the pressure) as we approach T_λ. This weak temperature dependence seems to give strong indirect evidence that the roton exists *above* as well as below T_λ (i.e., it is an *intrinsic* single-particle excitation of normal liquid ^{4}He), even though its weight in $S(\mathbf{Q}, \omega)$ vanishes above T_λ. In physical terms, we are saying that the roton is a single-particle excitation above T_λ, but it has zero overlap with the density fluctuation spectrum measured by neutron scattering.

In addition to the sharp quasiparticle peak in the superfluid phase, there is a broader distribution which can be identified with $\bar{\chi}^R_{nn}$ in (7.8). Within a simple "bubble diagram" approximation for $\bar{\chi}^R_{nn}$ as in (3.44), we expect a low-energy broad distribution associated with scattering from thermally excited rotons, described by an expression like the second term in (3.45). The associated particle–hole excitation spectrum $\omega = \omega_{p+Q} - \omega_p$ is described by a modified "Lindhard function" for rotons, and thus we expect that the intensity will be concentrated at low energies $\omega < \omega_Q$ (see Fig. 7.17). If this is due to thermally excited rotons (which are the dominant excitations at temperatures above 1.2 K), one would expect the scattering intensity to decrease as the temperature is lowered. This feature is consistent with data such as in Figs. 7.11 and 7.12.

In addition to the low-energy particle–hole contribution just described (which grows smoothly into the normal distribution above T_λ), we expect the usual multiparticle contribution to Im $\bar{\chi}^R_{nn}$ in (7.8) which is associated with creating a pair of excitations (see third line of (3.45)). If these excitations involve rotons and maxons, this multiparticle contribution will be peaked at very high energies, as in Fig. 1.6. It can be expected to lose weight as the temperature increases, since it depends on the Bose-induced hybridization. This high-energy component is clearly visible in the data, concentrated at around $\sim$ 2 meV up to about $Q = 2.3$ Å^{-1} (see Fig. 7.19).

The data in the region $2.2 \leq Q \leq 2.8$ Å^{-1} shown in Figs. 7.19 and 7.20 require some care in interpretation. What is happening is that the quasiparticle peak energy is steadily increasing; and when it becomes more or less degenerate with the multiparticle component (we estimate this to occur at $Q \sim 2.4$ Å^{-1} where the slope of the quasiparticle dispersion relation suddenly changes), we have the hybridization which occurs whenever two excitation branches cross. This is shown schematically in Fig. 7.21. The main point is that while at $Q = 2.3$ Å^{-1} the high-energy component is associated with the pair-excitation spectrum, by the time we reach $Q = 2.8$ Å^{-1} it should be interpreted as due to the p–h spectrum discussed above. At even higher Q, this goes over to the impulse-approximation line shape described in Chapter 4. It would be very useful to have data such as in Figs. 7.19 and 7.20 at a series of temperatures, including above T_λ. We expect the p–h spectrum to increase steadily with temperature below T_λ with a dependence related to the

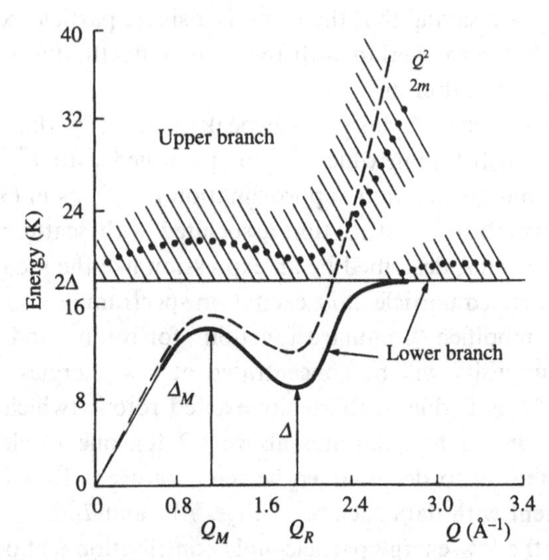

Fig. 7.21. The dashed line represents schematically the bare single-particle spectrum such as assumed in the analysis of Ruvalds and Zawadowski (1970). As discussed in Section 10.2, the condensate-induced hybridization of this sharp single-particle excitation with the broad pair-excitation branch (here represented by the two-roton continuum) leads to renormalized branches which transform into each other. The lower branch (solid line) is the (observed) maxon–roton dispersion relation, which goes over into the broad pair-excitation branch for $Q \gtrsim 2.5$ Å^{-1}. Similarly the upper pair-excitation branch transforms into the particle–hole branch, peaked at the recoil or single-particle energy $Q^2/2m$ at high Q. The shaded areas (centred at the black dots) are a schematic representation of this renormalized upper band [Source: adapted from Bedell, Pines and Zawadowski, 1984].

number of rotons present. In contrast, the weight of the pair-excitation spectra in $S(\mathbf{Q}, \omega)$ should decrease with the condensate density $n_0(T)$.

An adequate theory of the momentum region $2.3 < Q < 2.6$ Å^{-1} will involve extending the kind of analysis given by Glyde and Griffin (1990) to include the effects of the pair-excitation branch shown in Fig. 7.21, i.e., the Pitaevskii–Ruvalds–Zawadowski analysis discussed in Section 10.2. It follows that any discussion of this wavevector region is inadequate if it ignores the existence of the two-roton branch. In Fig. 9 of by Glyde (1992a) (see also Fig. 3 of Fåk and Andersen, 1991), the high-energy two-roton peak at $Q = 2.3$ Å^{-1} (shown in Fig. 7.19) is identified with the low-temperature remnant of the normal scattering intensity above T_λ. In Fig. 10 of Glyde (1992a), the low-energy two-roton peak at

$Q = 2.5$ Å^{-1}(shown in Fig. 7.20) is interpreted as the single-particle branch in the Glyde–Griffin picture.

It is clear that in our SP scenario, the expression in (7.8) is the analogue of the impulse approximation (4.2) which becomes valid at much higher momentum. We argue that the strong roton resonance which appears in $S(\mathbf{Q},\omega)$ as we go below T_λ is precisely the elusive "single-particle peak" which has been the object of so many experimental studies in the high-Q region (see Chapter 4). While the simple impulse approximation is not valid at values of Q as low as 2 Å^{-1}, (7.8) can be viewed as a generalized version of it. One reason that the roton resonance is so visible is that, as T goes below T_λ, it becomes extremely sharp. Moreover, we note that the instrumental line width typically increases linearly with Q and, as a result, the single-particle resonance no longer appears as a distinct peak sitting on a broad (Doppler-broadened) background. See Section 4.3 for further discussion.

A separate reason why the roton is so visible is that its weight $Z(Q_R)$ in $S(\mathbf{Q},\omega)$ is of order unity at low temperatures (Fig. 2.2). In contrast, at even larger values of $Q \gtrsim 3$ Å^{-1}, the single-particle peak in $S(\mathbf{Q},\omega)$ is predicted to have a weight given by n_0/n (see (4.2)). This reduction in the scattering intensity by a factor of 10 adds to the experimental difficulties of detecting the single-particle excitation in $S(\mathbf{Q},\omega)$ at larger wavevectors.

The essentials of this SP scenario for the roton data were first suggested in the classic paper by Miller, Pines and Nozières (MPN, 1962), although their analysis was at $T = 0$. Indeed, MPN already suggested that the weight of the roton resonance which appears in $S(\mathbf{Q},\omega)$ in the superfluid phase might be a *direct* measure of the condensate density n_0. (This idea was later extended by Hohenberg and Platzman, 1966, to much larger values of Q, as we discuss in Chapter 4.) MPN argued that the ratio

$$f(Q) = Z(Q)/S(Q) \tag{7.10}$$

should be a measure of the probability of exciting an atom "out of the condensate." At $Q \sim 2.5$ Å^{-1}, this expression should reduce to n_0/n, since at such large values of Q and ω, various Bose coherence factors should have disappeared (i.e., $u_Q = 1, v_Q = 0$). MPN suggested that the experimental value $f(Q) \sim 0.1$ at $Q \sim 2.5$ Å^{-1} (see Fig. 2.8) was consistent with the value of n_0/n at low temperatures predicted by Penrose and Onsager (1956). Strong hybridization effects with the two-roton spectra which start at about 2.4 Å^{-1} mean that the ratio in

(7.10) loses meaning at wavevectors larger than this (see Fig. 7.21 and Section 10.2).

A subject well worth further study is the rapid change in $Z(Q)$ shown in Fig. 2.1 as one goes from $Q = 1$ Å^{-1} to 2 Å^{-1}. In the scenario we have been discussing, based on the hybridization of single-particle excitations and density fluctuations, it is clear that the quasiparticle weight shown in Fig. 2.1 is describing a fairly complicated situation. Earlier, we noted that in the roton region, the quasiparticle weight $Z(Q)$ is related to the Bose broken-symmetry vertex function by (7.9). This relation is based on the validity of (7.8). In the maxon intermediate-wavevector region, the condensate-induced hybridization would seem to preclude the validity of any simple relation like (7.9).

Fig. 12 of Svensson (1991) shows a schematic decomposition of $Z(Q)$ into zero sound and single-particle components, based on the Glyde–Griffin picture. Relative to the free-atom energy $Q^2/2m$, the maxon is a high-energy excitation and the roton is a low-energy excitation. Griffin and Svensson (1990) have used these facts to argue that $Z(Q_M) \sim n_0/n$ while $Z(Q_R) \sim 1$ (which is consistent with the data in Fig. 2.1). However their arguments are qualitative and more detailed studies are needed. The main point to remember is that the Bose vertex functions play a crucial role in determining the strength of the single-particle excitations in $S(\mathbf{Q}, \omega)$ and this strength depends very much on both $n_0(T)$ and the wavevector region of interest.

Phonon–maxon region

In Section 7.1, we noted that the scattering intensity for intermediate values of $Q \sim 1$ Å^{-1} contained a broad distribution both above and below T_λ. We argued that this was partly associated with a strongly damped zero sound mode. This broad component was in addition to a sharp resonance which only appeared *below* T_λ. These data can be naturally understood on the basis of (7.6), the sharp resonance being associated with an intrinsic pole of $G_{\alpha\beta}(\mathbf{Q}, \omega)$ overlapping with a zero sound mode arising from a zero of the real part of $\epsilon^R(\mathbf{Q}, \omega)$ in (7.7). This is the remnant of the sharp zero sound phonon mode which dominates $S(\mathbf{Q}, \omega)$ at lower values of Q.

In our present scenario, the maxon or intermediate-Q region is especially interesting since it is a transition region between the ZS collective region at low Q (where both $G_{\alpha\beta}$ and χ_{nn} are dominated by a sharp zero sound ω_1 pole) and the single-particle region at high Q (where both

$G_{\alpha\beta}$ and χ_{nn} are dominated by a renormalized single-particle roton ω_2 excitation). We recall from the general discussion in Chapter 5 (for a summary, see Section 5.5) that if both $G_{\alpha\beta}$ and χ_{nn} have well defined poles above T_λ ($\bar{\omega}_2$ and $\bar{\omega}_1$, respectively), these excitations will be hybridized below T_λ and both will appear in $G_{\alpha\beta}$ and χ_{nn} with finite weight. Section 5.2 illustrates this with a dilute Bose gas calculation. The frequencies of these two renormalized modes (ω_1 and ω_2) are given by the zeros of the function $C(\mathbf{Q},\omega)$ in (5.22).

As is evident from the discussion in Section 7.1, a zero sound mode exists in the normal phase scattering intensity at low and intermediate Q values (estimated to be $\lesssim 1.3$ Å^{-1}). In this case, the two terms in (7.6) or (5.24) are strongly coupled through the Bose vertex functions since $\Lambda_\alpha = \bar{\Lambda}_\alpha/\epsilon_R$. It is then more useful to work with the alternative (but equivalent) form (see (5.2) or (5.76))

$$\chi_{nn}(\mathbf{Q},\omega) = \frac{\bar{\chi}_{nn}(\mathbf{Q},\omega)}{1 - V(Q)\bar{\chi}_{nn}(\mathbf{Q},\omega)} \equiv \frac{\bar{\chi}_{nn}(\mathbf{Q},\omega)}{\epsilon(\mathbf{Q},\omega)} , \tag{7.11}$$

where

$$\bar{\chi}_{nn}(\mathbf{Q},\omega) = \sum_{\alpha,\beta} \bar{\Lambda}_\alpha(\mathbf{Q},\omega)\bar{G}_{\alpha\beta}(\mathbf{Q},\omega)\bar{\Lambda}_\beta(\mathbf{Q},\omega) + \bar{\chi}^R_{nn}(\mathbf{Q},\omega) . \tag{7.12}$$

As discussed in Section 5.1, one can prove that if $n_0 \neq 0$, the poles of $G_{\alpha\beta}(\mathbf{Q},\omega)$ are also given by the zeros of $\epsilon(\mathbf{Q},\omega)$, with a weight governed by the Bose vertex functions $\bar{\Lambda}_\alpha$. Here $\bar{G}_{\alpha\beta}(\mathbf{Q},\omega)$ is the single-particle Green's function *without* the reducible self-energy contributions $\Sigma^c_{\alpha\beta} = V(Q)\bar{\Lambda}_\alpha\bar{\Lambda}_\beta/\epsilon_R$, as given in (5.75). Above T_λ, we have $G_{\alpha\beta} = \bar{G}_{\alpha\beta}$.

In physical terms, our scenario can be summarized as follows. The phonon and roton regions of the quasiparticle spectrum involve quite distinct excitation branches. The low-energy phonon is a zero sound mode involving the effective fields associated with the condensate as well as the thermally excited quasiparticles. The high-energy roton, in contrast, is viewed as a renormalized atomic-like excitation associated with the normal liquid. This picture inevitably implies that the intermediate maxon region will involve some sort of hybridization where the phonon and roton branches cross, producing the observed continuous quasiparticle spectrum. In order to illustrate this scenario (first suggested by Glyde and Griffin, 1990), we now consider a schematic parameterization of the regular functions in (7.11) and (7.12). Our interest is to introduce a simple model which incorporates the essential physics of the above scenario. Suitably generalized, such parameterizations may be useful as a basis for fits to the $S(\mathbf{Q},\omega)$ line-shape data.

In order to calculate $S(\mathbf{Q},\omega)$ using (7.11) and (7.12), we recall from (5.79) that Im χ_{nn} is directly proportional to Im $(1/\epsilon)$, where

$$\epsilon(\mathbf{Q},\omega) = 1 - V(\mathbf{Q})\bar{\chi}^{c}_{nn}(\mathbf{Q},\omega) - V(\mathbf{Q})\bar{\chi}^{R}_{nn}(\mathbf{Q},\omega) \ . \tag{7.13}$$

Thus the resonances in $S(\mathbf{Q},\omega)$ will be directly related to the zeros of $\mathrm{Re}\,\epsilon(\mathbf{Q},\omega)$. In order to proceed, we must introduce approximations for $\bar{G}_{\alpha\beta}$, $\bar{\Lambda}_{\alpha}$ and $\bar{\chi}^{R}_{nn}$ in (7.12) which are consistent with the underlying microscopic constraints discussed in Chapter 5 and Section 6.3.

We start with a two-pole ansatz for $\bar{G}_{\alpha\beta}$ in (7.12), given schematically by

$$\bar{G}_{\alpha\beta} = \frac{Z_1}{\omega^2 - \bar{\omega}^2_{s1}} + \frac{1 - Z_1}{\omega^2 - \bar{\omega}^2_{s2}} \ . \tag{7.14}$$

Here $\bar{\omega}_{s2}$ is a high-energy single-particle excitation which is assumed to be present in both the normal and superfluid phases of liquid ^{4}He. At large Q, this excitation is expected to be quite close to the observed maxon–roton quasiparticle frequency ω_Q. At lower $Q \lesssim 1$ Å^{-1}, nothing is really known about the dispersion relation of this high-energy excitation. The first term in (7.14) describes the low-energy, long-wavelength excitation $\bar{\omega}_{s1}$ in $\bar{G}_{\alpha\beta}$ which is explicitly associated with the Bose broken symmetry, as discussed in Section 6.3. This first term arises due to the appearance of a condensate mean field, which increasingly loses its effectiveness at larger wavevectors as well as higher temperatures. Little is known about the behaviour of the first term in (7.14) outside the low-Q, low-T region. Hypothetical dispersion relations of the single-particle branches $\bar{\omega}_{s1}$ and $\bar{\omega}_{s2}$ are sketched in Fig. 7.22.

Using the simple ansatz for $\bar{G}_{\alpha\beta}$ given in (7.14), one is led to parameterizing the condensate contribution $\bar{\chi}^{c}_{nn}$ in (7.12) and (7.13) by

$$V(\mathbf{Q})\bar{\chi}^{c}_{nn}(\mathbf{Q},\omega) = \frac{A_{c1}}{\omega^2 - \bar{\omega}^2_{s1}} + \frac{A_{c2}}{\omega^2 - \bar{\omega}^2_{s2}} \ . \tag{7.15}$$

The weights A_{c1} and A_{c2} are both proportional to the square of the Bose vertex function $\bar{\Lambda}_{\alpha}$ and hence vanish as $T \to T_{\lambda}$. For simplicity, we leave out any explicit reference to damping in (7.14) and (7.15). The contribution of the first term in (7.15) was not included by Glyde and Griffin (1990). It will be seen to be crucial in understanding why the phonon mode at low Q has a temperature-independent velocity in superfluid ^{4}He.

Finally, we turn to the second or "normal" contribution in (7.12) and

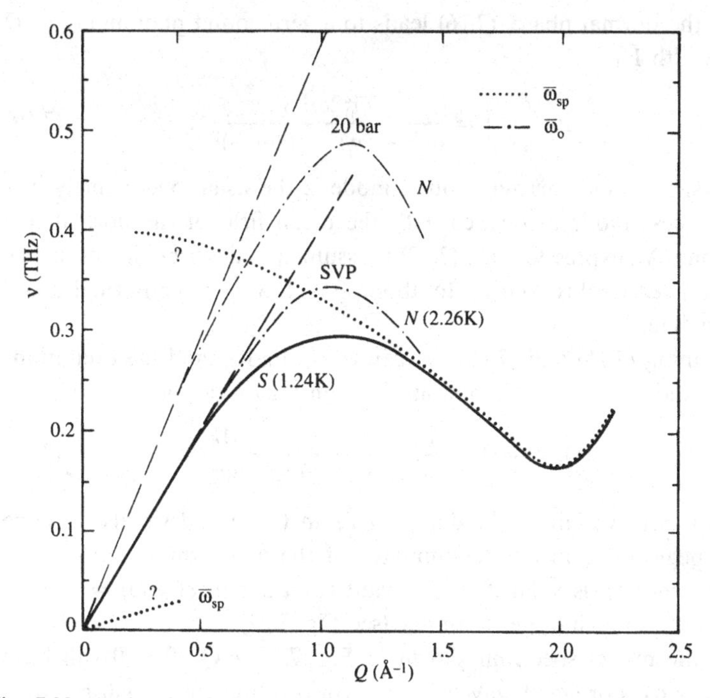

Fig. 7.22. A sketch of one model for the excitation branches associated with $G(\mathbf{Q},\omega)$ and $\chi_{nn}(\mathbf{Q},\omega)$ in the normal phase. The frequencies of the peaks in $S(\mathbf{Q},\omega)$ at 1.24 K and 2.26 K are taken from ILL data at SVP (Andersen, Stirling *et al.*, 1991). The normal distribution at 20 bar is schematic since the only data point is at 1.1 Å^{-1} (Talbot *et al.*, 1988). As indicated in Figs. 7.17–7.19, $S(\mathbf{Q},\omega)$ in the normal phase is quite broad. The high-Q maxon–roton single-particle excitation $\bar{\omega}_{s2}$ associated with the normal phase is extrapolated to low Q "by hand." The lower energy $\bar{\omega}_{s1}$ acoustic branch associated with the condensate is also indicated (see Section 6.3). The high-energy multiparticle branches (see Fig. 7.21) are not shown here.

introduce the traditional approximation used in discussing zero sound,

$$V(\mathbf{Q})\bar{\chi}^R_{nn}(\mathbf{Q},\omega) = \frac{A_R}{\omega^2 + 2i\omega\bar{\Gamma}_0} . \tag{7.16}$$

In the normal phase (where $\bar{\chi}^c_{nn} = 0$), the zero of (7.13) is given by

$$\omega^2 = A_R , \tag{7.17}$$

where the f-sum rule (see (2.24) and Section 8.1) gives

$$A_R = \frac{nV(Q)}{m}Q^2 \equiv c^2Q^2 . \tag{7.18}$$

Thus in the normal phase, (7.16) leads to a zero sound phonon $\bar{\omega}_0 = cQ$ of half-width $\bar{\Gamma}_0$,

$$-\operatorname{Im}\left(\frac{1}{\epsilon}\right) = \frac{\bar{\omega}_0^2 2\omega\bar{\Gamma}_0}{(\omega^2 - \bar{\omega}_0^2)^2 + (2\omega\bar{\Gamma}_0)^2} \ . \tag{7.19}$$

The physics behind this zero sound mode is the usual one, namely it is a collisionless mode associated with the mean field of the normal p–h excitations. An expression like (7.19) is assumed to describe the resonance in $S(\mathbf{Q}, \omega)$ reasonably well up to about $Q \sim 1$ Å^{-1} in the normal phase of liquid ^{4}He.

Combining (7.15) and (7.16), we can try to understand the excitations in superfluid ^{4}He in the phonon and maxon region on the basis of

$$\epsilon(\mathbf{Q}, \omega) = 1 - \frac{A_{c1}}{\omega^2} - \frac{A_{c2}}{\omega^2 - \bar{\omega}_{s2}^2} - \frac{A_R}{\omega^2} \ . \tag{7.20}$$

For simplicity, we ignore the damping $\bar{\Gamma}_0$ in (7.16) and we also assume we can neglect $\bar{\omega}_{s1}$ in the denominator of the first term in (7.15). This latter assumption is valid if the relevant solutions of $\epsilon(\mathbf{Q}, \omega) = 0$ occur at frequencies much larger than $\bar{\omega}_{s1}$ (see Fig. 7.22).

With the model spectrum shown in Fig. 7.22, $\epsilon(\mathbf{Q}, \omega) = 0$ will have two solutions. For small wavevectors, hybridization effects will be small and the excitation branches are given by

$$\left.\begin{aligned} \omega_0^2 &\simeq A_{c1} + A_R \ , \\ \omega_{sp}^2 &\simeq \bar{\omega}_{s2}^2 \ . \end{aligned}\right\} \tag{7.21}$$

While we expect A_{c1} and A_R to be individually quite temperature-dependent in the Bose-condensed phase, their sum should be relatively temperature-independent. This can be seen using the f-sum rule, which requires that

$$A_{c1} + A_R = c^2 Q^2 - A_{c2} \ , \tag{7.22}$$

where c is defined in (7.18). The observed phonon velocity is essentially constant in the superfluid and normal phases of superfluid ^{4}He. Within the dielectric formalism, this fact is naturally understood in terms of the phonon mode being zero sound arising from contributions from both the condensate ($\bar{\chi}_{nn}^c$) and regular ($\bar{\chi}_{nn}^R$) parts of $\bar{\chi}_{nn}$. As the temperature increases, the decreasing weight of the condensate associated with A_{c1} is compensated by the increasing strength of the normal fluid mean field A_R. We have already alluded to this picture in Section 5.2 (see (5.50)).

The results of Gavoret and Nozières (1964) reviewed in Section 6.3

imply that, at $T = 0$ and in the limit of small Q, both A_R and A_{c2} vanish. In this case, (7.22) reduces to $A_{c1} = c^2Q^2$ and the $\bar{\omega}_{s2}$ mode in (7.20) would have zero weight in $S(\mathbf{Q},\omega)$. Thus it would seem that the high-energy peak shown in Figs. 7.1 and 7.2 is not evidence for the $\bar{\omega}_{s2}$ mode in (7.20) but is rather related to a pair excitation (as discussed in Chapter 10). The precise nature of this high-energy peak at low Q deserves further study.

Within the context of the preceding discussion, one might attempt a unified analysis of the phonon and maxon line-shape data based on

$$\epsilon(\mathbf{Q},\omega) = 1 - \frac{A_{c2}}{\omega^2 - \bar{\omega}_{s2}^2 + i\omega\bar{\Gamma}_s} - \frac{\bar{\omega}_0^2 - A_{c2}}{\omega^2 + i\omega\bar{\Gamma}_0}\,, \tag{7.23}$$

where $\bar{\omega}_0$ and $\bar{\omega}_{s2}$ are sketched in Fig. 7.23. This kind of simplified expression was originally used by Griffin and Payne (1986) in the low-Q region and later by Glyde and Griffin (1990) to discuss hybridization effects in the maxon wavevector region. It is clear that in the cross-over region $\bar{\omega}_{s2} \sim \bar{\omega}_0$ there will be the usual hybridization and level repulsion. The lower-energy branch is interpreted as the observed phonon–maxon–roton dispersion relation. One cannot expect a very realistic treatment of the upper hybridized branch on the basis of the simplified expression in (7.23) since we have ignored the pair-excitation spectrum (see Chapter 10). This multiparticle spectrum would itself hybridize and strongly renormalize the upper branch predicted by (7.23), assuming the sort of input spectrum illustrated in Fig. 7.23.

Starting from the parameterized form (7.23), one finds

$$S(\mathbf{Q},\omega) \propto -\mathrm{Im}\left(\frac{1}{\epsilon}\right) = \frac{\tilde{\omega}_0^2 2\omega\bar{\Gamma}_0(\omega^2 - \bar{\omega}_{s2}^2)^2}{(\omega^2 - \omega_0^2)^2(\omega^2 - \omega_{s2}^2)^2 + [2\omega\bar{\Gamma}_0(\omega^2 - \tilde{\omega}_{s2}^2)]^2}\,. \tag{7.24}$$

Here ω_0 and ω_{s2} are the renormalized zero sound and single-particle excitations, given by the zeros of

$$\mathrm{Re}\,\epsilon(\mathbf{Q},\omega) = (\omega^2 - \tilde{\omega}_0^2)(\omega^2 - \tilde{\omega}_{s2}^2) - A_{c2}\tilde{\omega}_0^2 \tag{7.25}$$

and

$$\left.\begin{aligned} \tilde{\omega}_0^2 &\equiv \bar{\omega}_0^2 - A_{c2}\,, \\ \tilde{\omega}_{s2}^2 &\equiv \bar{\omega}_{s2}^2 + A_{c2}\,. \end{aligned}\right\} \tag{7.26}$$

This is a slightly generalized version of the expression introduced by Glyde and Griffin (1990). If the hybridization is weak ($\tilde{\omega}_0 \not\simeq \tilde{\omega}_{s2}$), the

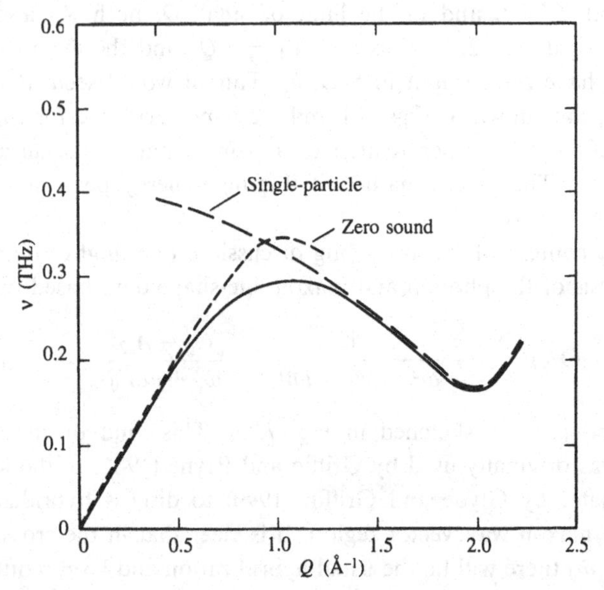

Fig. 7.23. The hybridized excitation spectrum in superfluid ^{4}He as given by (7.23)–(7.25), based on the general scenario introduced by Glyde and Griffin (1990). The uncoupled spectra ($T > T_\lambda$) for the zero sound ($\tilde{\omega}_0$) and the high-energy single-particle ($\tilde{\omega}_{s2}$) branches are shown. For simplicity, the widths are left out.

approximate solutions for the zeros of (7.25) are given by

$$\omega^2 = \begin{cases} \tilde{\omega}_0^2 + \dfrac{A_{c2}\tilde{\omega}_0^2}{\tilde{\omega}_0^2 - \tilde{\omega}_{s2}^2}, \\ \tilde{\omega}_{s2}^2 - \dfrac{A_{c2}\tilde{\omega}_0^2}{\tilde{\omega}_0^2 - \tilde{\omega}_{s2}^2}. \end{cases} \tag{7.27}$$

As long as A_{c2} is positive, we have $\omega_0 < \tilde{\omega}_0$ and $\omega_{s2} > \tilde{\omega}_{s2}$ for Q below the level-crossing wavevector Q^*; and $\omega_0 > \tilde{\omega}_0$ and $\omega_{s2} < \tilde{\omega}_{s2}$ for $Q > Q^*$. This shows the expected level repulsion and is illustrated in Fig. 7.23.

The precise line shape predicted by (7.24) is extremely dependent on what we take for the frequency and temperature dependence of Γ_0. Experimentally, the $S(\mathbf{Q}, \omega)$ data at low temperatures (say $T \lesssim 1.5$ K) indicate that the quasiparticle peak is extremely sharp. This is clear on the low-frequency side $\omega < \omega_Q$, where there is effectively no scattering intensity at low temperatures. At $\omega > \omega_Q$, the situation is more difficult to assess since, while the broad zero sound distribution present in the normal phase appears to be losing intensity as T decreases, at the same time there are multiparticle (pair) resonances which are predicted to develop increasing intensity in $S(\mathbf{Q},\omega)$ due to the Bose broken symmetry (see Chapter 10). We interpret the data shown in Fig. 7.7 as evidence that the zero sound scattering intensity is being suppressed at all frequencies as we go to lower temperatures in the superfluid phase. We have noted earlier that at $p = 20$ bar, the broad zero sound structure peak at 0.5 THz (see Fig. 7.8) seems to persist down at low temperatures. As we have noted earlier, however, we expect that multiparticle resonances would become evident in high pressure $S(\mathbf{Q},\omega)$ data if taken at higher resolution.

In summary, our basic scenario concerning how the $S(\mathbf{Q},\omega)$ line shape for $Q \sim 1$ Å^{-1} should change with decreasing temperature is as follows. In the normal phase, $S(\mathbf{Q},\omega)$ is dominated by a broad zero sound peak which is not very temperature-dependent. As one enters the superfluid phase, however, $S(\mathbf{Q},\omega)$ begins to exhibit new resonances associated with single-particle and pair-excitations, as a result of the condensate-induced hybridization. These new resonances increasingly dominate the spectral weight in $S(\mathbf{Q},\omega)$, at the expense of the original zero sound p–h mode. This is the essential content of (7.23), although we have not included the pair excitations in the latter expression. We view the condensate-induced hybridization of the low-energy zero sound branch as being completely analogous to the hybridization of the high-energy two-roton branch with the roton single-particle branch. The latter is illustrated in Fig. 7.21 and reviewed in Section 10.2. The analogue of the expression in (7.23) is given by (10.27) and (10.28).

A fruitful area of investigation is to see how well model expressions such as (7.23)–(7.26) can be fitted to the temperature-dependent $S(\mathbf{Q},\omega)$ line-shape data. Glyde (1992a) has done this starting from an expression for the dielectric function of the form (Δ is not the roton energy, as in the rest of this book)

$$\epsilon(\mathbf{Q},\omega) = 1 - \left(\frac{\Delta}{\omega^2 - \bar{\omega}_{\rm sp}^2 + i\,2\omega\bar{\Gamma}_{\rm sp}} + \frac{\alpha}{\omega^2 - \Omega^2 + i\,2\omega\bar{\Gamma}_0} \right) . \tag{7.28}$$

All six parameters in (7.28) depend on Q. Setting $\bar{\Gamma}_{\rm sp} = 0$, this gives a scattering intensity

$$\begin{aligned} S(\mathbf{Q},\omega) &\propto -{\rm Im}\,\chi_{nn}(\mathbf{Q},\omega) \\ &= -\frac{1}{V(\mathbf{Q})}{\rm Im}\frac{1}{\epsilon(\mathbf{Q},\omega)} \end{aligned}$$

$$= \frac{\alpha}{V(\mathbf{Q})} \frac{2\omega\bar{\Gamma}_0(\omega^2 - \bar{\omega}_{\rm sp}^2)^2}{\left[(\omega^2 - \tilde{\omega}_N^2)(\omega^2 - \tilde{\omega}_{\rm sp}^2) - \Delta\alpha\right]^2 + \left[2\omega\bar{\Gamma}_0(\omega^2 - \bar{\omega}_{\rm sp}^2)\right]^2}, \tag{7.29}$$

where

$$\left.\begin{aligned} \tilde{\omega}_{\rm sp}^2 &\equiv \bar{\omega}_{\rm sp}^2 + \Delta\,, \\ \tilde{\omega}_N^2 &\equiv \Omega^2 + \alpha\,. \end{aligned}\right\} \tag{7.30}$$

Above T_λ (where $\Delta = 0$), (7.29) simplifies to

$$S(\mathbf{Q},\omega) \propto -\frac{1}{V(\mathbf{Q})}\mathrm{Im}\frac{1}{\epsilon(\mathbf{Q},\omega)} = \frac{\alpha}{V(\mathbf{Q})}\frac{2\omega\bar{\Gamma}_0}{[\omega^2 - \tilde{\omega}_N^2]^2 + [2\omega\bar{\Gamma}_0]^2}\,. \tag{7.31}$$

If the mean energy Ω of the particle–hole excitations is set to zero, (7.29) reduces to (7.24).

The fitting procedure used by Glyde is based on the key simplifying assumption that only Δ changes with temperature (through its dependence on the condensate $n_0(T)$) in the superfluid phase. The other four parameters $\bar{\omega}_{\rm sp}$, Ω, α and $\bar{\Gamma}_0$ are assumed to be temperature-independent. In the normal phase ($T > T_\lambda$), line-shape fits to (7.31) determine $\tilde{\omega}_N$ and $\bar{\Gamma}_0$. The fact that the numerator of $S(\mathbf{Q},\omega)$ in (7.29) has a minimum at $\omega = \bar{\omega}_{\rm sp}$ can be used to estimate $\bar{\omega}_{\rm sp}$ from the low-temperature data. Finally, at low T, line-shape fits are used to determine the best values for Δ and α. At intermediate temperatures, $\Delta(T)$ is then allowed to vary in magnitude in order to simulate the condensate-induced hybridization. The results shown in Fig. 5 of Glyde (1992a) illustrate in a concrete way the scenario suggested by Glyde and Griffin (1990) in the maxon region $Q \sim 1.1$ Å^{-1} to explain data such as in Fig. 7.8.

Unfortunately, the simplifying assumption that only $\Delta(T)$ varies with temperature below T_λ appears to lead to unphysical values for some of the parameters in (7.28). In particular, for $Q = 1.1$ Å^{-1} at both SVP and 20 bar, the parameter fits which Glyde obtained imply that:

(a) α is negative. It is easily verified from (7.29) and (7.31) that this would require that $V(\mathbf{Q}) < 0$, otherwise $S(\mathbf{Q},\omega)$ would be negative. However, a negative $V(\mathbf{Q})$ in the maxon region is inconsistent with its positive value in the zero sound phonon region at slightly lower values of Q.

(b) The resonance frequency Ω is extremely large – in particular, $\Omega \simeq 1.1\tilde{\omega}_N$ and $\Omega > \bar{\omega}_{\rm sp}$. In contrast with the $\bar{\omega}_{\rm sp}$ pole of $\bar{\chi}_{nn}^c$ in (7.28), the physical origin of such a high-energy resonance in $\bar{\chi}_{nn}^R$ is unclear.

Moreover, in this parameterization, the peak in $S(\mathbf{Q},\omega)$ above T_λ (see (7.31)) would have its origin in this assumed resonance in $\bar{\chi}^R_{nn}$, rather than as a zero sound mode as in the Glyde–Griffin picture. This is most dramatically shown by the fact that at $T > T_\lambda$, the density fluctuation associated with $\bar{\chi}^R_{nn}$ by itself is given by

$$S_0(\mathbf{Q},\omega) \propto -\text{Im } \bar{\chi}^R_{nn}(\mathbf{Q},\omega) = \frac{\alpha}{V(\mathbf{Q})} \frac{2\omega\bar{\Gamma}_0}{(\omega^2 - \Omega^2)^2 + (2\omega\bar{\Gamma}_0)^2} \ . \quad (7.32)$$

When $\Omega \simeq 1.1\tilde{\omega}_N$, this is almost identical to the expression in (7.31).

The above discussion illustrates some of the difficulties in using (7.28) and (7.29) as the basis for a fit to experimental line shapes. It shows that one must allow other parameters besides $\Delta(T)$ to vary with temperature. In any fit to experimental data, one should also ensure that known constraints are satisfied:

(a) $S(\mathbf{Q},\omega)$ should satisfy the f-sum rule at all T, as the expression in (7.23) does.
(b) The static structure factor $S(\mathbf{Q})$, defined in (2.11), should be independent of temperature.
(c) At low T ($\sim$ 1 K), the intensity of the normal zero sound distribution appears to vanish, all of its spectral weight shifting into an extremely sharp single-particle mode at ω_Q (with $S(\mathbf{Q},\omega < \omega_Q) = 0$) as well as pair excitations at higher energies.

One simple way of proceeding would be to start with some reasonable assumption about the temperature dependence of the various parameters in (7.28), such as

$$\left.\begin{array}{lll} \Delta(T) = a(1-\tilde{N}) & , & \alpha(T) = \bar{\omega}_0^2 - \Delta(T) \ , \\ \bar{\Gamma}_0(T) = \Gamma_0\tilde{N} & , & \bar{\Gamma}_{\text{sp}} = \Gamma_{\text{sp}}\tilde{N} \ . \end{array}\right\} \quad (7.33)$$

Here the temperature-dependent parameter $\tilde{N}$ goes from $\tilde{N} = 1$ at $T \geq T_\lambda$ to $\tilde{N} = 0$ at $T = 0$. In an approximate way, one may think of $\tilde{N}$ as a measure of the "normal fluid" excitation density, with $\tilde{N} = 0.1$ corresponding to $T \sim 1$ K.

Further studies are clearly needed to find simple parameterized forms for $S(\mathbf{Q},\omega)$ such as (7.24) and (7.29) which fit the data and incorporate the general structure implied by the dielectric formalism. We note that it may be more advantageous to develop approximate forms for the regular longitudinal current response function $\bar{\chi}^{\ell}_{JJ}(\mathbf{Q},\omega)$ and use (5.81) to find $\epsilon(\mathbf{Q},\omega)$. The precise nature of the hybridization in the maxon wavevector region is complicated by the fact that we are dealing with strongly damped excitation branches at all but the lowest temperatures. In this connection, we note that the last term in (7.23) or (7.28) doesn't really describe Landau damping of normal zero sound (decay into two single-particle excitations). The width $\bar{\Gamma}_0$ in (7.28) describes the half-width of particle–hole excitations and Landau damping will still occur even if $\bar{\Gamma}_0 = 0$.

The scattering intensity in the roton region does not exhibit any evidence (see Fig. 7.12) for the kind of asymmetrical line shape associated with hybridization which is evident in the maxon region. This is consistent with our previous argument that in the roton region, (7.8) is a better starting point than (7.11) or (7.12). By the same token, expressions like (7.23) and (7.28) do not appear to be the appropriate starting points for understanding the line shapes in the roton wavevector region.

In our analysis, the simple expression (7.23) is introduced not as an *ad hoc* formula for data analysis but within the general conceptual picture summarized in Sections 5.5 and 6.3. A key role is played by the poles of $\bar{G}_{\alpha\beta}$ in (7.14). In our analysis, we were naturally led to assume that there may be two $\bar{\omega}_{\rm sp}$ branches: a long-wavelength, low-energy mode $\bar{\omega}_{s1}$ associated with the condensate (see Section 6.3), and a short-wavelength, high-energy mode $\bar{\omega}_{s2}$ already present in normal liquid ^{4}He and identified with the maxon–roton. Both branches are sketched in Fig. 7.22.

Line-shape results predicted by (7.23) clearly ignore any low-energy structure associated with the low-energy symmetry-restoring $\bar{\omega}_{s1}$ mode. As we discuss in Section 6.3, the precise dispersion relation of this mode and its continuation outside the long-wavelength two-fluid region is not that well understood at present, especially at finite temperatures ($T > 1$ K). Further theoretical studies are needed. In this connection, we note that Svensson and Tennant (1987) have looked for scattering intensity on the low-frequency side of the maxon resonance at low temperatures. Although they could have detected scattering as low as 10^{-3} of the maxon intensity at $Q = 1.1$ Å^{-1}, they found nothing. Svensson and Tennant have also noted that $S(\mathbf{Q},\omega)$ can exhibit *spurious* peaks in this low-frequency region at the free ^{4}He atom energy. These originate from higher-order Bragg-scattering contamination when the energy of the scattered neutron is measured.

Summary

To summarize this chapter, we have shown how the neutron-scattering line shape exhibited by superfluid ^{4}He over a wide range of Q, ω and T can be naturally understood in terms of a zero sound phonon at low Q and a single-particle maxon–roton at high Q, the latter appearing in $S(\mathbf{Q},\omega)$ through the hybridizing effect of the Bose condensate. At some intermediate Q value, say $Q_c \sim 0.7 - 0.8$ Å^{-1}, we predict that the sharp zero sound mode at low Q will go smoothly over into the maxon–roton branch, with the usual level repulsion and cross-over behaviour characteristic of hybridization when two modes cross (see Fig. 7.23). The dielectric formalism thus provides a natural microscopic basis for the *continuous* phonon–maxon–roton dispersion curve first postulated by Landau (1947), a curve which has always been puzzling given that the physical nature of phonons and rotons has been long recognized as being quite different (see also Section 12.1).

The work of Glyde and Griffin (1990) provides a new scenario built on microscopic theory which anchors the phonon–maxon–roton quasiparticle to the existence of a Bose condensate. In this picture, however, the maxon–roton excitation is viewed as essentially a normal phase elementary excitation. In addition, the zero sound oscillations in the non-condensate atoms play a crucial role in renormalizing the condensate fluctuations, leading to a temperature-independent phonon velocity. In an alternative scenario (see Sections 6.3 and 12.1), the entire phonon–maxon–roton quasiparticle spectrum in superfluid ^{4}He is viewed as a single branch associated with the oscillations of the condensate. While theoretically possible, this more traditional scenario does not seem to be compatible with the observed line-shape changes in superfluid ^{4}He as the temperature increases towards T_λ.

We have also introduced a simple model (7.23) which illustrates some of the features expected from the microscopic field-theoretic analysis. This model expression was constructed keeping the experimental data on $S(\mathbf{Q},\omega)$ in mind, especially as concerns the temperature-dependent changes. Clearly there are many ramifications of this new picture, which will require much further experimental and theoretical work to both confirm and/or elucidate. In particular, the simple parameterization in (7.23) of the two terms in (7.13) is meant only as an illustration (see Fig. 7.23) of the kind of temperature-dependent structure we might expect in $S(\mathbf{Q},\omega)$ due to the condensate coupling of SP and ZS modes. We emphasize, however, that this kind of hybridization is an inevitable consequence of the Bose broken symmetry and the dielectric formalism, as summarized in Section 5.5. Moreover, a proper parameterization at finite temperatures most naturally starts with the two terms in $\bar{\chi}_{nn}$ given by (7.13). Attempting to parameterize directly the two contributions in (7.6) is more difficult, since the physics of the hybridization is much less clearly exhibited. The Woods–Svensson (1978) parameterization (summarized in Section 8.2) suffers from this difficulty as it is based on (7.6) rather than (7.11) and (7.12). Thus it is incapable of reproducing the hybridization in the maxon region predicted by (7.24) and illustrated in Fig. 7.23.

In further studies, a fruitful avenue of research would be to develop simple parameterizations such as given in (7.24) and (7.29) in order to see how many of the features exhibited by $S(\mathbf{Q}, \omega)$ can be fitted. One needed generalization at higher energies would be to include a contribution to $\bar{\chi}^R_{nn}$ in (7.12) from the two-excitation continuum discussed in Chapter 10. In particular, this latter extension is crucial before one can discuss the line shapes in the interesting region around 2.4 Å^{-1} (see Figs. 7.19–7.21).

8

Sum-rule analysis of the different contributions to $S(\mathbf{Q}, \omega)$

The dielectric formalism gives the general structure of various correlation functions in a Bose-condensed fluid. However, quantitative calculations of the "regular" functions discussed in Chapter 5 are difficult even for a dilute Bose gas, let alone a Bose liquid like superfluid ^{4}He. As an aid to analysing neutron data and also in developing parameterizations of the dielectric formalism expressions (such as we discussed at the end of Section 7.2), frequency-moment sum rules specific to a Bose-condensed fluid are of considerable interest.

In Section 8.1, we discuss several rigorous f-sum rules for the condensate and non-condensate parts of $\chi_{nn}(\mathbf{Q}, \omega)$ based on (5.24). These were originally derived by Wagner (1966), Hohenberg (1967) and Wong and Gould (1974) for soft-core interatomic potentials. In Section 8.2, we critically review the phenomenological two-fluid formula for $S(\mathbf{Q}, \omega)$ of Woods and Svensson (1978). This was the first attempt to describe the fact that the maxon–roton quasiparticle peak appeared to disappear at T_λ. In Section 8.3, we use the long-wavelength, zero-frequency limit of the longitudinal current-response function $\chi^{\ell}_{JJ}(\mathbf{Q}, \omega)$ to derive what one might call "superfluid" and "normal fluid" f-sum rules. The usefulness of these results and those in Section 8.1 is briefly discussed.

8.1 The Wagner–Wong–Gould f-sum rules

The Placzek or longitudinal f-sum rule (2.24) is a direct consequence of the high-frequency behaviour given by (2.26),

$$\lim_{\omega \to \infty} \chi_{nn}(\mathbf{Q}, \omega) = \lim_{\omega \to \infty} \bar{\chi}_{nn}(\mathbf{Q}, \omega) = \frac{nQ^2}{m\omega^2} + \dots \, . \tag{8.1}$$

This leads to the first frequency moment

$$-\frac{1}{n}\int_{-\infty}^{\infty}\frac{d\omega}{2\pi}\omega \text{ Im } \chi_{nn}(\mathbf{Q},\omega) = \int_{-\infty}^{\infty} d\omega\, \omega S(\mathbf{Q},\omega) = \varepsilon_Q \ , \tag{8.2}$$

valid for arbitrary wavevectors. We will now show that in any Bose-condensed fluid, the two components of $S(\mathbf{Q},\omega)$ which are given by (5.24) each satisfy their own "f-sum rule."

A sum rule can be derived for S_1 in (3.18) by using the kind of result first discussed by Wagner (1966). Expanding $G_{\alpha\beta}$ in (3.29) in powers of $1/\omega$, one finds

$$\lim_{\omega\to\infty}\sum_{\alpha,\beta} G_{\alpha\beta}(\mathbf{Q},\omega) = \lim_{\omega\to\infty}\sum_{\alpha,\beta}\bar{G}_{\alpha\beta} = \frac{2}{\omega^2}[\varepsilon_Q + B_\infty(Q)] \ , \tag{8.3}$$

where we have defined

$$B_\infty(Q) \equiv \Sigma_{11}(Q,\omega\to\infty) - \Sigma_{12}(Q,\omega\to\infty) - \mu \ . \tag{8.4}$$

This means that $S_1(\mathbf{Q},\omega)$ satisfies the high-frequency sum rule

$$\int_{-\infty}^{\infty} d\omega\, \omega S_1(\mathbf{Q},\omega) = \frac{n_0}{n}\varepsilon_Q + \frac{n_0}{n}B_\infty(Q) \ . \tag{8.5}$$

In the $\omega\to\infty$ limit, the *only* self-energy diagrams which are left in (8.4) are those which are frequency-independent. For a "soft" potential such that $V(\mathbf{q})$ is non-singular, these are the Hartree–Fock diagrams given by the first terms in (5.52) and (5.53). In this case, (8.4) reduces to

$$B_\infty(Q) = B_{HF}(Q) \equiv nV(\mathbf{q}=0) - \mu + \int\frac{d\mathbf{p}}{(2\pi)^3}V(\mathbf{p}+\mathbf{Q})(\tilde{n}_p - \tilde{m}_p) \ . \tag{8.6}$$

The sum rule (8.5) is closely related to the one given in (4.5).

Both of these sum rules can be generalized to potentials with a hard core, by introducing the t-matrix for free-atom scattering (Griffin, 1984). For an arbitrary interatomic potential, one can show that

$$\lim_{\omega\to\infty}\Sigma_{11}(\mathbf{Q},\omega) = \int\frac{d\mathbf{p}}{(2\pi)^3}n_p\Big[\Gamma_\infty\big(\tfrac{1}{2}(\mathbf{Q}-\mathbf{p}),\tfrac{1}{2}(\mathbf{Q}-\mathbf{p})\big) + \Gamma_\infty\big(-\tfrac{1}{2}(\mathbf{Q}-\mathbf{p}),\tfrac{1}{2}(\mathbf{Q}-\mathbf{p})\big)\Big] \ , \tag{8.7}$$

where the vertex function defined by

$$m\Gamma_\infty(\mathbf{q},\mathbf{q}') \equiv \tilde{f}(\mathbf{q},\mathbf{q}') + \int\frac{d\mathbf{k}'}{(2\pi)^3}\frac{\tilde{f}(\mathbf{q},\mathbf{k}')\tilde{f}^*(\mathbf{q}',\mathbf{k}')}{k'^2 - q'^2 - i0^+} \tag{8.8}$$

is the high-frequency limit of Eq. (22.3) of Fetter and Walecka (FW, 1971). The scattering amplitude $\tilde{f}(\mathbf{q},\mathbf{q}')$ for two ^{4}He atoms in free space is given by the solution of the integral equation in Eq. (11.14) of FW. The scattering amplitude $\tilde{f}$ is well defined *even* for a potential with a hard core and hence so is Γ_∞ in (8.8) (in Eqs. (17) and (19) of Griffin, 1984, one should replace $\tilde{f}(\pm\mathbf{q},\mathbf{q})$ by $m\Gamma_\infty(\pm\mathbf{q},\mathbf{q})$). It is interesting to note that for a non-singular potential with a well defined Fourier transform, one can show that $\Gamma_\infty(\mathbf{q},\mathbf{q}')$ as defined in (8.8) reduces to the *bare* potential $V(\mathbf{q}-\mathbf{q}')$. This follows from the third equation on p.131 of FW. Thus we see (somewhat surprisingly) that in this case (8.7) does reduce to the expression which led to (4.5), namely

$$\lim_{\omega\to\infty} \Sigma_{11}(\mathbf{Q},\omega) = \int \frac{d\mathbf{p}}{(2\pi)^3} n_p [V(\mathbf{q}=0) + V(\mathbf{p}+\mathbf{Q})] \ . \tag{8.9}$$

Suitably generalized using (8.7), one can view $B_\infty(Q)$ in (8.5) as, in principle, known. One can numerically calculate the diagonal and off-diagonal momentum distributions $\tilde{n}_p$ and $\tilde{m}_p$ for realistic interatomic potentials (see, for example, McMillan, 1965). It would be useful to have $B_\infty(Q)$ evaluated as a function of Q and tabulated in the literature.

An additional f-sum rule can be derived using the exact Ward identities (5.17)–(5.19). In the high-frequency limit, only the lowest-order diagram in (5.55) contributes to $\bar{\Lambda}_\alpha^\ell$. Using this in (5.17), one obtains to leading order

$$\lim_{\omega\to\infty} \bar{\Lambda}_\alpha(\mathbf{Q},\omega) = \sqrt{n_0}\left[1 - \operatorname{sgn}\alpha \frac{B_\infty(Q)}{\omega}\right] \ . \tag{8.10}$$

Using results analogous to (8.3) in combination with (8.10) gives (Wong and Gould, 1974; Talbot and Griffin, 1983)

$$\begin{aligned}\lim_{\omega\to\infty} \chi^c_{nn}(\mathbf{Q},\omega) &= \lim_{\omega\to\infty} \bar{\chi}^c_{nn} = \sum_{\alpha,\beta} \bar{\Lambda}_\alpha \bar{G}_{\alpha\beta} \bar{\Lambda}_\beta \\ &= \frac{2}{\omega^2}[n_0\varepsilon_Q - n_0 B_\infty(Q)] \ .\end{aligned} \tag{8.11}$$

Thus one finds

$$\int_{-\infty}^{\infty} d\omega\, \omega S_c(\mathbf{Q},\omega) = \frac{n_0}{n}\varepsilon_Q - \frac{n_0}{n} B_\infty(Q) \ , \tag{8.12}$$

where S_c is the dynamic structure factor associated with $\chi^c_{nn} = \Lambda G \Lambda$ in (5.24). An immediate consequence of (8.1) and (8.12) is that the *second* term in (5.24) satisfies

$$\int_{-\infty}^{\infty} d\omega\, \omega S_R(\mathbf{Q},\omega) = \frac{\tilde{n}}{n}\varepsilon_Q + \frac{n_0}{n} B_\infty(Q) \ , \tag{8.13}$$

where $\tilde{n} \equiv n - n_0$. In summary, when $n_0 \neq 0$, the usual f-sum rule (8.2) splits into two separate sum rules (8.12) and (8.13). These are rigorous results, valid at all wavevectors. The same sum rules also hold for $\bar{\chi}^c_{nn}$ and $\bar{\chi}^R_{nn}$.

As we have mentioned, $B_\infty(Q)$ should be calculable by Monte Carlo numerical methods. For large enough Q, it is clear from (8.6) (or its equivalent for hard-core potentials) that $B_\infty(Q)$ will become Q-independent. In particular, it will become negligibly small compared to the free-atom energy $\varepsilon_Q = Q^2/2m$. In this large-Q limit, the Bose fluid f-sum rules (8.12) and (8.13) reduce to

$$\left.\begin{aligned} \lim_{Q\to\infty}\int_{-\infty}^{\infty} d\omega\, \omega S_c(\mathbf{Q},\omega) &= \frac{n_0}{n}\varepsilon_Q\ , \\ \lim_{Q\to\infty}\int_{-\infty}^{\infty} d\omega\, \omega S_R(\mathbf{Q},\omega) &= \frac{\tilde{n}}{n}\varepsilon_Q\ . \end{aligned}\right\} \tag{8.14}$$

The two components of $S(\mathbf{Q},\omega)$ given by the high-Q impulse approximation in (4.2) satisfy these sum rules.

The Wagner–Wong–Gould sum rules can be recast into analogous ones for cross-correlation functions such as $\langle\hat{\rho}(\mathbf{Q},t)\hat{a}_{-\mathbf{Q}}(0)\rangle$. These are of special interest since they give a direct measure of how the density and single-particle fluctuations are mixed via the condensate. Recalling the formal definitions in (5.4) and (5.8), we have (here $\hat{a}$ represents the ^{4}He atom destruction operator)

$$\bar{\chi}_{n,a}(\mathbf{Q},\omega) \equiv \bar{\chi}_{n,2}(\mathbf{Q},\omega) = \bar{\Lambda}_1\bar{G}_{12} + \bar{\Lambda}_2\bar{G}_{22}\ . \tag{8.15}$$

Using (5.6), (8.10) and the high-frequency expansions of $\bar{G}_{\alpha\beta}$ (see also (8.3)),

$$\bar{G}_{22}(\mathbf{Q},\omega) = -\frac{1}{\omega} + \frac{\varepsilon_Q - \mu + \bar{\Sigma}_{11}(\mathbf{Q},\infty)}{\omega^2}\ ,\quad \bar{G}_{12}(\mathbf{Q},\omega) = -\frac{\bar{\Sigma}_{12}(\mathbf{Q},\infty)}{\omega^2}\ , \tag{8.16}$$

we finally obtain the remarkably simple high-frequency expansion

$$\lim_{\omega\to\infty}\chi_{n,a}(\mathbf{Q},\omega) = \lim_{\omega\to\infty}\bar{\chi}_{n,a}(\mathbf{Q},\omega) = -\frac{\sqrt{n_0}}{\omega} + \frac{\sqrt{n_0}}{\omega^2}\varepsilon_Q\ . \tag{8.17}$$

Recalling the spectral representation such as in (2.25), (8.17) immediately gives the frequency-moment sum rules (compare with (8.2))

$$\left.\begin{aligned} \int_{-\infty}^{\infty}\frac{d\omega}{\pi}\mathrm{Im}\ \chi_{n,a}(\mathbf{Q},\omega) &= \sqrt{n_0(T)}\ , \\ -\int_{-\infty}^{\infty}\frac{d\omega}{\pi}\omega\ \mathrm{Im}\ \chi_{n,a}(\mathbf{Q},\omega) &= \sqrt{n_0(T)}\frac{Q^2}{2m}\ . \end{aligned}\right\} \tag{8.18}$$

In contrast with (8.5) and (8.12), these sum rules do not involve $B_\infty(Q)$

on the right hand side. This difference can be traced to the fact that $\chi_{n,a}$ involves the *total* density $\hat{\rho}(\mathbf{Q})$ rather than $\tilde{\rho}_{\mathbf{Q}}$ (see (3.13)).

Stringari (1991, 1992) and Giorgini and Stringari (1990) have recently given an elementary derivation of various frequency moment sum rules for response functions involving combinations of $\hat{a}_{\mathbf{Q}}, \hat{a}^{+}_{\mathbf{Q}}$ and $\hat{\rho}_{\mathbf{Q}}$, using equal-time commutation relations. These authors have emphasized the usefulness of particle density sum rules such as (8.18) as a way of determining the overlap between single-particle and density fluctuations (see also remarks at the end of Section 9.1).

Stringari (1992) has also derived additional sum rules analogous to those of Wagner in (4.5) and Wong and Gould in (8.12) but with the detailed balance factor $[N(\omega)+1]$ in the frequency integral. This factor effectively gives more weight to the low-frequency region and thus these new sum rules can be useful in studying low-energy, long-wavelength excitations. In contrast, the sum rules (4.5) and (8.12) give more weight to high frequencies. This spectral region is much more dependent on the details of the hard-core potential (see also Wong and Gould, 1974). On the other hand, it is the high-energy spectral region of various response functions which is of most interest in relation to the predicted hybridization effects resulting from the maxon excitation overlapping the zero sound particle–hole mode (see discussion at end of Section 7.2). In addition, at around $Q \sim 2.4$ Å^{-1}, we expect similar hybridization effects when the roton branch crosses the two-roton branch (see Section 10.2).

High-frequency tail of $S(\mathbf{Q},\omega)$

The f-sum rules (8.5) and (8.12) are direct consequences of the high-frequency behaviour of the various terms contributing to χ_{nn} in (5.24). A related subject is the high-frequency behaviour of $S(\mathbf{Q},\omega)$ itself. This has been the subject of several studies related to the so-called "deep inelastic scattering" behaviour of quantum fluids. It is argued that at high enough energy transfers, $S(\mathbf{Q},\omega)$ should be *independent* of the nature of the elementary excitations, quantum statistics, etc. This deep inelastic region is sometimes defined as when the energy transfer is much larger than the free-particle energy ($\omega \gg \varepsilon_Q$).

This deep inelastic region has been mainly studied at $T = 0$ (Wong and Gould, 1974; Bartley and Wong, 1975; Family, 1975; Wong, 1977). One finds the asymptotic behaviour is given by (Kirkpatrick, 1984)

$$S(\mathbf{Q},\omega) \sim \frac{(Qa)^n}{\omega^{7/2}} \tag{8.19}$$

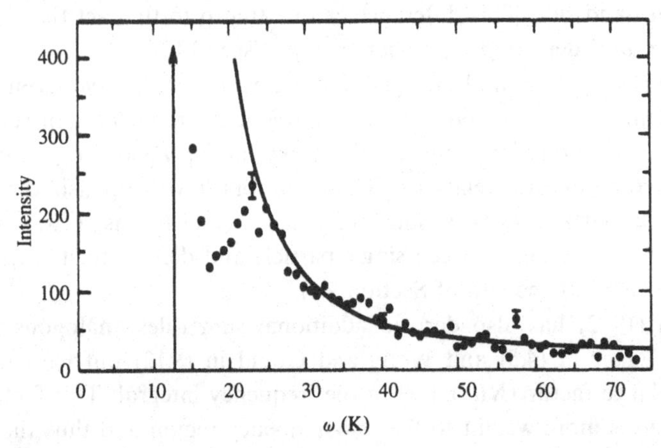

Fig. 8.1. Fit of the high-frequency tail of $S(\mathbf{Q},\omega)$ to the predicted $\omega^{-7/2}$ line shape, for $Q = 0.8$ Å^{-1}, $T = 1.2$ K. The neutron data are from Woods *et al.*, 1972 [Source: Wong, 1977].

for a quantum fluid of hard spheres with diameter a (where $n = 4$ for $Qa \ll 1$; $n = 2$ for $Qa \gg 1$). The coefficient in (8.19) is proportional to the second derivative of the pair distribution function $g''(r = a)$. In Fig. 8.1, we show a fit of the high-energy tail to $\omega^{-7/2}$ for $Q = 0.8$ Å^{-1} and $T = 1.2$ K (Wong, 1977). It would be useful to have more systematic studies of the high-frequency tail of $S(\mathbf{Q},\omega)$ over a wide spectrum of Q values, as well as theoretical predictions based on a more realistic interatomic potential.

While no calculations at $T \neq 0$ have been reported, it is expected that the high-frequency tail of $S(\mathbf{Q},\omega)$ will be almost temperature-independent since it apparently only depends on the pair distribution function. As noted by Talbot and Griffin (1984a) and Talbot *et al.* (1988), this temperature independence is in agreement with the $S(\mathbf{Q},\omega)$ data in the range $0.8 \lesssim Q \lesssim 2$ Å^{-1}. As shown by the examples in Figs. 1.5 and 2.3, the high-frequency tail appears to be the same at all of temperatures for $\omega \gtrsim 0.6$ THz (30 K). For further discussion, see Svensson (1991).

8.2 Woods–Svensson two-fluid ansatz for $S(\mathbf{Q},\omega)$

The first systematic high-resolution study of $S(\mathbf{Q},\omega)$ as a function of the temperature was by Woods and Svensson (1978). Their results strongly indicated that the intensity of the maxon–roton peak went to zero as one approached T_λ. In an attempt to explain their observations, Woods and Svensson introduced an *ad hoc* two-component expression which divided $S(\mathbf{Q}, \omega)$ into a "superfluid" part and a "normal" part, allowing one to extract out a "quasiparticle" resonance from the former. As with any parameterization, there are two questions about the WS formula: (a) Does it give a reasonable, consistent fit to the data? (b) Does the parameterization have a well defined microscopic basis? Both aspects have been extensively discussed in the literature (Glyde and Svensson, 1987; Griffin, 1987; Talbot, Glyde, Stirling and Svensson (TGSS), 1988; Svensson, 1989). On the experimental question as to how good the fit is, the detailed analysis by TGSS at $p = 20$ bar suggests serious problems. We also recall from Chapter 7 that the temperature dependence of the $S(\mathbf{Q}, \omega)$ line shape is qualitatively different in the phonon, maxon and roton wavevector regions. Even in the $Q \gtrsim 0.8$ Å^{-1} region, the analysis presented in Section 7.2 gives no reason to expect that there will be a single "universal formula" for $S(\mathbf{Q}, \omega)$ which describes *both* the maxon and roton regions. However, we think it is still worthwhile to review the WS scenario, as historically the first attempt at the kind of finite-temperature parameterization we are looking for. Moreover, a similar parameterization of $\bar{\chi}_{nn}$ instead of χ_{nn} may be a promising direction in analysing data, as we discuss in Sections 7.2 and 8.3.

The WS ansatz divides $S(\mathbf{Q}, \omega)$ below T_λ into two parts,

$$S = S_S + S_N \ , \tag{8.20}$$

where the "superfluid" component is

$$S_S(\mathbf{Q}, \omega) = Z(Q, T)[N(\omega)+1]A(\mathbf{Q}, \omega; T) + \frac{\rho_S}{\rho} S_M(\mathbf{Q}, \omega; T = 1 \text{ K}) \tag{8.21}$$

and the "normal" component is

$$S_N(\mathbf{Q}, \omega) = \frac{\rho_N}{\rho} S(\mathbf{Q}, \omega; T_\lambda) \ . \tag{8.22}$$

Effectively, $S(\mathbf{Q}, \omega)$ is assumed to be a weighted sum of three separate contributions: a quasiparticle Lorentzian peak (of frequency ω_Q and width Γ_Q) described by $A(\mathbf{Q}, \omega)$ as in (7.4); a multiparticle part S_M; and a normal component S_N. Within this scheme, one makes the key simplifying assumptions that S_M can be described by its low-temperature value (~ 1 K) with a weighting factor ρ_S/ρ; and S_N can be approximated by its value just above T_λ, weighted by ρ_N/ρ. Using (8.20)–(8.22), one can then isolate the quasiparticle contribution in the $S(\mathbf{Q}, \omega)$ data and determine the temperature dependence of $Z(Q, T)$, ω_Q and Γ_Q. According to WS, this procedure leads to the prediction that

$$Z(Q,T) \propto \rho_S/\rho\ , \tag{8.23}$$

$$\Gamma_Q(T) \propto \rho_N/\rho\ , \tag{8.24}$$

for SVP data in the range $0.8 \lesssim Q \lesssim 2$ Å^{-1}. However, analysis of high-resolution data at $p = 20$ bar by TGSS led to significant deviations from (8.23) above 1.7 K (we recall that $T_\lambda = 1.9$ K at this pressure).

The WS expressions in (8.21) and (8.22) are equivalent to experimentally *defining* the quasiparticle contribution as

$$S_{qp}(\mathbf{Q},\omega) \equiv S(\mathbf{Q},\omega) - \frac{\rho_S}{\rho}S_M(\mathbf{Q},\omega;1\text{ K}) - \frac{\rho_N}{\rho}S(\mathbf{Q},\omega;T_\lambda)\ . \tag{8.25}$$

Since $S(\mathbf{Q})$ is essentially the same above and below T_λ, it follows that the *integrated* quasiparticle intensity is

$$\begin{aligned}\int_{-\infty}^{\infty} d\omega S_{qp}(\mathbf{Q},\omega) &\equiv S_{qp}(\mathbf{Q}) \\ &= S(\mathbf{Q}) - \frac{\rho_S}{\rho}\int_{-\infty}^{\infty} d\omega S_M(\mathbf{Q},\omega;1\text{ K}) - \frac{\rho_N}{\rho}S(\mathbf{Q}) \\ &= \frac{\rho_S}{\rho}\left[S(\mathbf{Q}) - S_M(\mathbf{Q},1\text{ K})\right]\ .\end{aligned} \tag{8.26}$$

As TGSS have emphasized (following Mineev, 1980), this means the *integrated* intensity $S_{qp}(\mathbf{Q})$ as defined by (8.26) must necessarily be proportional to ρ_S if we use (8.25). In contrast, TGSS have determined the temperature dependence of the quasiparticle intensity $Z(Q)$ in (8.21) by directly fitting $S_{qp}(\mathbf{Q},\omega)$ as defined in (8.25) to the Lorentzian line shape in (7.4).

The first frequency moments of (8.21) and (8.22) are easily calculated to give

$$\left.\begin{aligned}\int_{-\infty}^{\infty} d\omega\ \omega S_S(\mathbf{Q},\omega) &= \frac{\rho_S}{\rho}\{\varepsilon_Q + Z(Q)[\omega_Q(T) - \omega_Q(1\text{ K})]\}\ , \\ \int_{-\infty}^{\infty} d\omega\ \omega S_N(\mathbf{Q},\omega) &= \frac{\rho_N}{\rho}\varepsilon_Q\ .\end{aligned}\right\} \tag{8.27}$$

Thus the WS expressions are consistent with the f-sum rule (8.2) to the extent that the quasiparticle energy ω_Q is taken to be temperature-independent, which is a pretty good first approximation.

From a theoretical point of view, a key requirement of any parameterization of $S(\mathbf{Q},\omega)$ is that it be consistent with the general structure implied by the dielectric formalism as summarized in Section 5.5. In the low-Q phonon region, which is dominated by a single sharp zero sound mode at *all* temperatures, the WS ansatz would not appear to be an appropriate starting point (Griffin, 1987). On the face of it, the WS ansatz might be useful as an approximation in the roton region to the extent this region can be described by (7.8).

The most serious deficiency of the WS ansatz (8.21) in the maxon region $Q \sim 1$ Å^{-1} is that it cannot describe the characteristic effects associated with the condensate-induced hybridization of zero sound with a high-energy maxon (see discussion at the end of Section 7.2). From the point of view of the dielectric formulation, as we remarked at the end of Section 5.5, it is $\bar{\chi}_{nn}$ in (5.76) which is most naturally split into "superfluid" and "normal" components, rather than the full χ_{nn} as Woods and Svensson tried to do. It seems clear that one should try to develop some sort of WS parameterization for $\bar{\chi}^{c}_{nn}$ and $\bar{\chi}^{R}_{nn}$. One may view the simple model approximation (7.23) introduced in Section 7.2 as an attempt in this direction.

8.3 Superfluid and normal fluid f-sum rules

In Section 8.1, we derived f-sum rules for the two components of χ_{nn} given by (5.24) or (3.47) by considering the high-frequency limit. The condensate parts satisfy

$$-\int_{-\infty}^{\infty} \frac{d\omega}{2\pi} \omega \operatorname{Im} \chi^{c}_{nn}(\mathbf{Q},\omega) = n_0[\varepsilon_Q - B_\infty(Q)] \ , \tag{8.28a}$$

$$-\int_{-\infty}^{\infty} \frac{d\omega}{2\pi} \omega \operatorname{Im} \bar{\chi}^{c}_{nn}(\mathbf{Q},\omega) = n_0[\varepsilon_Q - B_\infty(Q)] \ , \tag{8.28b}$$

which are equivalent to (8.12). Analogous results for Im χ^{R}_{nn} and Im $\bar{\chi}^{R}_{nn}$ can be written down, as in (8.13).

Sum rules such as (8.28*b*) are useful as a constraint on parameterized expressions for $\bar{\chi}^{c}_{nn}$. The question of how useful sum rules such as (8.28*a*) are in analysing $S(\mathbf{Q},\omega)$ data is more complicated. We have emphasized in Chapters 5–7 that the two parts of (5.24) may not have much physical significance as separate entities in the intermediate maxon region, where the expression (5.76) is more appropriate. On the other hand, we argued in Section 7.2 that in the roton region $Q \gtrsim 1.5$ Å^{-1}, the two-component form (7.6) is useful and hence so should be sum rules like (8.28*a*).

However there is the complication that the high-frequency multiparticle contributions (usually peaked at ~ 25 K) can arise from *both* the vertex functions $\bar{\Lambda}_\alpha$ and the self-energies $\bar{\Sigma}_{\alpha\beta}$ in the first term in (7.8), in addition to contributions from $\bar{\chi}^{R}_{nn}$ in the second term. There seems no

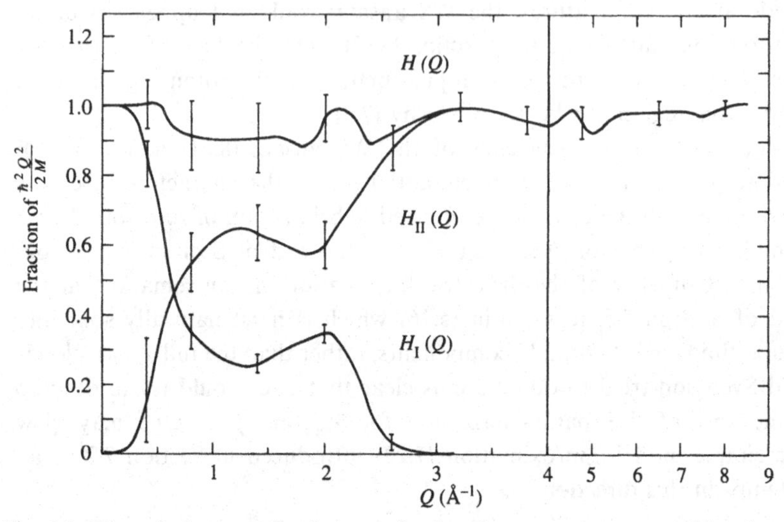

Fig. 8.2. The relative contributions to the first frequency moment arising from the quasiparticle peak and the multiphonon distribution at $T = 1.1$ K [Source: Cowley and Woods, 1971].

easy way to disentangle the contribution each makes to the observed multiparticle spectral weight. The same problem arises with the particle–hole excitations (as well as the broad zero sound mode which appears to be present in $S(\mathbf{Q},\omega)$ in the intermediate-Q region ~ 1 Å^{-1}, as discussed in Section 7.2). Such contributions can come from *both* terms in (7.8).

What we would really like to have is a sum rule for S_{I} in (2.34), i.e., the part of $S(\mathbf{Q},\omega)$ which only describes the quasiparticle resonance at ω_Q. In Fig. 8.2, we show the relative contribution of S_{I} and S_{II} to the ordinary f-sum rule at T=1.1 K (Cowley and Woods, 1971). As noted by Wong (1979), the sum rules for S_c and S_R are analogous to the Ambegaokar–Conway–Baym (ACB) sum rules in anharmonic crystals. As we discuss in Section 11.2, while useful, the ACB f-sum rules in quantum crystals still do not allow one to cleanly isolate the phonon peak in $S(\mathbf{Q},\omega)$ data on solid Helium at large Q, where interference effects are significant.

We now derive another set of Bose f-sum rules, based on the low-frequency, long-wavelength limit of longitudinal current-response functions. We recall that, using (5.16*a*) and (5.16*b*),

$$\left.\begin{aligned} \operatorname{Im} \chi_{nn}(\mathbf{Q},\omega) &= \frac{Q^2}{m^2\omega^2}\operatorname{Im} \chi^{\ell}_{JJ}(\mathbf{Q},\omega) \ , \\ \operatorname{Im} \bar{\chi}_{nn}(\mathbf{Q},\omega) &= \frac{Q^2}{m^2\omega^2}\operatorname{Im} \bar{\chi}^{\ell}_{JJ}(\mathbf{Q},\omega) \ . \end{aligned}\right\} \tag{8.29}$$

As with χ_{nn}, one can split χ^{ℓ}_{JJ} into the sum of two terms, one of which vanishes in the normal phase (Talbot and Griffin, 1984b)

$$\chi^{\ell}_{JJ} = \chi^{\ell c}_{JJ} + \chi^{\ell R}_{JJ} \ , \tag{8.30}$$

with

$$\left.\begin{aligned} \chi^{\ell c}_{JJ} &= \sum_{\alpha,\beta} \tilde{\Lambda}^{\ell}_{\alpha} G_{\alpha\beta} \tilde{\Lambda}^{\ell}_{\beta} \ , \\ \chi^{\ell R}_{JJ} &= \bar{\chi}^{\ell R}_{JJ} + \bar{\chi}^{\ell R}_{Jn} \frac{V(Q)}{\epsilon^R} \bar{\chi}^{\ell R}_{Jn} \ , \end{aligned}\right\} \tag{8.31}$$

where we find it convenient to define a new vertex function

$$\tilde{\Lambda}^{\ell}_{\alpha} \equiv \bar{\Lambda}^{\ell}_{\alpha} + \bar{\Lambda}_{\alpha} \frac{V(Q)}{\epsilon^R} \bar{\chi}^{\ell R}_{Jn} \ . \tag{8.32}$$

The correlation function $\bar{\chi}^{\ell R}_{Jn}$ is defined in (5.13*b*) and (5.6). Similarly, the regular part $\bar{\chi}^{\ell}_{JJ}$ is given by (see also (5.13*b*))

$$\bar{\chi}^{\ell}_{JJ} = \bar{\chi}^{\ell c}_{JJ} + \bar{\chi}^{\ell R}_{JJ} \ , \tag{8.33}$$

with the proper condensate contribution

$$\bar{\chi}^{\ell c}_{JJ} = \sum_{\alpha,\beta} \bar{\Lambda}^{\ell}_{\alpha} \bar{G}_{\alpha\beta} \bar{\Lambda}^{\ell}_{\beta} \ . \tag{8.34}$$

The zero-frequency limit of $\chi^{\ell}_{JJ}(\mathbf{Q},\omega)$ is given by (6.2),

$$-\chi^{\ell}_{JJ}(\mathbf{Q},\omega=0) = -\int_{-\infty}^{\infty} \frac{d\omega}{\pi} \frac{\operatorname{Im} \chi^{\ell}_{JJ}(\mathbf{Q},\omega)}{\omega} = \rho \ , \tag{8.35}$$

valid for arbitrary wavevectors. The zero-frequency, long-wavelength limits of the condensate and regular contributions in (8.30)–(8.34) can be summarized by

$$\left.\begin{aligned} \lim_{Q\to 0} -\chi^{\ell R}_{JJ}(\mathbf{Q},\omega=0) &= -\bar{\chi}^{\ell R}_{JJ}(\mathbf{Q},\omega=0) = \rho_N(T) \ , \\ \lim_{Q\to 0} -\chi^{\ell c}_{JJ}(\mathbf{Q},\omega=0) &= -\bar{\chi}^{\ell c}_{JJ}(Q,\omega=0) = \rho_S(T) \ . \end{aligned}\right\} \tag{8.36}$$

These results follow from (6.2), (6.4) and the fact that $\bar{\chi}^{\ell R}_{Jn}(\mathbf{Q},\omega=0) = 0$.

Writing (8.36) in the form of spectral representations as in (8.35), we immediately obtain

$$\left.\begin{aligned} \lim_{Q\to 0} -\int_{-\infty}^{\infty} \frac{d\omega}{\pi} \frac{\text{Im}\,\chi_{JJ}^{\ell c}(\mathbf{Q},\omega)}{\omega} &= \rho_S(T)\ , \\ \lim_{Q\to 0} -\int_{-\infty}^{\infty} \frac{d\omega}{\pi} \frac{\text{Im}\,\chi_{JJ}^{\ell R}(\mathbf{Q},\omega)}{\omega} &= \rho_N(T)\ , \end{aligned}\right\} \tag{8.37}$$

and the analogous expressions for $\overline{\chi}_{JJ}^{\ell c}$ and $\overline{\chi}_{JJ}^{\ell R}$.

Results such as (8.37) immediately suggest that at low Q, it is useful to divide $S(\mathbf{Q},\omega)$ into "superfluid" and "normal" parts as follows:

$$\left.\begin{aligned} S_S(\mathbf{Q},\omega) &\equiv -\frac{[N(\omega)+1]}{\pi n}\frac{Q^2}{m^2\omega^2}\text{Im}\,\chi_{JJ}^{\ell c}(\mathbf{Q},\omega)\ , \\ S_N(\mathbf{Q},\omega) &\equiv -\frac{[N(\omega)+1]}{\pi n}\frac{Q^2}{m^2\omega^2}\text{Im}\,\chi_{JJ}^{\ell R}(\mathbf{Q},\omega)\ . \end{aligned}\right\} \tag{8.38}$$

With these *definitions*, the sum rules in (8.37) lead to the following new long-wavelength f-sum rules:

$$\lim_{Q\to 0}\int_{-\infty}^{\infty} d\omega\ \omega S_S(\mathbf{Q},\omega) = \frac{\rho_S}{\rho}\varepsilon_Q\ , \tag{8.39a}$$

$$\lim_{Q\to 0}\int_{-\infty}^{\infty} d\omega\ \omega S_N(\mathbf{Q},\omega) = \frac{\rho_N}{\rho}\varepsilon_Q\ . \tag{8.39b}$$

Such sum rules are discussed and applied in Chapter 7 of Nozières and Pines (1964, 1990) as well as in the review by Pines (1965). More recently, they have been used by Griffin (1979b) and Griffin and Talbot (1981). We emphasize that they are only valid in the $Q \to 0$ limit and do not offer a microscopic basis for the Woods–Svensson two-component ansatz introduced to explain line-shape data in the region $Q \gtrsim 0.8$ Å^{-1} (see Section 8.2).

Several comments are needed concerning the significance of the results in (8.39). First of all, we note that these superfluid and normal fluid f-sum rules are a consequence of taking the $Q,\omega \to 0$ limit of the components of the longitudinal current-response function. In contrast, the f-sum rules in (8.12) and (8.13) involve the high-frequency limit of the components of the density response function and are, formally, valid at arbitrary values of Q. The results we have found seem physically reasonable, namely that the high-frequency limit f-sum rules of Section 8.1 depend on n_0 and $\tilde{n}$ while the low-frequency limit sum rules of this section involve ρ_S and ρ_N. The sum rules in (8.12) and (8.13) are thus not equivalent to those in (8.39). In particular, one cannot make a naïve identification based on (8.29) and assume that

$$\text{Im}\, \chi^c_{nn} = \frac{Q^2}{m^2\omega^2}\text{Im}\, \chi^{\ell c}_{JJ} \qquad \text{[wrong]}\ . \tag{8.40}$$

This would lead to the equalities $S_S = S_c$ and $S_N = S_R$, which are not satisfied in general (the right hand sides of (8.12) and (8.39*a*) are not equal). For further discussion of these differences, see Wong (1979) and Talbot and Griffin (1983).

The preceding analysis also shows that there are two equally valid ways of splitting Im χ_{nn} into two components, one component being proportional to the single-particle Green's function. Depending on whether we start from (5.24) for χ_{nn} or (8.30) for χ^{ℓ}_{JJ}, the part of χ_{nn} containing $G_{\alpha\beta}$ as a separate factor involves quite different Bose broken-symmetry vertex functions. This means that the particle–hole and multiparticle spectral contributions (both contained in Λ_α and $\tilde{\Lambda}^{\ell}_\alpha$) would modulate how the structure of $G_{\alpha\beta}$ appears in $S(\mathbf{Q},\omega)$ in different ways. As a result, the apparent weight of the quasiparticle peak could be different in the two formulations. A partial resolution of this apparent arbitrariness lies in the fact that there may be cancellations between the two components of expressions such as (8.30) and (5.24). It was for this reason that we argued in Section 7.2 that (5.24) and (3.47) are only useful in the large-Q limit ($\gtrsim 1.5$ Å^{-1}).

In conclusion, while we have derived formally exact two-component expressions for χ_{nn} and χ^{ℓ}_{JJ}, there does not appear to be a simple way of experimentally distinguishing the resulting two components in $S(\mathbf{Q},\omega)$ data. The same difficulty occurs in solid ^{4}He, as discussed in Section 11.2. On the other hand, the analogous two-component forms for $\overline{\chi}_{nn}$ and $\overline{\chi}^{\ell}_{JJ}$, and the various f-sum rules they satisfy, should be useful in developing simple parameterized expressions for these regular functions. The preliminary analysis given at the end of Section 7.2 worked in terms of $\overline{\chi}^c_{nn}$ and $\overline{\chi}^R_{nn}$. An improved analysis might start with simple approximations for $\overline{\chi}^{\ell c}_{JJ}$ and $\overline{\chi}^{\ell R}_{JJ}$ for use in (5.81).

The application of the Bose sum rules of the kind discussed in this chapter clearly is in its infancy and further studies are needed. For a systematic but elementary discussion of rigorous frequency-moment sum rules associated with Bose broken symmetry, we recommend the recent papers by Pitaevskii and Stringari (1991) and Stringari (1992).

9

Variational and parameterized approaches

In Chapter 5, we argue that in an interacting fluid having a Bose condensate, a sharp peak in $S(\mathbf{Q}, \omega)$ could have two origins. It could be either a collective zero sound (ZS) density fluctuation or a renormalized single-particle (SP) excitation. In Section 7.2, we use this scenario to argue that the low-Q phonon peak observed in superfluid ^{4}He was a collective ZS mode but the high-Q maxon–roton peak was associated with an SP excitation. However, we also saw that while this qualitative picture is plausible in terms of the microscopic theory for a Bose-condensed fluid (see summary in Section 5.5 and discussion in Section 7.2), at the present time we have few quantitative calculations of the regular functions $\bar{\chi}^R_{nn}$, $\bar{\Lambda}_\alpha$ and $\bar{\Sigma}_{\alpha\beta}$ which are relevant to liquid ^{4}He. These include results at arbitrary temperatures in the case of a dilute Bose gas (see Sections 5.2 and 5.3) and the $T = 0$ calculations of Gavoret and Nozières (1964) in the long-wavelength limit (discussed in Section 6.3). In addition, we have rigorous limits at very low and very high frequencies (see Chapters 6 and 8). In this chapter, we discuss some alternative approaches which have been used to calculate (mainly at $T = 0$) the phonon–maxon–roton dispersion curve as well as the $S(\mathbf{Q}, \omega)$ line shape.

In Section 9.1, we review what is perhaps the most successful method, that based on direct numerical calculations of the many-body wavefunctions and energies using variational techniques. The ground-state wavefunction is described by a Jastrow–Feenberg type form which can incorporate the strong short-range correlations present in a quantum liquid. (The same kind of wavefunction is also used in quantum solids, as we discuss in Chapter 11.) The variational excited states are then built up from this ground state by considering density fluctuations coupled into backflow processes, as first done by Feynman and Cohen (1956). These methods (often referred to as the correlated-basis-function or CBF approach) are now able to obtain impressive results for the phonon–maxon–roton quasiparticle dispersion relation at $T = 0$, as well as for its weight in $S(\mathbf{Q}, \omega)$ using a direct evaluation of the expression (2.14). While these methods can be used to evaluate the single-particle density matrix (and hence the ^{4}He atomic momentum distribution and condensate fraction, as discussed in Chapter 4), they do not deal with the single-particle Green's functions. Consequently, the *dynamical* role of the condensate is not very clear. In addition, using variational methods to deal with the effects of finite temperature and damping on time-dependent correlation functions is always difficult and somewhat *ad hoc*.

In Section 9.2, we review the polarization potential theory introduced by Aldrich and Pines (1976). This approach (which so far has been worked out only at $T = 0$) may be viewed as a phenomenological parameterization of the density response function $\chi_{nn}(\mathbf{Q}, \omega)$, which attempts to include the correct many-body physics. We discuss its consistency with the general structure of $\chi_{nn}(\mathbf{Q}, \omega)$ implied by the dielectric formalism (as summarized in Section 5.5). The memory function formalism sketched in Section 9.3 has many of the same goals as polarization potential theory. In existing formulations, however, we do not believe that broken symmetry *unique* to a Bose-condensed fluid has been adequately incorporated.

As a general comment on these alternative methods, we note that they all focus on the density-response function $\chi_{nn}(\mathbf{Q}, \omega)$ directly, rather than on the more fundamental single-particle Green's function $G_{\alpha\beta}(\mathbf{Q}, \omega)$ which lies at the basis of more complicated two-particle correlation functions like χ_{nn}. Consequently these approaches, useful as they are, are quite different from the field-theoretic methods used in the rest of this book.

9.1 Variational theories in coordinate space

Beginning with the pioneering work of Bijl (1940) and Feynman (1954), there have been many studies of superfluid ^{4}He based on *ab initio* calculations of the many-body wavefunctions using a variational approach. To the extent that one can find good approximations to the ground state and low-lying excited states, one can directly obtain the excitation energies and also evaluate $S(\mathbf{Q}, \omega)$. In this approach, the variational wavefunctions are given in a coordinate-space representation, depending on the positions of all the atoms $(\mathbf{r}_1, \mathbf{r}_2, \ldots, \mathbf{r}_N)$. A key requirement for Bose fluids is that these wavefunctions be symmetric with respect to the interchange of any two particles. Matrix elements taken with respect to these wavefunctions can then be reduced to quantities which are dependent on various static distribution functions (the radial two-body distribution function $g(\mathbf{r})$ in (2.10) being the simplest one).

A nice introduction to this approach is given in Chapter 10 of Mahan (1990). Extensive reviews of this method have been given by Feenberg (1969), Woo (1976) and Campbell (1978). More recent developments are summarized by Clark and Ristig (1989) and Campbell and Clements (1989). The work of Manousakis and Pandharipande (1984, 1986) gives the results of state-of-the-art calculations using correlated-basis-function methods for both the quasiparticle excitation energy ω_Q and the dynamic structure factor $S(\mathbf{Q}, \omega)$. Up to the present time, these variational methods have been developed mainly at $T = 0$ (see, however, Senger *et al.*, 1992). Campbell and Clements (1989) review attempts to extend this approach to finite temperatures, building on the pioneering work by Penrose (1958) and Reatto and Chester (1967).

Typically, trial variational approximations to excited states are given in terms of coordinate-space operators acting on the ground state $|\Phi_0\rangle$. The Jastrow-type ansatz

$$|\Phi_0\rangle = \Phi(\mathbf{r}_1, \mathbf{r}_2, \ldots, \mathbf{r}_N) = \prod_{i<j} f(r_{ij}) \tag{9.1}$$

incorporates the short-range two-body correlations ($r_{ij} = |\mathbf{r}_i - \mathbf{r}_j|$) and the conditions which the ground-state wavefunction for a Bose system must satisfy. Since two atoms cannot get closer than the hard-core diameter a, $f(r)$ must vanish for $r < a$. One usually takes

$$f(r) = e^{-u(r)/2} \quad , \tag{9.2}$$

with $u(r) = (b/r)^5$, where b is determined variationally. The long-range correlations induced by the condensate discussed in Section 4.1 can be shown to imply that

$$\lim_{r \to \infty} f(r) = 1 - \frac{m^2 c}{2\pi^2 \hbar \rho} \frac{1}{r^2} \quad . \tag{9.3}$$

A generalized Jastrow–Feenberg ansatz for $|\Phi_0\rangle$ which includes three-body correlations (“triplets”) gives a ground-state energy $\langle \Phi_0 | \hat{H} | \Phi_0 \rangle = E_0$, which is within a few per cent of the value obtained by direct Green’s function Monte Carlo (GFMC) methods. The associated atomic momentum distribution n_p and condensate fraction can also be calculated from $|\Phi_0\rangle$ using the single-particle density matrix (Clark and Ristig, 1989)

$$\rho_1(\mathbf{r}-\mathbf{r}') = N \int d\mathbf{r}_2 \int d\mathbf{r}_3 \cdots \int d\mathbf{r}_N \Phi_0^*(\mathbf{r}, \mathbf{r}_2, \cdots \mathbf{r}_N)\Phi_0(\mathbf{r}', \mathbf{r}_2, \cdots \mathbf{r}_N) \ , \quad (9.4)$$

with

$$n_p = \int d\mathbf{r} e^{i\mathbf{p}\cdot\mathbf{r}} \rho_1(\mathbf{r}) \ . \quad (9.5)$$

As we discussed in Section 1.2, the well known Feynman–Bijl approximation to the normalized excited state is given by (1.4) and (1.5),

$$|\Phi_{FB}\rangle = \frac{\hat{\rho}^+(\mathbf{Q})|\Phi_0\rangle}{\sqrt{NS(Q)}} \ . \quad (9.6)$$

Within this approximation, we note that such excited states are *pure* density fluctuations of wavevector Q. The excitation energy defined by

$$\langle \Phi_{FB}(Q)|\hat{H} - E_0|\Phi_{FB}(Q)\rangle \equiv \omega_Q^{FB} \quad (9.7)$$

can be expressed in terms of the radial distribution function $g(\mathbf{r})$ or, more precisely, its Fourier transform $S(\mathbf{Q})$. As indicated by (1.7), the spectral weight of $S(\mathbf{Q}, \omega)$ is exhausted by this mode, which has a weight given by $S(\mathbf{Q})$.

Feynman and Cohen (1956) introduced an improved approximation for the excited states in an attempt to include “backflow” processes. In their original coordinate-space description, the FC wavefunction is given by

$$|\Phi_{FC}\rangle = F_Q^+|\Phi_0\rangle \ , \quad (9.8)$$

with the generator given by

$$F_Q^+ = \sum_j e^{i\mathbf{Q}\cdot\mathbf{r}_j} \left[1 + i \sum_{i \neq j} \eta(r_{ij}) \mathbf{Q} \cdot \mathbf{r}_{ij}\right] \ . \quad (9.9)$$

At *large* r, it can be shown that $\eta(r) = A/r^3$, which describes dipolar backflow. In recent variational calculations based on (9.8) and (9.9), however, $\eta(r)$ is determined variationally so as to minimize the energy of the state (9.8). In momentum space, one can rewrite (9.9) in the form (Miller, Pines and Nozières, 1962)

$$F_Q^+ = \hat{\rho}^+(\mathbf{Q}) + \sum_{\mathbf{q} \neq \mathbf{Q}} A_{q,Q} \hat{\rho}^+(\mathbf{q}) \hat{\rho}^+(\mathbf{Q} - \mathbf{q}) \ . \quad (9.10)$$

One finds that $A_{q,Q}$ vanishes as $Q \to 0$ and hence F_Q^+ generates a pure density fluctuation. However the Feynman–Cohen excited state (9.10) increasingly deviates from a *pure* density fluctuation as Q increases.

While they were originally introduced to describe backflow processes, the wavefunctions in (9.9) and (9.10) are much more general when η and A are determined variationally. In the original Feynman–Bijl approximation, it is implicitly assumed that all the important correlations between atoms are already contained in the ground-state wavefunction $|\Phi_0\rangle$. The excited states do not involve any *new* correlations and thus $|\Phi_{FB}\rangle$ simply involves multiplying $|\Phi_0\rangle$ by a one-body operator, as in (1.4) or (9.6). This is *not* the case with $|\Phi_{FC}\rangle$ in (9.8).

Manousakis and Pandharipande (MP, 1984, 1986) use second-order perturbation theory to renormalize the FC state by coupling it to the 2-FC excitation subspace. Defining non-orthogonal correlated basis functions (CBF)

$$\left.\begin{aligned} |1\rangle &\equiv F_Q^+|\Phi_0\rangle \ , \\ |2\rangle &\equiv F_q^+F_{q'}^+|\Phi_0\rangle \ , \end{aligned}\right\} \tag{9.11}$$

they introduce the orthogonal CBF basis

$$\left.\begin{aligned} |Q\rangle &\equiv \frac{|1\rangle}{\sqrt{N_{11}}} \ , \\ |q,q'\rangle &\equiv \Big[|2\rangle - \frac{N_{12}}{N_{11}}|1\rangle\Big]/\sqrt{N_{22}} \ , \end{aligned}\right\} \tag{9.12}$$

with $\langle i|j\rangle = N_{ij}$. The approach of MP makes a significant improvement over earlier CBF work (see Feenberg, 1969), which was based on the Feynman–Bijl one- and two-excitation states (i.e., F_Q^+ in (9.11) is approximated by the first term in (9.10)). Carrying out this calculation, MP found that the renormalized FC excitation dispersion relation was in good agreement with the measured spectrum, even in the roton region. This is shown by the comparison in Fig. 9.1. However, as we noted earlier, at large wavevectors $Q \sim 2$ Å^{-1}, one can no longer simply interpret the second term in (9.10) in terms of backflow processes. Thus, while the observed roton spectrum can now be calculated quite accurately, we believe that the physical content of the starting variational wavefunction (9.10) still needs further elucidation (see remarks at end of this section).

Within their variational scheme, Manousakis and Pandharipande (MP, 1986) have also evaluated $S(\mathbf{Q},\omega)$ at $T = 0$ starting directly from (2.14),

$$S(\mathbf{Q},\omega) = \frac{1}{N}\sum_m \left|\langle m|\hat{\rho}^+(\mathbf{Q})|0\rangle\right|^2 \delta(\omega - \omega_{m0}) \ , \tag{9.13}$$

where $\hbar\omega_{m0} = E_m - E_0$ and $|0\rangle$ represents the ground state. Their final result for the density-response function can be written in the following

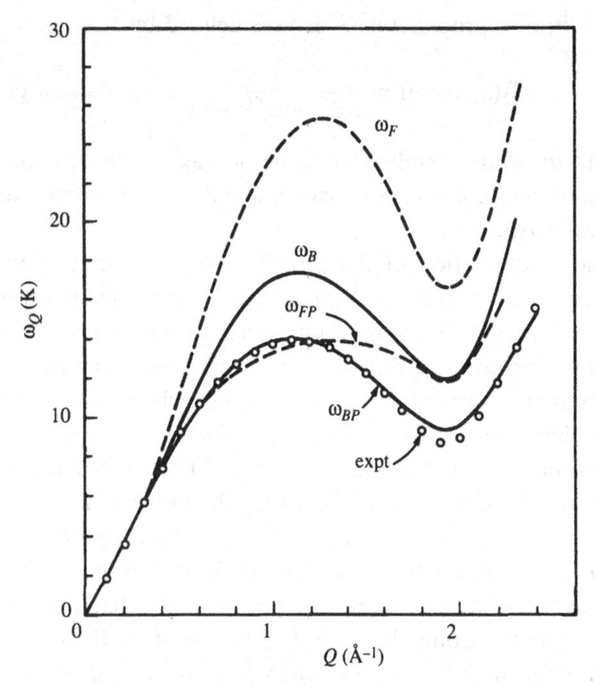

Fig. 9.1. Excitation spectra predicted by various variational-perturbative calculations, using the experimental value of the pair distribution function. Feynman, ω_F; Feynman–Cohen including backflow, ω_B; Brillouin–Wigner perturbation theory with Feynman basis, ω_{FP}; Brillouin–Wigner with Feynman–Cohen basis, ω_{BP} [Source: Manousakis and Pandharipande, 1984].

form (using the basis (9.12)):

$$n^{-1}\chi_{nn}(\mathbf{Q},\omega) = \frac{\Lambda^2_{MP}(Q)}{\omega_Q^{FC} + \Sigma_0(Q,\omega) - \omega} + \frac{1}{2}\sum_{q_1,q_2}\left|\left\langle q_1,q_2\middle|\hat{\rho}^+(\mathbf{Q})\middle|0\right\rangle\right|^2 G_2^0(q_1,q_2,\omega) \ , \tag{9.14a}$$

where the second-order excitation self-energy is

$$\Sigma_0(\mathbf{Q},\omega) = -\frac{1}{2}\sum_{q_1,q_2}\left|\left\langle q_1,q_2\middle|\hat{H} - E_0\middle|Q\right\rangle\right|^2 G_2^0(q_1,q_2,\omega) \ , \tag{9.14b}$$

while the single-excitation spectral weight is given by

$$\Lambda_{MP}(Q) = \langle 0|\hat{\rho}(\mathbf{Q})|Q\rangle - \frac{1}{2}\sum_{q_1,q_2}\langle 0|\hat{\rho}(\mathbf{Q})|q_1,q_2\rangle G_2^0(q_1,q_2,\omega)\langle q_1,q_2|\hat{H}-E_0|0\rangle \ . \tag{9.14c}$$

The two-FC excitation propagator G_2^0 is here defined by

$$G_2^0(q_1, q_2, \omega) = \frac{1}{\omega_{q_1}^{FC} + \omega_{q_2}^{FC} - \omega} \,. \tag{9.14d}$$

MP also extend the above results to include propagator renormalization in G_2^0, but ignore vertex corrections which should be considered at the same level of approximation.

For the detailed derivation of the preceding results, we refer to MP (1986). However, the general structure of (9.14*a*) is easily interpreted: the first term describes the one-excitation spectrum and the second term describes the two-excitation spectrum. The detailed two-excitation spectrum exhibits structure arising from $G_2(q_1, q_2, \omega)$, which can be related to q_1 and q_2 being (maxon, roton), (maxon, maxon) and (roton, roton) due to the high density of states at q_M and q_R. For further discussion of pair excitations, we refer to Chapter 10. In the numerical evaluations involved in (9.14), MP replace the sharp resonances arising from (9.14*d*) by a distribution of width $\sim$ 0.2 K. Earlier work by Jackson (1973) was similar in spirit except that he used the Feynman–Bijl basis instead of the improved Feynman–Cohen basis. In Fig. 9.2, we show the theoretical predictions for $S(\mathbf{Q}, \omega)$ of MP at $Q = 0.825$ Å^{-1} and $T = 0$.

The results are significantly altered in going from G_2^0 to the renormalized G_2 in (9.14). Including the associated vertex corrections may also produce comparable changes. Indeed, such vertex corrections in the ladder approximation are what give rise to bound-state resonances involving roton–roton, maxon–roton and maxon–maxon pairs (see Chapter 10). A serious weakness of the MP work is that it does not consider the effect of interactions on the pair-excitation contributions, such as considered by Ruvalds and Zawadowski (1970).

The CBF approach to liquid ^{4}He has always had the clear goal of developing quantitative numerical methods for dealing with the effects of the hard-core interaction in order to calculate (a) the ground-state energy and static properties, such as the pair distribution function $g(\mathbf{r})$ and the momentum distribution function $n(\mathbf{p})$, and (b) the observed phonon–maxon–roton dispersion curve using similar variational methods, as well as the two-excitation spectrum at higher energies. Since the late 1960's, the CBF approach has been energetically pursued by many workers, with increasingly impressive agreement with experimental data (for a brief review, see Campbell and Clements, 1989). As with Green's function and path-integral Monte Carlo simulation methods, the test of the correlated-basis-function programme is in its final numerical results,

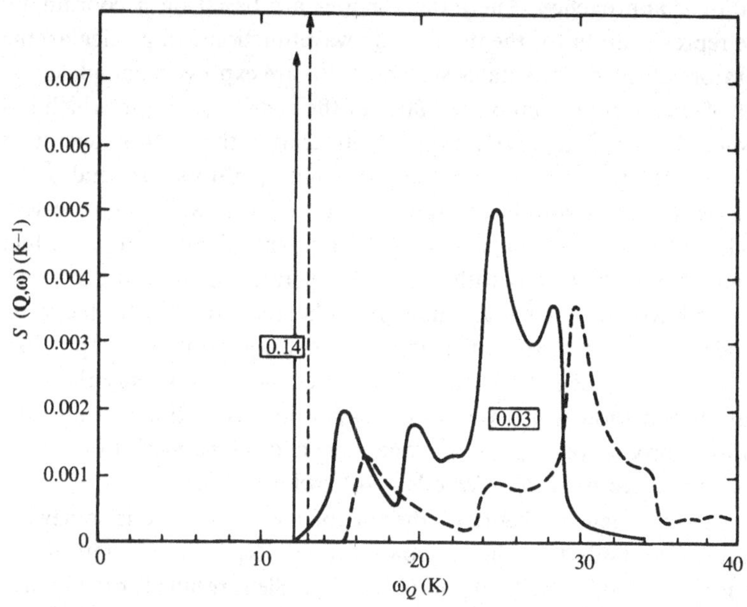

Fig. 9.2. Theoretical results for $S(\mathbf{Q},\omega)$ at $Q = 0.825$ Å^{-1}, showing the single-quasiparticle peak and the multiparticle structure at $T = 0$. These results are based on (9.13)–(9.14). The dashed (full) lines are based on using two-excitation propagator G_2^0 (G_2). The numbers in the boxes give the relative contribution to the static structure factor $S(\mathbf{Q}) = 0.21$ [Source: Manousakis and Pandharipande, 1986].

not in a particular microscopic model of superfluid ^{4}He (such as we have been emphasizing in this book). In spite of these different goals, it is still interesting to formulate the physical picture described in Section 5.5 in terms of some equivalent variational ansatz for the excited states. In turn, this may lead to an improved understanding of the physics behind expressions such as the FC generator in (9.9) or (9.10). We conclude this section with some remarks concerning this connection.

First of all, we recall that the key role of the Bose condensate is made very clear in the dielectric formalism of Chapter 5. The order parameter associated with the condensate gives rise to a coupling of the single-particle excitations and the density fluctuations. Indeed, in the field-theoretic microscopic theory, the coupling of SP excitations to both the particle–hole (density) and two-particle spectra are *both* connected with the existence of a Bose broken symmetry. The analogue of this condensate-induced hybridization is, however, only hinted at in variational approaches. The latter theories are based on a coordinate-space representation for the many-body wavefunctions. In particular, the generators of the excited states such as (9.10) are expressed completely in terms of the Fourier components $\hat{\rho}(\mathbf{Q})$ of the density operator, which has an especially simple representation (1.5) in terms of the position operators of the atoms. In contrast, the quantum field operators $\hat{\psi}(\mathbf{r})$ and $\hat{\psi}^+(\mathbf{r})$ are not easily dealt with in coordinate space. In their well known review, Woods and Cowley (1973, see pp. 1215ff) classify theories of superfluid ^{4}He into those which start with basic field operators $\hat{\psi}^+(\mathbf{r})$ and $\hat{\psi}(\mathbf{r})$ (such as the dielectric formalism and most field-theoretic approaches based on a condensate) and those which treat the density fluctuation operators $\hat{\rho}(\mathbf{r})$ as the fundamental variables. The latter include not only the CBF variational method which we are discussing in this section but also the kind of approach referred to as "quantum hydrodynamics" (for a brief introduction, see p. 1216 of Woods and Cowley, 1973).

In all such "density theories", the role of the condensate is somewhat obscure. One finds that the Jastrow–Feenberg-type ground-state wavefunctions do lead (using (9.4) and (9.5)) to excellent results for the atomic momentum distribution and the Bose condensate density n_0. On the other hand, the phase coherence and the resulting correlations implied by a Bose broken symmetry (over and above a Bose condensate) are not very transparent. That wavefunctions such as (9.1) imply a finite value of n_0 is not surprising, since we know that the ground state of a *free* Bose gas already has all the atoms in the condensate ($n_0 = n$) and that the short-range correlations induced by the hard cores have the effect of reducing but not destroying this condensate. One of the problems of dealing in coordinate space with the role of the condensate atoms is that spatially they are completely delocalized.

We have already noted that almost all modern variational approaches for the excited states start with the Feynman–Cohen wavefunctions (9.9). As (9.10) makes clear, this wavefunction does not describe a *pure* density fluctuation. What insight we have into the physical meaning of this kind of wavefunction is still nicely summarized by the pioneering paper of Miller, Pines and Nozières (1962), based on a detailed analysis of the wavefunction describing a "tagged" or "impurity" atom moving in a Bose liquid. The effect of the backflow around the atom (of momentum $\mathbf{Q}$) is to modify its wavefunction to

$$\phi_Q(\mathbf{r}) = e^{i\mathbf{Q}\cdot\mathbf{r}}\left[1 + \int \frac{d\mathbf{q}}{(2\pi)^3} A(\mathbf{q})\hat{\rho}^+(\mathbf{q})e^{-i\mathbf{q}\cdot\mathbf{r}}\right] . \tag{9.15}$$

The form of this wavefunction emphasizes that the origin of the second term in (9.10) lies in the coupling of the single-particle motion with the density fluctuations, a feature which is somewhat hidden in the symmetrized form of the many-body wavefunction given by (9.8) and (9.9). Moreover, we recall the important observation (see p. 65 of Chester, 1963) that at large Q (i.e., in the roton region), the symmetrization in (9.9) has little effect and can be ignored. Within the context of the dielectric formalism, backflow has been discussed in terms of the virtual excitation of two quasiparticles and their coupling to a single quasiparticle (see p. 292 of Wong and Gould, 1974)

For small $Q \lesssim 0.7$ Å^{-1}, the excited states essentially correspond to zero sound density fluctuations in the dielectric formalism, with relatively minor renormalization effects due to their coupling to underlying single-particle (SP) excitations. In contrast, at large $Q \sim 2$ Å^{-1}, the dominant roton peak in $S(\mathbf{Q}, \omega)$ is predicted to be SP in nature. We note that this is precisely the short-wavelength region where the variational wavefunction (9.10) is significantly different from that describing a *pure* density fluctuation.

The recent work of Stringari (1991, 1992) is a promising beginning in our understanding the relative role of single-particle and collective modes in Bose-condensed liquids in a quantitative manner, generalizing earlier work which concentrated almost entirely on the Feynman and Feynman–Cohen states. As we noted in Section 8.1, Stringari (1992) has given a detailed derivation and discussion of various frequency-moment sum rules in Bose-condensed fluids involving the single-particle operators $\hat{a}_{\mathbf{Q}}, \hat{a}^+_{\mathbf{Q}}$ and the density operator $\hat{\rho}_{\mathbf{Q}}$. In this work, he has emphasized the distinction between the Feynman–Bijl state $|\mathrm{FB}\rangle$ in (9.6) generated by the density operator and the normalized single-particle state defined by

$$|\mathrm{SP}\rangle = \frac{1}{\sqrt{n(\mathbf{Q})}}\hat{a}_{-\mathbf{Q}}|\Phi_0\rangle \ . \tag{9.16}$$

Here $n(\mathbf{Q})$ is the atomic momentum distribution discussed in Chapter 4. In general, neither $|\mathrm{FB}\rangle$ or $|\mathrm{SP}\rangle$ coincides with an exact excited state of the many-body system, although both will have significant overlap with such a state in a Bose-condensed fluid. The exception is a dilute, weakly interacting gas, where $|\mathrm{FB}\rangle$ and $|\mathrm{SP}\rangle$ do indeed describe an excited state of momentum Q.

Stringari defines a single-particle energy (the analogue of (9.7)) by

$$\varepsilon_{\mathrm{SP}}(\mathbf{Q}) = \frac{\langle \mathrm{SP}|\hat{H} - \mu\hat{N}|\mathrm{SP}\rangle}{\langle \mathrm{SP}|\mathrm{SP}\rangle} \ . \tag{9.17}$$

At $T = 0$, this SP energy can be evaluated using various exact frequency-moment sum rules (such as in Section 8.1). These involve the off-diagonal components of the $T = 0$ two-particle density matrix (Clark and Ristig, 1989)

$$\rho^{(2)}(\mathbf{r}_1,\mathbf{r}_2;\mathbf{r}_1',\mathbf{r}_2') = N(N-1)\int d\mathbf{r}_3\ldots\int d\mathbf{r}_N \times\Phi_0(\mathbf{r}_1,\mathbf{r}_2,\mathbf{r}_3\ldots\mathbf{r}_N)\Phi_0(\mathbf{r}_1',\mathbf{r}_2',\mathbf{r}_3,\ldots\mathbf{r}_N)\ , \quad (9.18)$$

which should be compared with (9.4). The pair distribution function $g(\mathbf{r}_1 - \mathbf{r}_2)$ in (2.10) is given by the diagonal components $\rho^{(2)}(\mathbf{r}_1,\mathbf{r}_2;\mathbf{r}_1,\mathbf{r}_2)$. Stringari (1992) gives some preliminary estimates of the single-particle energy $\varepsilon_{SP}(\mathbf{Q})$ in the roton region and compares it with the Feynman energy excitation (1.6). The analysis initiated by Stringari and coworkers based on the precise definition of a single-particle state (9.16) seems to offer a powerful way of extracting information about the relative role of single-particle vs. collective excitations in Bose-condensed liquids.

9.2 Polarization potential theory

In an effort to give a *unified* description of the dynamics of *all* the Helium liquids in the $\mathbf{Q}$–ω regions studied by neutron scattering, Pines and coworkers have developed a phenomenological theory which attempts to include the important physics in a simple way. This polarization potential (PP) theory is built on mean-field theories (such as the RPA and Landau Fermi liquid theory) but also incorporates the correct static ($\omega = 0$) properties, various frequency-moment sum rules, and a physical picture of the short-range correlations. The latter picture is based on the observation that the restoring forces which lead to the existence of zero-sound-like density oscillations arise mainly from the strong short-range correlations, which are expected to be almost the same in both liquid ^{3}He and ^{4}He. The quantum-statistical and zero-point-energy correlations are much less important. PP theory was first developed for superfluid ^{4}He by Aldrich and Pines (1976), building on the pioneering work of Pines (1966). It was later extended to liquid ^{3}He (Aldrich, Pethick and Pines, 1976; Aldrich and Pines, 1978; Hess and Pines, 1988) and ^{3}He–^{4}He mixtures (Hsu, Pines and Aldrich, 1985). Excellent summaries of PP theory have been given by Pines (1985, 1987). In particular, we call attention to the history of its development (and the theory of quantum liquids in general) in Section 2.1 of Pines (1985). In this section, we concentrate on the PP theory for superfluid ^{4}He.

The essential insight is that the restoring forces which give rise to collective density oscillations in the Helium liquids are dominated by two phenomena:

(a) Short-range positional correlations between atoms, arising from the repulsive interatomic potential. These are incorporated in PP theory through a frequency-independent but momentum-dependent effective interaction f^s_Q, as in the RPA and Landau Fermi liquid theory.
(b) Backflow effects. As an atom moves around, it induces longitudinal currents which modify its motion by changing its effective mass. This backflow is described through an effective mass and a frequency-dependent self-consistent polarization field, parameterized by f^v_Q.

The polarization potential expression for the linear response function χ_{nn} is derived by introducing self-consistent scalar and vector potentials to describe the effects induced by an external scalar potential

$$\begin{aligned}\delta\rho(\mathbf{Q},\omega) &= \chi_{nn}(\mathbf{Q},\omega)\delta\phi^0(\mathbf{Q},\omega)\\ &\simeq \chi^{sc}_{nn}(\mathbf{Q},\omega)\left[\delta\phi^0(\mathbf{Q},\omega)+\delta\phi^{\mathrm{ind}}(\mathbf{Q},\omega)\right]\\ &\qquad +\frac{1}{m}\chi^{sc}_{nJ}(Q,\omega)\cdot\delta\mathbf{A}^{\mathrm{ind}}(\mathbf{Q},\omega)\ .\end{aligned} \tag{9.19}$$

The usual scalar polarization potential is given by

$$\delta\phi^{\mathrm{ind}}(\mathbf{Q},\omega) = f^s_Q\delta\rho(\mathbf{Q},\omega)\ , \tag{9.20}$$

while the new induced vector polarization field is

$$\delta\mathbf{A}^{\mathrm{ind}}(\mathbf{Q},\omega) = \frac{f^v_Q}{m}\delta\mathbf{J}(\mathbf{Q},\omega)\ , \tag{9.21}$$

the polarization strengths being denoted by f^s_Q and f^v_Q. The backflow effects enter through the last term in (9.19), which involves a screened density current-response function of the kind defined in (5.4) and (5.5). Using (5.5), (5.14) and (5.15b), the longitudinal components are

$$\left.\begin{aligned}\chi^{\ell sc}_{nJ}(\mathbf{Q},\omega) &= \frac{m\omega}{Q}\chi^{sc}_{nn}(\mathbf{Q},\omega)\ ,\\ \delta J^{\ell}(\mathbf{Q},\omega) &= \frac{m\omega}{Q}\delta\rho(\mathbf{Q},\omega)\ .\end{aligned}\right\} \tag{9.22}$$

Thus the backflow term in (9.19) reduces to $(\omega/Q)^2 f^v_Q\delta\rho(\mathbf{Q},\omega)$ and hence it can be solved to give

$$\chi_{nn}(\mathbf{Q},\omega) = \frac{\chi^{sc}_{nn}(\mathbf{Q},\omega)}{1-(f^s_Q+\frac{\omega^2}{Q^2}f^v_Q)\chi^{sc}_{nn}(\mathbf{Q},\omega)}\ . \tag{9.23}$$

This result is the formal starting point of the PP theory. All temperature-dependent effects (and quantum statistics) enter through the "screened" density-response function χ_{nn}^{sc}. To quote p. 587 of Pines (1985):

The expression in [(9.23)] provides a formal basis for a unified theory of normal liquid ^{3}He and superfluid ^{4}He. The physical basis is that if it is the strong interaction between the particles rather than their quantum statistics which plays a dominant role, then the influence of statistical correlations (or temperature) should be a minor one, so that the strength of the polarization potentials, f_Q^s and f_Q^v, should be very nearly the same for ^{3}He and ^{4}He at the *same* density. In other words, effects of statistics or temperature come in only through $\chi^{sc}(\mathbf{Q}, \omega)$.

Starting with the PP expression in (9.23), one may easily verify that the f-sum rule (see (2.23) and (2.26)) imposes the requirement

$$\lim_{\omega\to\infty} \chi_{nn}^{sc}(\mathbf{Q}, \omega) = \frac{nQ^2}{m_Q^* \omega^2} , \tag{9.24}$$

with the momentum-dependent effective mass being defined by $m_Q^* \equiv m + nf_Q^v$ (Aldrich and Pines, 1976). With (9.24), the zero sound dispersion relation given by the zero of the denominator of (9.23) reduces to

$$\omega^2 = \left(\frac{nf_Q^s}{m}\right) Q^2 . \tag{9.25}$$

This result implies that the sound velocity is directly related to the long-wavelength limit of f_Q^s,

$$\lim_{Q\to 0} f_Q^s = \frac{mc^2}{n} . \tag{9.26}$$

It is convenient to define the scalar potential f_Q^s in terms of a real-space soft potential $f^s(\mathbf{r})$. For $r > r_c$ (the core radius), $f^s(\mathbf{r})$ is assumed to be identical to the attractive long-range potential of the bare ^{4}He–^{4}He interaction. For $r < r_c$, $f^s(\mathbf{r})$ is parameterized in PP theory by the ansatz

$$f^s(\mathbf{r}) = a\left[1 - \left(\frac{r}{r_c}\right)^8\right] \theta(r_c - r) . \tag{9.27}$$

In liquid ^{4}He, it is found that the core radius is a constant $r_c \sim 2.68$ Å, at all densities. The only remaining parameter in (9.27) is the soft-core strength a and this is uniquely determined through (9.27) and (9.26) by the (pressure-dependent) sound velocity c. It is found that $a = 49$ K in (9.27) from a fit to SVP data, and increases with pressure. We note that the parameterized $f^s(\mathbf{r})$ leads to a f_Q^s which goes through zero at

$Q \simeq 1.85$ Å^{-1} at *all* pressures. The determination of f_Q^v is somewhat less direct, but a procedure for obtaining it will be discussed shortly.

In applying (9.23) to superfluid ^{4}He at $T = 0$, Aldrich and Pines (1976) and subsequent papers (Pines, 1985, 1987) have been based on the model approximation for the screened response function

$$\chi_{nn}^{sc}(\mathbf{Q}, \omega) = \alpha_Q \frac{nQ^2/m_Q^*}{\omega^2 - \varepsilon_Q^{*2}} + \chi_{\text{mp}}(\mathbf{Q}, \omega) \ . \tag{9.28}$$

Physically the first term (of weight α_Q) describes the density fluctuation produced by exciting a single quasiparticle $\varepsilon_Q^* \equiv Q^2/2m_Q^*$ out of the condensate. The second or "multiparticle" term describes the excitation of two (or more) such quasiparticles out of the condensate and becomes increasingly important as Q increases. This contribution is peaked at high energies (Hess and Pines, 1988).

As we have noted above, PP theory is based on separating χ_{nn} into two parts, the polarization fields (whose strength is determined by f_Q^s and f_Q^v) and the screened responses to these fields. The main effect of pressure is expected to be on renormalizing the values of f_Q^s and f_Q^v, rather than on changes in χ_{nn}^{sc}. Confirmation of this basic idea was obtained by Aldrich and Pines (1976) in the following way. They fitted the observed dispersion curve ω_Q at $T \sim 1$ K and SVP to the PP expression given by (9.23),

$$1 = \left(f_Q^s + \frac{\omega^2}{Q^2} f_Q^v \right) \chi_{nn}^{sc}(Q, \omega) \ . \tag{9.29}$$

This fit was based on the value of f_Q^s discussed above in conjunction with the ansatz (9.28), the multiparticle part being approximated by its zero-frequency limit, $-nA_Q$. The values of α_Q, A_Q and f_Q^v are then adjusted to give the best fit to the SVP data for ω_Q up to $Q \sim 2$ Å^{-1}. The resulting values of α_Q and A_Q are assumed to be independent of pressure and are shown in Fig. 9.3. Since ω_Q does not change with pressure at $Q = 1.85$ Å^{-1}, AP use (9.29) in conjunction with the ansatz (9.28) to determine the f_Q^v at this *particular* value of Q for different pressures (assuming that α_Q and A_Q are pressure-independent). The value of f_Q^v at $Q = 0$ can be obtained from the pressure-dependent effective mass m_0^* in liquid ^{3}He at the same density (at SVP, this gives $nf_0^v \simeq 3.2m$). Finally, f_Q^v at *intermediate* values of Q is determined by a smooth extrapolation between $Q = 0$ and 1.85 Å^{-1}.

Armed with the pressure-dependent values of f_Q^s and f_Q^v (the SVP values are shown in Fig. 9.4) and the pressure-independent values of α_Q and A_Q, Aldrich and Pines (1976) then used (9.28) and (9.29) to predict

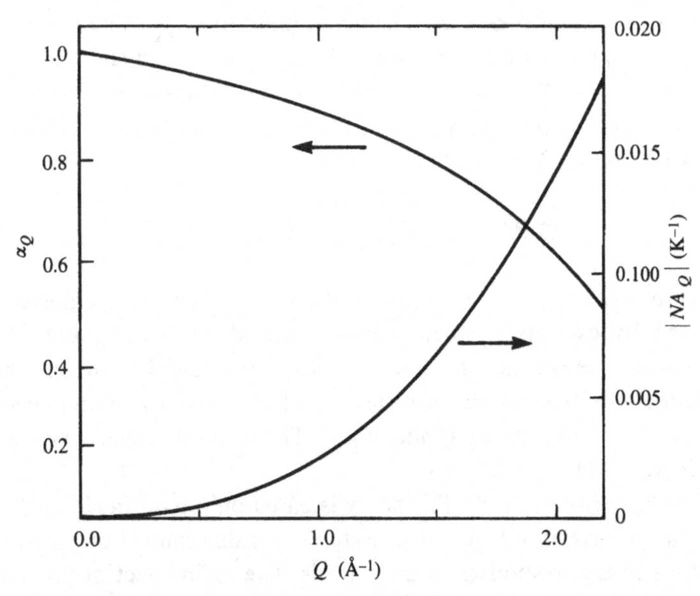

Fig. 9.3. The momentum dependence of the relative weight of the single-particle (α_Q) and the multiparticle ($nA_Q \equiv -\chi_{\text{mp}}(\mathbf{Q}, \omega \simeq 0)$) contributions to the screened density-response function given by the ansatz in (9.28) [Source: Aldrich and Pines, 1976; Pines, 1985].

how ω_Q changes with pressure. They found excellent agreement up to the maximum possible pressure of 25 bar. We view this agreement as justification of the idea that the *main* pressure dependence of $\chi_{nn}(\mathbf{Q}, \omega)$ comes through f^s_Q and f^v_Q, rather than the precise form of χ^{sc}_{nn} given by (9.28).

In trying to relate the PP expression (9.23) to the general results of the dielectric formalism of Chapter 5, we encounter a problem: the soft potential f^s_Q corresponds to some sort of t-matrix in the many-body analysis. By analogy with similar discussions in liquid ^{3}He, where multiple scattering has been extensively discussed, $V(Q)$ in (5.2) can be replaced by a soft-core potential such as f^s_Q, i.e.

$$\chi_{nn}(\mathbf{Q}, \omega) = \frac{\bar{\chi}_{nn}(\mathbf{Q}, \omega)}{1 - f^s_Q \bar{\chi}_{nn}(\mathbf{Q}, \omega)} . \tag{9.30}$$

If we compare this with the PP expression (9.23), we obtain

$$\bar{\chi}_{nn}(\mathbf{Q}, \omega) = \frac{\chi^{sc}_{nn}(\mathbf{Q}, \omega)}{1 - f^v_Q \frac{\omega^2}{Q^2} \chi^{sc}_{nn}(\mathbf{Q}, \omega)} . \tag{9.31}$$

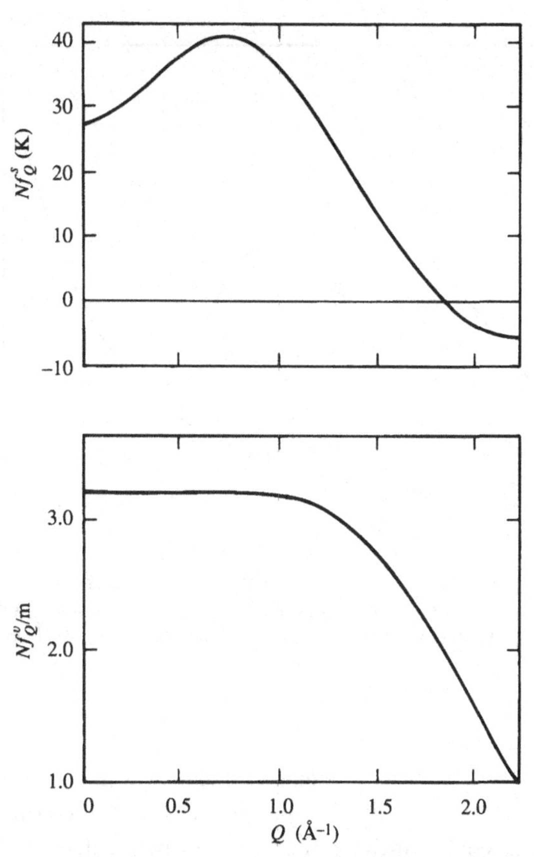

Fig. 9.4. The momentum dependence of the polarization potential scalar and vector (backflow) pseudopotentials at SVP [Source: Aldrich and Pines, 1976, 1978].

The two-component form (5.11b) for $\bar{\chi}_{nn} = \bar{\chi}^c_{nn} + \bar{\chi}^R_{nn}$ which results from the existence of the Bose condensate will be reflected in the ansatz one makes for χ^{sc}_{nn}. The PP expression given in (9.28) may be viewed as appropriate at $T = 0$, the first term being an approximation to $\bar{\chi}^c_{nn} = \bar{\Lambda}_\alpha \bar{G}_{\alpha\beta} \bar{\Lambda}_\beta$ in the dielectric formalism. In this context, (5.50) suggests a natural extension of (9.28) to finite temperatures which includes the effect of the "normal fluid" associated with thermally excited quasiparticles. For completeness, in Fig. 9.5 we plot the single-particle excitation energy ε^*_Q used in the polarization potential theory (see (9.28)). It would be useful to have further studies of ways to formulate the dielectric formalism results in terms of a suitably parameterized PP-type theory.

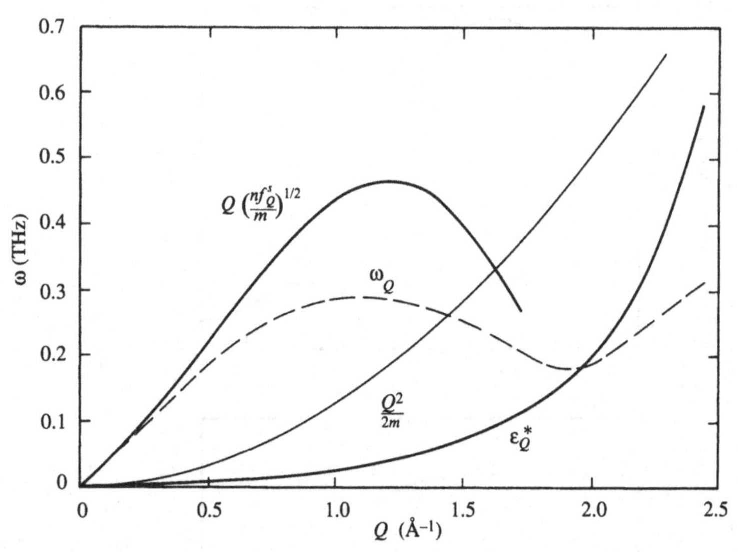

Fig. 9.5. The dispersion relation of the bare single-particle energy $\varepsilon^*_Q = Q^2/2m^*_Q$ used in the PP expression (9.28). The effective mass is $m^*_Q \equiv m + nf^v_Q$, where the vector polarization potential is given in Fig. 9.4. The "zero sound" frequency is defined by (9.25).

In Section 3.4, we outlined a mean-field derivation of the *simplest* version of the dielectric formalism results. It is a straightforward process to extend that analysis to include the linear response due to a vector potential. This generalization gives a version of the polarization potential theory which explicitly includes the dynamical effect of the Bose order parameter. One is thus able to exhibit the effects of backflow, which have been left out of the correlation functions given in (3.52)–(3.55). After a somewhat lengthy calculation, one finds that χ_{nn} is once again given by (9.23) (and hence $\bar{\chi}_{nn}$ is given by (9.31)), with $\chi^{sc}_{nn}(\mathbf{Q},\omega)$ being equal to the density-response function $\chi^0_{nn}(\mathbf{Q},\omega)$ of a non-interacting Bose gas. Thus the key PP expression (9.23) is seen to be valid even when one explicitly includes the effect of the Bose order parameter. In this extended PP theory, one can also discuss how backflow modifies the single-particle Green's functions $G_{\alpha\beta}(\mathbf{Q},\omega)$ due to Bose broken symmetry (Griffin, Wu and Lambert, 1992).

The arguments leading to (9.23) are also valid for liquid ^{3}He (Pines, 1985). Equally important is that, to a good first approximation, χ^{sc}_{nn} is given by the Lindhard function for a quasiparticle gas with an effective mass $m_0^* \simeq 3m$ – in contrast with superfluid ^{4}He, where we as yet do not have a simple model expression for χ_{nn}^{sc}.

9.3 Memory function formalism

Another method of evaluating the density response function $\chi_{nn}(\mathbf{Q}, \omega)$ is to use the memory function formalism. The method can be expressed in terms of projection operators which lead to a systematic (continued fraction) expansion for the memory (or polarization) function $M(\mathbf{Q}, \omega)$ defined below. (For a general introduction to this method, see Chapter 9 of Hansen and McDonald, 1986.) In most applications to quantum liquids, various simplified forms are introduced for $M(\mathbf{Q}, \omega)$. As with the polarization potential method discussed in Section 9.2, the basic physical assumptions concerning the nature of the excitations are then contained in the choice of model approximation for $M(\mathbf{Q}, \omega)$. To some extent, "what goes in is what comes out."

An advantage of the memory function approach is that one can treat both classical and quantum liquids in the same way. In addition, the hydrodynamic low-ω limit can be properly described, which is not the case with the variational approach discussed in Section 9.1. The application of the memory function formalism to the calculation of $S(\mathbf{Q}, \omega)$ in classical liquids is reviewed by Copley and Lovesey (1975). For a discussion of quantum liquids using the memory formalism, we recommend Götze and Lücke (1976) and Lücke (1980).

In the memory formalism, the usual "starting" expression for the density-response function or susceptibility is an expression of the kind

$$\chi_{nn}(\mathbf{Q}, \omega) = \frac{nQ^2/m}{\omega^2 - \Omega^2(Q) + \omega M(\mathbf{Q}, \omega)} , \tag{9.32}$$

where the characteristic frequency $\Omega(Q)$ is defined as

$$\Omega^2(Q) \equiv \frac{nQ^2}{m\chi(Q)} . \tag{9.33}$$

This definition ensures that $\chi_{nn}(\mathbf{Q}, \omega)$ automatically has the correct $\omega = 0$ limit, namely $\chi_{nn}(\mathbf{Q}, \omega = 0) = -\chi(Q)$, where $\chi(Q)$ is the static susceptibility discussed in Section 2.2 and satisfies the compressibility sum rule (2.28). In (9.32), $M(\mathbf{Q}, \omega)$ plays a role somewhat analogous to the self-energy function in the single-particle Green's function. In the derivation of (9.32), projector techniques are used to separate the motions into a coherent one (characterized by the frequency $\Omega(Q)$) and incoherent fluctuations described by $M(\mathbf{Q}, \omega)$. To the extent that this separation can be done, $M(\mathbf{Q},\omega)$ will be a fairly smooth function of ω; it is thus a better candidate for approximations than $\chi_{nn}(\mathbf{Q},\omega)$.

One may also treat (9.32) as a *definition* of $M(\mathbf{Q},\omega)$, since $\Omega(Q)$ in (9.33) is given in terms of an experimentally determined quantity $\chi(Q)$. At such a phenomenological level, one may attempt to extract $M(\mathbf{Q},\omega)$ directly from neutron scattering data. To the extent that $M(\mathbf{Q},\omega)$ is non-singular, this is a very convenient way of presenting experimental data. For example, expanding (9.32) for large ω and using (2.26), we find

$$\lim_{\omega\to\infty} \omega M(\mathbf{Q},\omega) = \frac{2m}{Q^2}\langle\omega^3\rangle - \Omega^2(Q) \ . \tag{9.34}$$

Both $\Omega(Q)$ in (9.33) and the frequency moment $\langle\omega^3\rangle$ given by (2.26) can be determined directly from the neutron-scattering data, for each value of Q. Another constraint is that the third frequency moment $\langle\omega^3\rangle$ satisfies a sum rule (Puff, 1965; Pines, 1985).

The simplest approximation is to set $M(\mathbf{Q},\omega)$ to zero. This effectively assumes that a single collective mode of frequency $\Omega_0(Q)$ dominates the dynamics, i.e., it saturates the various sum rules for the frequency moments ($n = -1$, 0 and 1) defined by (2.23). Computing the zeroth frequency moment of χ_{nn} using (9.32) with $M = 0$, we find $\Omega_0(Q) = \varepsilon_Q/S_0(\mathbf{Q})$. This is consistent with the Feynman expression (1.6). An equivalent way of writing this result is $\chi_0(Q) = 2nS_0(\mathbf{Q})/\Omega_0(Q)$, where $S_0(\mathbf{Q})$ is the static structure factor.

In using the density-response expression (9.32), one often introduces the response function of a model or "reference system", defined by

$$\chi^r(\mathbf{Q},\omega) \equiv \frac{nQ^2/m}{\omega^2 - \Omega_r^2(Q) - \omega M^r(Q,\omega)} \ , \tag{9.35}$$

with

$$\Omega_r^2(Q) \equiv \frac{nQ^2}{m\chi^r(Q)} \tag{9.36}$$

and $\chi^r(Q) \equiv -\chi^r(Q,\omega = 0)$. One may easily verify that (9.32) can be written in terms of χ^r as follows:

$$\chi_{nn}(\mathbf{Q},\omega) = \frac{\chi^r(\mathbf{Q},\omega)}{1 - V(\mathbf{Q},\omega)\chi^r(\mathbf{Q},\omega)} \ , \tag{9.37}$$

where a frequency-dependent potential has been introduced:

$$\begin{aligned} V(\mathbf{Q},\omega) &\equiv \frac{1}{\chi^r(\mathbf{Q},\omega)} - \frac{1}{\chi(\mathbf{Q},\omega)} \\ &= \frac{1}{\chi(\mathbf{Q})} - \frac{1}{\chi^r(\mathbf{Q})} + \frac{m\omega}{nQ^2}[M(\mathbf{Q},\omega) - M^r(\mathbf{Q},\omega)] \ . \end{aligned} \tag{9.38}$$

(In the literature – see, for example, Yoshida and Takeno, 1987 – such a reference system is often denoted by a "zero" superscript.) An approach based on (9.37) is only useful if one can incorporate enough of the dynamics into the reference system $\chi^r(\mathbf{Q},\omega)$, leaving $V(\mathbf{Q},\omega)$ in (9.38) to play the role of a relatively simple "effective" interaction. The similarity of (9.37) to the polarization potential expression (9.23) is clear. In normal liquid ^{3}He, the natural choice of χ^r is that of a non-interacting Fermi gas χ^0_{nn}, i.e., the Lindhard function. The appropriate form for the "model response function" of superfluid ^{4}He is not so obvious, as we noted in Section 9.2.

Götze and Lücke (1976) and Lücke (1980) have given a detailed analysis of $S(\mathbf{Q},\omega)$ for superfluid ^{4}He at $T = 0$ based on the assumption that the coherent mode $\Omega(Q)$ is self-consistently coupled into the two-mode spectrum. In their mode–mode coupling approximation model, the imaginary part of the memory function is (we refer to the original paper for details)

$$M''(Q,\omega) = \frac{m}{Q^2}\int\frac{d\mathbf{p}}{(2\pi)^3}\int\frac{d\omega'}{2\pi}\phi^2(\mathbf{Q},\mathbf{p}) \times\frac{1}{2}[\mathrm{sgn}\,\omega' + \mathrm{sgn}(\omega-\omega')]\chi''_{nn}(\mathbf{p},\omega)\chi''_{nn}(\mathbf{Q}-\mathbf{p},\omega-\omega')\ , \tag{9.39}$$

where the double primes denote the imaginary parts. It is important that the pair modes are described *self-consistently* by true spectral densities χ''_{nn} associated with the response functions in (9.32). They find the interaction vertex function in (9.39) to be given by

$$\phi(\mathbf{Q},\mathbf{p}) = \frac{1}{2m}[\Omega_0(Q)+\Omega_0(p)+\Omega_0(|\mathbf{Q}-\mathbf{p}|)] \times\left[\frac{\mathbf{Q}\cdot\mathbf{p}}{S(\mathbf{p})}+\frac{\mathbf{Q}\cdot(\mathbf{Q}-\mathbf{p})}{S(\mathbf{Q}-\mathbf{p})}-Q^2\right]\ , \tag{9.40}$$

where the static structure factors $S(\mathbf{p})$ and $S(\mathbf{Q}-\mathbf{p})$ are taken as inputs. As Götze and Lücke note, the memory function analysis based on (9.32) and (9.39) is reminiscent of the work carried out earlier by Jackson (1969, 1973) within a variational scheme where one- and two-density fluctuations are coupled (see Section 9.1). One should be able to work out an improved version of the Götze–Lücke theory which is built on the Feynman–Cohen states. It would also be useful to have the Manousakis–Pandharipande variational calculation (see Section 9.1) formulated via the Götze–Lücke approach.

We have seen in earlier chapters that the existence of a condensate has the crucial effect of coupling the density fluctuations of a normal Bose liquid to the single-particle excitations. The resulting hybridization leads to the characteristic phonon–maxon–roton spectrum (see Section 7.2) observed in $S(\mathbf{Q},\omega)$. A similar condensate-induced hybridization couples the one- and two-quasiparticle spectrum, as discussed in Section 10.1. The memory formalism with its continued-fraction expansion of $M(\mathbf{Q},\omega)$ is general enough to incorporate *explicitly* these effects of the Bose broken symmetry. It would be useful to extend the work of Götze and Lücke (1976) in this direction.

Finally, we sketch the bare bones of the continued-fraction representation of $\chi_{nn}(\mathbf{Q},\omega)$. Working with the Laplace transform

$$\chi_{nn}(Q,z) \equiv \int_0^\infty dt\, e^{-zt}\chi_{nn}(Q,t) \ , \tag{9.41}$$

one finds

$$\chi_{nn}(Q,z) = \Delta_0(Q)[z\overline{M}_0(Q,z) - 1] \ , \tag{9.42}$$

with the recursion relation

$$\overline{M}_n(\mathbf{Q},z) = \frac{1}{z + \Delta_{n+1}(Q)\overline{M}_{n+1}(\mathbf{Q},z)} \ . \tag{9.43}$$

The wavevector-dependent functions $\Delta_n(Q)$ are defined in terms of frequency moments of $\chi_{nn}(\mathbf{Q},\omega)$. For further details of how one derives these results, we refer to the literature (pp. 64ff of Boon and Yip, 1980; p. 311 of Yoshida and Takeno, 1989; Chapter 9 of Hansen and McDonald, 1986). Using (9.43), one can express $\overline{M}_0(\mathbf{Q},z)$ in (9.42) as a continued fraction as follows:

$$\overline{M}_0(\mathbf{Q},z) = \cfrac{1}{z + \cfrac{\Delta_1(Q)}{z + \Delta_2(Q)\overline{M}_2(\mathbf{Q},z)}} \tag{9.44}$$

$$= \cfrac{1}{z + \cfrac{\Delta_1(Q)}{z + \cfrac{\Delta_2(Q)}{z + \Delta_3(Q)\overline{M}_3(\mathbf{Q},z)}}} \ . \tag{9.45}$$

Thus one can formally express $\chi_{nn}(\mathbf{Q},z)$, in terms of the frequency moments $\Delta_n(Q)$, in the form of an infinite-order continued fraction.

One may verify using (9.42) and (9.44) that

$$\chi_{nn}(\mathbf{Q},z) = \frac{-\Delta_0(Q)\Delta_1(Q)}{z^2 + \Delta_1(Q) + \Delta_2(Q)z\overline{M}_2(\mathbf{Q},z)} \ . \tag{9.46}$$

This is the analogue of (9.32), with $z = i\omega$. Using the fact that $z\overline{M}_2(Q, z) = 0$ for $z = 0$, one concludes from (9.46) that $\Delta_0(Q) = \chi(Q)$. Similarly the high-frequency limit

$$\lim_{z\to\infty} \chi_{nn}(\mathbf{Q}, z) = -\frac{\Delta_0(Q)\Delta_1(Q)}{z^2}, \tag{9.47}$$

in conjunction with the f-sum rule, shows that $\Delta_0(Q)\Delta_1(Q) = nQ^2/m$. Thus we find $\Delta_1(Q) = \Omega^2(Q)$, where $\Omega(Q)$ is defined in (9.33).

Approximations enter when we truncate the expansion in (9.45) at some *finite* order. One can show that keeping terms up to order n in (9.45) means that $\overline{M}_0(\mathbf{Q}, t)$ will be given correctly up to order t^{2n}. If we truncate the expansion in (9.45) using

$$z\overline{M}_2(Q, z) = \frac{z^2}{z^2 + \Delta_3(Q)z\overline{M}_3(Q, z)} \simeq \frac{z^2}{z^2 + \Delta_3(Q)}, \tag{9.48}$$

then (9.46) reduces to ($z = i\omega$)

$$\chi_{nn}(\mathbf{Q}, \omega) \simeq \frac{nQ^2/m}{\omega^2 - \Omega^2(Q) - \dfrac{\Delta_2(Q)\omega^2}{\omega^2 - \Delta_3(Q)}}. \tag{9.49}$$

This result is equivalent to that obtained by Yoshida and Takeno (1987) for the time dependence of χ_{nn}. The poles are given by the solution of Eq. (3.10) of YT. Clearly this form can be interpreted as a density mode of frequency $\Omega(Q) = \sqrt{\Delta_1(Q)}$ being coupled (or hybridized) via $\Delta_2(Q)$ with another mode of frequency $\Omega_3(Q) = \sqrt{\Delta_3(Q)}$.

From this point of view, we conclude that the peculiar oscillatory time dependence of the memory function obtained by YT (1987) is most naturally interpreted as a condensate-induced hybridization of quasiparticles with the two-quasiparticle continuum. (Note, however, that the ideal Bose gas response function used by YT as a "reference system" is not valid for superfluid ^{4}He.) This same problem was originally studied in frequency space by Ruvalds and Zawadowski (1970) using Green's function techniques (see Section 10.2). It would be interesting to use (9.49), or something similar, to treat the hybridization of zero sound and the maxon–roton excitations which is discussed in Section 7.2.

A phenomenon similar to the hybridization of zero sound and the single-particle maxon–roton excitations in pure superfluid ^{4}He also occurs in ^{3}He–^{4}He mixtures. As discussed by Lücke and Szprynger (1982), when we use the memory function formalism generalized to deal with a two-component ^{3}He–^{4}He mixture, the ^{4}He quasiparticle memory function is found to have a resonance associated with the ^{3}He particle–hole modes. The resulting hybridization of the ^{4}He quasiparticle with the ^{3}He p–h states is often referred to as mode–mode coupling (for further discussion, see Hsu, Pines and Aldrich, 1985). The dielectric formalism for ^{3}He–^{4}He mixtures is briefly reviewed in Section 12.2.

10

Two-particle spectrum in Bose-condensed fluids

At many points in this book, we have mentioned the high-frequency scattering intensity which appears in the $S(\mathbf{Q}, \omega)$ data. This high-frequency component (see Fig. 1.6) is usually identified with the spectrum of two excitations (with total momentum Q) and is thus referred to as the multiphonon or multiparticle component. In addition to inelastic neutron scattering, this two-excitation spectrum can be more directly probed by inelastic Raman light scattering, but only at $Q = 0$. In this chapter, we briefly review the microscopic theory of such pair excitations and how they show up in both neutron and Raman scattering cross-sections.

Raman light scattering in superfluid ^{4}He has been extensively studied both theoretically and experimentally, especially with regard to the possible formation of bound states involving roton–roton, roton–maxon and maxon–maxon pairs (Ruvalds and Zawadowski, 1970; Iwamoto, 1970). High-resolution Raman experiments over a wide range of pressure and temperature are reviewed by Greytak (1978) and more recently by Ohbayashi (1989, 1991). An excellent theoretical introduction at a phenomenological level is given by Stephen (1976). Our emphasis will be on the role of the Bose broken symmetry.

In earlier chapters, we have carefully distinguished the single-particle Green's function $G_1(\mathbf{Q}, \omega)$ (which may be a 2×2 matrix) and the density-response function $\chi_{nn}(\mathbf{Q}, \omega)$. The latter gives the dynamic structure factor measured by neutron scattering. In the present chapter, we introduce several additional correlation functions which are needed to describe the pair-excitation spectrum and Raman scattering. The two-particle Green's function $G_2(\mathbf{Q}, \omega)$ describing the propagation of two atoms of total momentum $\hbar Q$ and energy $\hbar\omega$ is discussed in Section 10.1. We follow the analysis of Pitaevskii (1959) as well as Ruvalds and Zawadowski (1970). We also review how the pair spectrum is hybridized into the single-particle spectrum described by $G_1(\mathbf{Q}, \omega)$. This mixing occurs through the effect of the Bose condensate and is analogous to the hybridization of the p–h spectrum of $\chi_{nn}(\mathbf{Q}, \omega)$ with $G_1(\mathbf{Q}, \omega)$, as described in Chapters 5–7. In Section 10.2, we discuss the pair-excitation spectrum as exhibited in $S(\mathbf{Q}, \omega)$.

In Section 10.3, we briefly consider the Raman light-scattering cross-section $h(\mathbf{Q} \simeq 0, \omega)$ and its relation to a correlation function involving four density operators (in contrast to the two density operators in $S(\mathbf{Q}, \omega)$ involved in the Brillouin light scattering and in neutron scattering). In the usual analysis of this four-point correlation function, the density fluctuations are treated as elementary excitations and hence $h(\mathbf{Q} \simeq 0, \omega)$ is viewed as describing the propagation and interaction of two such density fluctuations. In contrast, our analysis views the pair excitation in the context of the field-theoretic dielectric formalism of Chapter 5, in which the quantum field operators (rather than the density fluctuation operators) are the starting point. We indicate the crucial role of the Bose broken symmetry in coupling the pair-excitation spectrum into both $G_{\alpha\beta}$ and χ_{nn}.

10.1 Two-particle Green's function

We first outline the $T = 0$ calculations of Ruvalds and Zawadowski (1970) and Zawadowski, Ruvalds and Solana (ZRS, 1972), which are a development of those by Pitaevskii (1959). In contrast to more phenomenological approaches to be discussed in later sections, the work of ZRS is grounded in a field-theoretic analysis of Bose-condensed liquid, such as we have used in earlier chapters. The pair-excitation spectrum is discussed in terms of the two-particle Green's function $G_2(\mathbf{Q}, \omega)$, the Fourier transform of

$$G_2(\mathbf{r} - \mathbf{r}', t - t') = -\langle T\tilde{\psi}(\mathbf{r}, t)\tilde{\psi}(\mathbf{r}, t)\tilde{\psi}^+(\mathbf{r}', t')\tilde{\psi}^+(\mathbf{r}', t')\rangle \ . \qquad (10.1)$$

Here the quantum field operators $\tilde{\psi}^+, \tilde{\psi}$ are defined as in (3.10) and describe the creation and destruction of non-condensate ^{4}He atoms. Clearly $G_2(\mathbf{Q}, \omega)$ describes the propagation of two atoms with total (centre-of-mass) momentum $\hbar\mathbf{Q}$ and total energy $\hbar\omega$. ZRS calculate $G_2(\mathbf{Q}, \omega)$ in the standard ladder approximation used for discussing bound states, which is described by a Bethe–Salpeter integral equation. The existence of a bound state (for an attractive interaction) shows up as a resonance below the two-particle continuum of two non-interacting excitations.

Before considering the approximate calculations of ZRS and Pitaevskii, it is useful to recall the general analysis of two-particle Green's functions $K^{\gamma\delta}_{\alpha\beta}(p', p; Q)$ given by Gavoret and Nozières (1964) and summarized in Section 5.4. Using the notation of that section, we found that $K^{\gamma\delta}_{\alpha\beta}$ as defined in (5.60) is given by the sum of a condensate term (5.63) (see Fig. 5.12)

$$^{c}K^{\gamma\delta}_{\alpha\beta}(p', p; Q) = \sum_{\rho,\sigma} Q^{\rho}_{\alpha\beta}(p', Q) G_{\rho\sigma}(Q) Q^{\gamma\delta}_{\sigma}(p, Q) \tag{10.2}$$

and a regular term (5.65), written schematically as (see Fig. 5.13)

$$^{R}K = G_1 G_1 + G_1 G_1 + G_1 G_1 \Gamma G_1 G_1 \ . \tag{10.3}$$

Finally the generalized interaction vertex Γ is given by the Bethe–Salpeter equation (5.68), written schematically as (see Fig. 5.14)

$$\Gamma = I + \frac{1}{2} I G_1 G_1 \Gamma \ . \tag{10.4}$$

These are all matrix equations.

We note that $K^{\gamma\delta}_{\alpha\beta}$ in (5.59) is defined in terms of the total field operators $\hat{\psi}, \hat{\psi}^{+}$ and not just the non-condensate parts $\tilde{\psi}, \tilde{\psi}^{+}$. Thus G_2 in (10.1) is associated with the regular part $^{R}K^{++}_{++}(p', p; Q)$, as given by (10.3) and (10.4). We see, however, that because the G_1's are 2×2 matrix propagators, the Bethe–Salpeter equation (10.4) for $\Gamma^{++}_{++}(p', p; Q)$ will be coupled into several of the other 15 functions $\Gamma^{\gamma\delta}_{\alpha\beta}$. The functions $K^{\gamma\delta}_{\alpha\beta}$ and $\Gamma^{\gamma\delta}_{\alpha\beta}$ satisfy various symmetry relations (see GN) which can be used to simplify the resulting coupled integral equations. Such calculations are illustrated by the analysis of Cheung and Griffin (1971b) at $T \neq 0$ and also Nepomnyashchii and Nepomnyashchii (1974) at $T = 0$. At this basic level, of course, the single-particle excitations, the particle–hole excitations and the two-particle excitations are all coupled into each other due to the Bose condensate. Apart from the asymptotic region of $Q, \omega \to 0$ (see Section 6.3), there are essentially no theoretical studies which work out the details of such a complete microscopic theory (while ensuring that all correlation functions share the same poles).

Rather than carry out a full calculation of the type sketched in the preceding paragraph, we make use of a simplified scheme following ZRS and Pitaevskii. First of all, we note that all components of $G_{\alpha\beta}$ share the same poles in a Bose-condensed fluid. Moreover, if we are interested in the spectrum associated with creating two excitations, we can concentrate on the positive-energy poles of $G_{\alpha\beta}$ and effectively schematize the structure of the equations of motion (Pitaevskii, 1959; pp. 237ff of Abrikosov, Gor'kov and Dzyaloshinskii, 1963). This is especially the case when one is considering a region (Q_c, ω_c) where a single excitation can decay into two excitations, both of which come from (Q, ω) regions far removed from (Q_c, ω_c). More specifically, one can also argue that at large wavevectors the numerator of the anomalous propagator $G_{12}(\mathbf{Q}, \omega)$ is much smaller than that of $G_{11}(\mathbf{Q}, \omega)$ because of the decreased importance of the Bose coherence factors (this is illustrated by the Bogoliubov approximation results in Section 3.2, with $u_Q \to 1$ and $v_Q \to 0$ at large Q).

All these remarks set the stage for the ZRS calculations, based on solving (5.65) and (5.68) for ${}^RK^{++}_{++}$ keeping only the G_{11} (or G_{++}) diagonal component of the 2×2 matrix Green's functions. Moreover, the latter is approximated by

$$G_{11}(\mathbf{Q}, \omega) \simeq \frac{1}{\omega - E_Q} \, , \tag{10.5}$$

where the "unhybridized" single-particle excitation E_Q is assumed to be given approximately by the observed maxon–roton dispersion relation in the region $1 \lesssim Q \lesssim 2.4$ Å^{-1}. As Q increases, E_Q goes over smoothly to $Q^2/2m$ (see the dashed line in Fig. 7.21). Following Ruvalds and Zawadowski (1970), we distinguish E_Q from the observed spectrum ω_Q since the former will be strongly hybridized when it overlaps with the pair-excitation spectrum.

As we have noted above, this approach attempts to isolate the problem of finding the pair spectrum and ignores the tricky self-consistency problem which requires that all two-particle correlation functions $K^{\gamma\delta}_{\alpha\beta}$ and one-particle functions $G_{\alpha\beta}$ share the same poles. This kind of procedure may still give reasonable results for the two-particle energy spectrum. We recall, for example, that a conserving approximation for χ_{nn} may be built on the Hartree–Fock single-particle Green's functions (Cheung and Griffin, 1971b). In Section 10.2, we discuss how the two-particle spectrum we obtain here shows up in the spectrum of $G_{\alpha\beta}$ and χ_{nn} using the dielectric formalism results of Chapter 5.

In the context of the above kind of simplified analysis, we now discuss the evaluation of $G_2(\mathbf{Q}, \omega)$ in (10.1) in a slightly more direct manner. It is convenient to introduce a general four-point two-particle Green's function for Bosons:

$$G_2(1, 2; 1', 2') = -\langle T\tilde{\psi}(1)\tilde{\psi}(2)\tilde{\psi}^+(2')\tilde{\psi}^+(1')\rangle \, . \tag{10.6}$$

In the ladder-diagram approximation equivalent to (5.68), G_2 satisfies the equation of motion

$$G_2(12;1'2') = G_1(1,1')G_1(2,2') + G_1(1,2')G_1(2,1') + \int d\bar{1} \int d\bar{2}\, G_1(1,\bar{1})G_1(2,\bar{2})I(\bar{1}-\bar{2})G_2(\bar{1}\bar{2};1'2') \; . \quad (10.7)$$

Here we represent the space-time coordinates as $1 \equiv \mathbf{r}, \tau$, etc. with $0 < \tau < \beta\hbar$; barred coordinates are integrated over (for more details concerning (10.7), see, for example, Chapter 13 of Kadanoff and Baym, 1962). In the case of interest, we set 1=2 and $1' = 2'$ and then (10.7) reduces to

$$\begin{aligned} G_2(11;1'1') &\equiv G_2(1-1') \\ &= 2G_1(1,1')G_1(1,1') \\ &\quad + \int d\bar{1} \int d\bar{2}\, G_1(1,\bar{1})G_1(1,\bar{2})I(\bar{1}-\bar{2})G_2(\bar{1}\bar{2};11') \; . \end{aligned} \quad (10.8)$$

If we introduce an effective short-range interaction $I(1-2) = g_4\delta(1-2)$, (10.8) simplifies to

$$G_2(1-1') = 2G_1(1,1')G_1(1,1') + g_4 \int d\bar{1}\, G_1(1,\bar{1})G_2(\bar{1}-1') \; . \quad (10.9)$$

Eq. (10.9) is easily solved by Fourier transformation to give

$$G_2(\mathbf{Q},\omega) = \frac{2F_0(\mathbf{Q},\omega)}{1 - g_4(Q)F_0(\mathbf{Q},\omega)} \; , \quad (10.10)$$

where

$$F_0(\mathbf{Q}, i\omega_n) = -\frac{1}{\beta}\sum_{\omega_\ell} \int \frac{d\mathbf{k}}{(2\pi)^3} G_1(\mathbf{k}, i\omega_\ell)G_1(\mathbf{Q}-\mathbf{k}, i\omega_n - i\omega_\ell) \; . \quad (10.11)$$

In (10.10), we have written the effective interaction as $g_4(Q)$ to emphasize that the most appropriate value may be quite different depending on the centre-of-mass momentum Q. One may think of the function F_0 as the particle–particle propagator (p–p) describing two non-interacting excitations, as compared with a particle–hole (p–h) propagator. Using the simple expression given by (10.5) in (10.11) and carrying out the Bose Matsubara frequency sums as in Section 3.3, we obtain the well known result

$$F_0(\mathbf{Q},\omega) = \int \frac{d\mathbf{k}}{(2\pi)^3} \frac{1 + N(E_{Q-k}) + N(E_k)}{\omega - (E_{Q-k} + E_k)} C(Q,k) \; . \quad (10.12)$$

Here the function $C(Q,k)$ has been introduced once just as a reminder of the various Bose coherence factors which have been ignored. This type of approximation for the pair-excitation spectrum was first used by

Pitaevskii (1959) at $T = 0$, where the Bose distributions $N(E)$ in (10.12) vanish.

The spectral density associated with $G_2(\mathbf{Q}, \omega)$ in (10.10) is conveniently defined as

$$\begin{aligned}\rho_2(\mathbf{Q}, \omega) &\equiv -\frac{1}{4\pi}\text{Im } G_2(\mathbf{Q}, \omega + i\eta) \\ &= -\frac{1}{2\pi}\frac{\text{Im } F_0(\mathbf{Q}, \omega)}{[1 - g_4 \text{ Re } F_0]^2 + [g_4 \text{ Im } F_0]^2} .\end{aligned} \tag{10.13}$$

The spectral density associated with two non-interacting excitations described by $G_2^0 = 2F_0$ is given by (at $T = 0$)

$$\rho_2^0(\mathbf{Q}, \omega) = \frac{1}{2}\int \frac{d\mathbf{k}}{(2\pi)^3}\delta(\omega - [E_{Q-k} + E_k]) . \tag{10.14}$$

This function ρ_2^0 has been evaluated by ZRS using a simple ansatz for the important roton and maxon wavevector regions (because of the high density of states, these dominate the integral in (10.14))

$$\left.\begin{aligned} E_Q &= \Delta + (Q - Q_R)^2/2\mu_R \quad , \quad Q \sim Q_R , \\ E_Q &= \Delta_M - (Q - Q_M)^2/2\mu_M \quad , \quad Q \sim Q_M . \end{aligned}\right\} \tag{10.15}$$

The resulting pair-excitation spectrum depends very much on the value of the centre-of-mass momentum $\mathbf{Q}$. The $\mathbf{Q} = 0$ region is probed by Raman scattering (Section 10.3), while relatively large values of $\mathbf{Q}$ are involved in neutron scattering (Section 10.2).

We first discuss the case $Q = 0$, which has been examined in most detail in the literature. One finds using (10.14) that

$$\begin{aligned}\rho_2^0(\mathbf{Q} = 0, \omega) &= -\frac{1}{2\pi}\text{Im } F_0(\mathbf{Q} = 0, \omega) \\ &= \left(\frac{Q_R}{2\pi}\right)^2\left(\frac{\mu_R}{\omega - 2\Delta}\right)^{\frac{1}{2}} + \left(\frac{Q_M}{2\pi}\right)^2\left(\frac{\mu_M}{2\Delta_M - \omega}\right)^{\frac{1}{2}}\end{aligned} \tag{10.16}$$

in the region $2\Delta \leq \omega \leq 2\Delta_M$ and zero elsewhere. Strictly speaking, these results are only correct close to the square root singularities at $\omega \gtrsim 2\Delta$ and $\omega \lesssim 2\Delta_M$. So far, we have not really addressed the question of how we determine the paramaterized (roton–roton) interaction g_4 in (10.9) and (10.10). This is a complicated problem in its own right and we refer to Bedell, Pines and Zawadowski (1984) for further discussion. As we shall see in Section 10.3, the Raman-scattering data indicate that there arises a d-like two-roton bound state, which requires the $\ell = 2$ angular momentum component of the roton–roton interaction to be attractive. For the particular case of $\mathbf{Q} = 0$ (where both rotons have a wavevector of magnitude Q_R), the Bethe–Salpeter equation (10.8) can be solved for each angular momentum channel. One finds (10.10) to be valid for the pair excitations in the ℓ-th angular momentum channel; the only change is that g_4 is now replaced by g_4^ℓ, but $F_0(\mathbf{Q} = 0, \omega)$ is the same for all values of ℓ. This is not the case when $\mathbf{Q} \neq 0$.

In the region below the two-roton continuum $\omega < 2\Delta$, (10.13) may exhibit a bound state if $g_4 < 0$. Integrating (10.12) at $T = 0$, one obtains

$$F_0(\mathbf{Q} = 0, \omega < 2\Delta) = -4\left(\frac{Q_R}{2\pi}\right)^2 \sqrt{\mu_R}\frac{1}{\sqrt{|E|}} \arctan\left(\left|\frac{2D}{E}\right|^{\frac{1}{2}}\right) , \quad (10.17)$$

where $E \equiv \omega - 2\Delta$ and $D \equiv \Delta_M - \Delta$ is an energy cutoff. Using this result in (10.13) gives the pair spectral density

$$\rho_2(\mathbf{Q} = 0, \omega < 2\Delta) = \frac{1}{2g_4}\delta[1 - g_4 \text{ Re } F_0(\mathbf{Q} = 0, \omega)] . \quad (10.18)$$

The pair continuum spectral density just *above* 2Δ is easily found from

$$F_0(\mathbf{Q} = 0, \omega > 2\Delta) = 2\left(\frac{Q_R}{2\pi}\right)^2 \sqrt{\mu_R}\frac{1}{\sqrt{E}}\left(\ln\left|\frac{\sqrt{E} + \sqrt{(2D)}}{\sqrt{E} - \sqrt{(2D)}}\right| - i\pi\right) . \quad (10.19)$$

For further details, we refer to ZRS. These results for $\mathbf{Q} = 0$ are relevant to Raman-scattering experiments on superfluid ^{4}He (see Section 10.3 for further discussion).

In neutron-scattering studies, in contrast, the pair-excitation spectrum with *finite* centre-of-mass momentum Q is probed. Ignoring the roton lifetime (width) for simplicity, one finds (ZRS)

$$\text{Re } F_0(\mathbf{Q}, \omega) = 2\rho_0(Q) \ln\left(\frac{|E|}{2D}\right) , \quad (10.20)$$

$$\text{Im } F_0(\mathbf{Q}, \omega) = \begin{cases} 0^+, & \text{for } E < 0 , \\ -2\pi\rho_0(Q), & \text{for } E > 0 , \end{cases} \quad (10.21)$$

where $\rho_0(Q) \equiv \mu_R Q_R^2/4\pi Q$ (and again we recall that $E \equiv \omega - 2\Delta$). These results are only valid for intermediate values of Q. As with the $Q = 0$ case discussed above, we see that (10.13) reduces to ($g_4' \equiv g_4\rho_0(Q)$)

$$\rho_2(\mathbf{Q}, \omega) = \frac{1}{2g_4'}\delta\left(1 - 2g_4' \ln \frac{|E|}{2D}\right) \quad (10.22)$$

for $\omega < 2\Delta$. Thus, as long as g_4 is attractive (< 0), we have a roton bound state with energy

$$\omega_B = 2\Delta - 2D \exp\left(-\frac{1}{2|g_4'|}\right) . \quad (10.23)$$

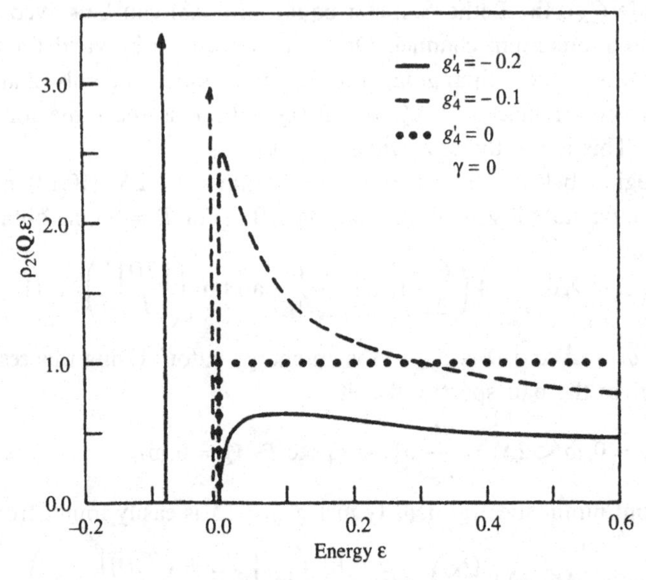

Fig. 10.1. Calculated two-particle density of states $\rho_2(\mathbf{Q}, \varepsilon)$ for non-zero total momentum Q, plotted as a function of the dimensionless energy $\varepsilon \equiv (\omega - 2\Delta)/2D$. ρ_2 is normalized with respect to $\rho_0(\mathbf{Q})$. Results are shown for three different values of the roton interaction $g_4' \equiv g_4\rho_0(\mathbf{Q})$. The bound state at $\omega < 2\Delta$ is shown by an arrow [Source: Zawadowski, Ruvalds and Solana, 1972].

In the continuum region $\omega > 2\Delta$, the interaction also modifies the pair spectrum, with (10.13) giving

$$\rho_2(\mathbf{Q}, \omega) = \frac{\rho_0(Q)}{\left[1 - 2g_4\rho_0(Q)\ \ln\left(\frac{E}{2D}\right)\right]^2 + [2\pi g_4\rho_0(Q)]^2} \tag{10.24}$$

in the frequency region not too far from the two-roton threshold at 2Δ. For illustration purposes, in Fig. 10.1 we show the two-roton spectral density predicted by (10.22) and (10.24). Including a finite roton lifetime broadens the bound-state singularity considerably, as shown in Fig. 10.2. For further details concerning such calculations, we refer to ZRS (1972), Zawadowski (1978) and Bedell, Pines and Zawadowski (BPZ, 1984).

It should be emphasized that the bound-pair spectrum reviewed in this section is mainly dependent on there being an attractive interaction. In no sense does the Bose condensate play any direct role. Indeed, the approximate non-self-consistent calculations of the pair spectrum we have been discussing are completely analogous to those given for bound states in normal systems. In principle, the bound roton state discussed

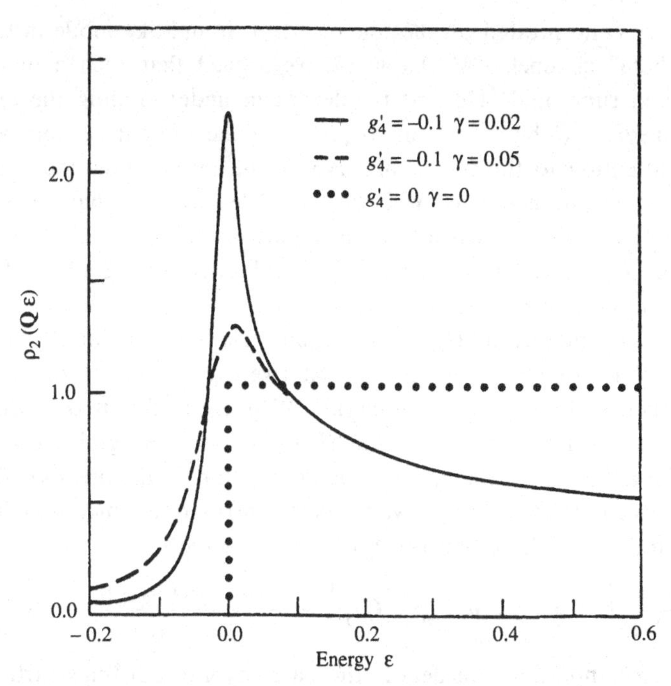

Fig. 10.2. The two-particle density of states $\rho_2(\mathbf{Q}, \varepsilon)$ at finite $\mathbf{Q}$ is plotted as a function of the dimensionless energy ε (same notation as in Fig. 10.1). The results for three different roton line widths (γ) are shown. The finite lifetime smears out of the sharp bound states shown in Fig. 10.1 [Source: Zawadowski, Ruvalds and Solana, 1972].

here could exist in normal liquid ^{4}He (ignoring the inevitable broadening effects which appear at increasing temperature).

Pitaevskii and Fomin (1974) have given a detailed analysis of the dependence of the two-roton bound-state energy on the centre-of-mass net momentum $\mathbf{Q}$, for different values of the total angular momentum ℓ and the azimuthal quantum number m. Raman-scattering experiments excite only states with even ℓ. In contrast, neutron scattering involves s-wave scattering and only excites the $m = 0$ state at large values of Q (see also BPZ, 1984, and Zawadowski, 1978).

10.2 Evidence for the two-excitation spectrum in neutron scattering

In earlier chapters we emphasized the inevitable hybridization of the single-particle and density fluctuation spectra through the action of the broken-symmetry vertex functions $\bar{\Lambda}_\alpha(\mathbf{Q}, \omega)$. In Chapters 5 and

7, we mainly concentrated on the role of a zero-sound-like mode in the "particle–hole" channel. In Chapter 7, we argued that such a mode does exist in superfluid ^{4}He and is relevant in understanding the low and intermediate Q behaviour of $S(\mathbf{Q},\omega)$. In the present section, we shift our attention to the region of *large* ω and discuss how the pair-excitation spectrum associated with the "particle–particle" channel (see Section 10.1) will also appear in the single-particle ($G_{\alpha\beta}$) and the density fluctuation (χ_{nn}) correlation function, due to the hybridizing effect of a Bose condensate (see Fig. 12.1).

The precise structure of $G_{\alpha\beta}$ in superfluid ^{4}He at large wavevectors ($Q \gtrsim 1$ Å^{-1}) is not known in any quantitative detail. For reasons discussed earlier, Ruvalds and Zawadowski (1970) assumed that $G_{11}(\mathbf{Q},\omega)$ could be approximated by (10.5) and $G_{12}(\mathbf{Q},\omega)$ was negligible. Using this kind of ansatz for the single-particle propagators in the one-loop diagrams of Section 5.3, the only two-particle contribution which remains comes from irreducible self-energy $\bar{\Sigma}_{11}$ in (5.52), namely

$$n_0 \int \frac{d\mathbf{p}}{(2\pi)^3}\left[V^2(\mathbf{p}) + V(\mathbf{p})V(\mathbf{p}+\mathbf{Q})\right]\frac{1+N(\omega_p)+N(\omega_{p+Q})}{\omega-(\omega_p+\omega_{p+Q})} . \qquad (10.25)$$

In our present notation, we denote the bare (hybridized) quasiparticle energies ω_p, ω_{p+Q} in (10.25) by E_p and E_{p+Q}. Generalizing this kind of contribution to include higher-order particle–particle ladder diagrams, one generates the self-energy terms Ruvalds and Zawadowski (1970) considered, namely (see Fig. 10.3)

$$\bar{\Sigma}_{11}(\mathbf{Q},\omega) = n_0 g_4^2(Q) G_2(\mathbf{Q},\omega) , \qquad (10.26)$$

where $G_2(\mathbf{Q},\omega)$ is given by (10.10) and $F_0(\mathbf{Q},\omega)$ is the propagator (10.11) describing two non-interacting rotons. As we discuss in Section 10.1, G_2 will exhibit a two-roton bound-state pole with centre-of-mass momentum $\mathbf{Q}$ if the roton–roton pseudopotential $g_4(\mathbf{Q})$ is attractive at that wavevector.

One also sees from (10.26) that the renormalized single-particle Green's function G_{11} will be given by (approximately)

$$G_{11}(\mathbf{Q},\omega) = \frac{1}{\omega - E_Q - n_0 g_4^2 G_2(\mathbf{Q},\omega)} , \qquad (10.27)$$

where

$$G_2(\mathbf{Q},\omega) = \frac{2\rho_0(Q)\left[\ln\frac{|\omega-2\Delta|}{2D} - i\pi\theta(\omega-2\Delta)\right]}{1-g_4^2\rho_0(Q)\left[\ln\frac{|\omega-2\Delta|}{2D} - i\pi\theta(\omega-2\Delta)\right]} . \qquad (10.28)$$

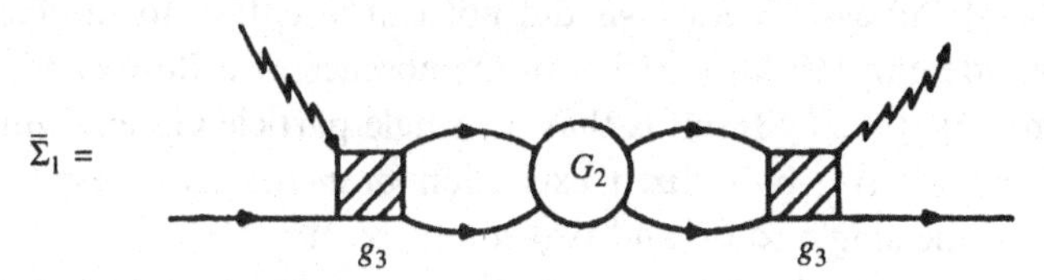

Fig. 10.3. The irreducible self-energy approximation used in the Pitaevskii–Ruvalds–Zawadowski theory. The two-particle propagator G_2 is coupled into the single-particle self-energy via the condensate n_0. Here $g_3 \equiv \sqrt{n_0}g_4$ and a jagged line represents an atom in the condensate. This self-energy is equivalent to that given in Figs. 79 and 80 of Abrikosov *et al.* (1963).

Here $\rho_0(Q)$ is defined after (10.21) and $\theta(\omega)$ is the step function. The zeros of (10.27) can give rise to two branches, a renormalized roton quasiparticle and (possibly) a renormalized two-roton resonance. These two branches can be expected to be strongly hybridized when they are degenerate, i.e., when $E_Q \simeq 2\Delta$. We emphasize that such hybridization effects occur *only* because of the Bose condensate. Needless to say, the dependence on n_0 exhibited in (10.27) is simply a consequence of the crude approximation we have made for the anomalous g_3-vertex function in Fig. 10.3.

In essentials, the Green's function analysis of RZ is already contained in the *tour de force* work of Pitaevskii (1959). Pitaevskii's work specifically addressed the question of the instability of the quasiparticle spectrum at large wavevectors, arising from the spontaneous decay into two rotons which occurs at some threshold wavevector Q_c. Not knowing the precise form of the numerators of the single-particle Green's functions, Pitaevskii concentrated on the singular behaviour of $F_0(Q,\omega)$ in (10.12) at $T = 0$ near Q_c, which is found to be given by

$$F(Q_c,\omega) \sim \ln\left(\frac{1}{2\Delta - \omega}\right) . \tag{10.29}$$

All other functions near this threshold can be treated as constants whose precise magnitude requires a (difficult) microscopic calculation. Ignoring the possibility of a two-roton bound state at such large wavevectors, Pitaevskii obtained (for $Q \sim Q_c$, $\omega \sim 2\Delta$)

$$G^{-1}(\mathbf{Q},\omega) \sim b + c\left[\ln\left(\frac{a}{2\Delta - \omega}\right)\right]^{-1} \tag{10.30}$$

(see §35 of Lifshitz and Pitaevskii, 1980). While it is implicit in the class of self-energy diagrams which were used to derive this result, the generality of Pitaevskii's analysis did not call attention to the fact that the amplitude c in (10.30) *vanishes* in the absence of a Bose condensate.

In summary, the key result is that the single-particle Green's functions $G_{\alpha\beta}$ will exhibit *two* hybridized excitation branches as a result of the coupling of the single-roton and two-roton spectra.

These results can be immediately used in connection with neutron-scattering studies since we know quite generally from Chapter 5 that $G_{\alpha\beta}$ and χ_{nn} are inter-related in the superfluid state. It is customary in the older literature (RZ, ZRS, BPZ) simply to assume that

$$S(\mathbf{Q},\omega) \propto n_0 \operatorname{Im} G_{11}(\mathbf{Q},\omega) \ , \tag{10.31}$$

where G_{11} is given by (10.27) and (10.28). As we discussed at some length in Section 7.2, this kind of approximation is partly justified microscopically in the high-Q maxon–roton wavevector region, where (7.6) is a useful starting point. In spite of the criticism of Griffin (1972), it captures some of the physics involved in the high-energy pair-excitation spectrum in $S(\mathbf{Q},\omega)$. More precisely, when there is no well defined zero sound mode present, (7.6) can be approximated by (7.8). On the other hand, the pair-excitation spectrum will show up in χ_{nn} as structure in $G_{\alpha\beta}$ not only via the self-energy $\bar{\Sigma}_{\alpha\beta}$ (as discussed above) but also via $\bar{\chi}^R_{nn}$ and $\bar{\Lambda}_\alpha$ in (7.8). This can already be seen from the one-loop diagrams in (5.52)–(5.55). Thus while the pair-excitation spectrum will appear with finite weight in $S(\mathbf{Q},\omega)$ due to the condensate, it cannot be simply identified as coming only from $G_{\alpha\beta}$ in (7.8) – an implicit assumption in using (10.31).

More generally, a proper treatment of the pair spectrum in $S(\mathbf{Q},\omega)$ must include the properly weighted contributions from both components in (3.47) or (7.6), with careful attention to the interference effects arising from the Bose-vertex functions $\bar{\Lambda}_\alpha$. It is clear that the pair contribution to $\bar{\chi}^R_{nn}$ in (7.8) can be directly related to $G_2(\mathbf{Q},\omega)$, given by (10.10) and (10.11) in the ladder-diagram approximation.

In this language, the pair-excitation spectrum in (3.45) corresponds to using $G_2 = 2F_0$, where F_0 is defined in (10.11). Examining (3.45), one sees that the pair excitation has its origin in the fact that the single-particle Green's functions have poles at $\pm\omega_Q$, the negative-energy pole having a finite weight due to the Bose broken symmetry (see (3.36)). The resulting cross-terms in (3.44) lead to the two-particle contributions of energy $\omega_p + \omega_{p+Q}$. This two-particle continuum contribution of energy $S(\mathbf{Q},\omega)$ has a coherence factor involving products of the u_p and v_p amplitudes and hence vanishes if v_p does (i.e., if $n_0 = 0$). Needless to say, (3.45) involves

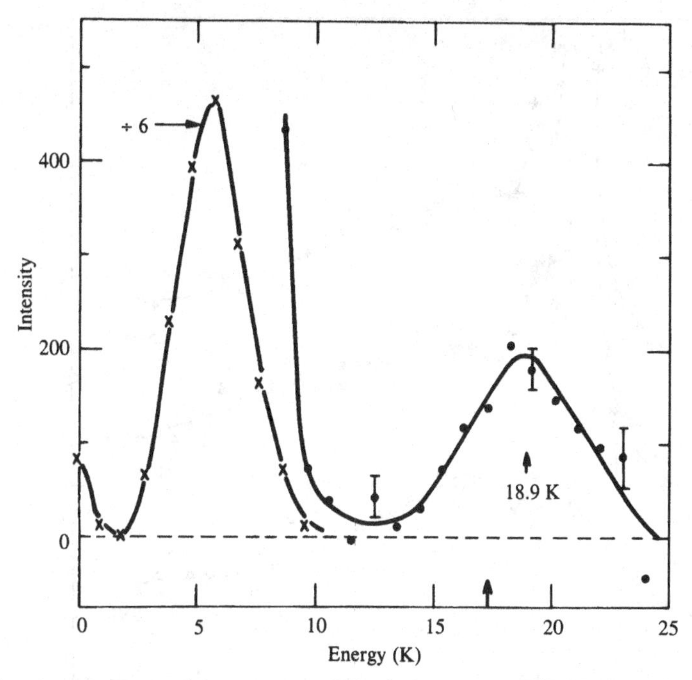

Fig. 10.4. Scattering intensity vs. frequency for $Q = 0.3$ Å^{-1}and $T = 1.2$ K. The unlabelled arrow is at $2\Delta = 17.3$ K. The lines are only a guide to the eye [Source: Woods, Svensson and Martel, 1972].

the same kind of integration as discussed in Section 10.1. The dominant contributions come from the roton and maxon regions because of high density of states. This multiparticle contribution should not, however, be thought of as the excitation of two atoms out of the condensate. Rather it arises from the fact that in a Bose-condensed system, creating (or destroying) an atom with finite momentum is a coherent process involving both creation and destruction of quasiparticles of that momentum.

In an analogous way, $S(\mathbf{Q}, \omega)$ in the Gavoret–Nozières formalism of Section 5.4 is the sum of the two contributions ${}^{c}K^{+-}_{-+}$ and ${}^{R}K^{+-}_{-+}$ given by (10.2) and (10.3), respectively. Fukushima and Iseki (1988) have made a careful analysis of the pair spectrum (including the maxon–roton part left out of (10.28)) based on the GN formalism, but only consider the ${}^{R}K^{+-}_{-+}$ contribution, as given by (5.65) and (5.68). The contribution given by the analogue of (10.31) was not included.

The first detailed study of the multiparticle distribution for *low Q* values (0.3 Å^{-1} and 0.8 Å^{-1}) was by Woods, Svensson and Martel

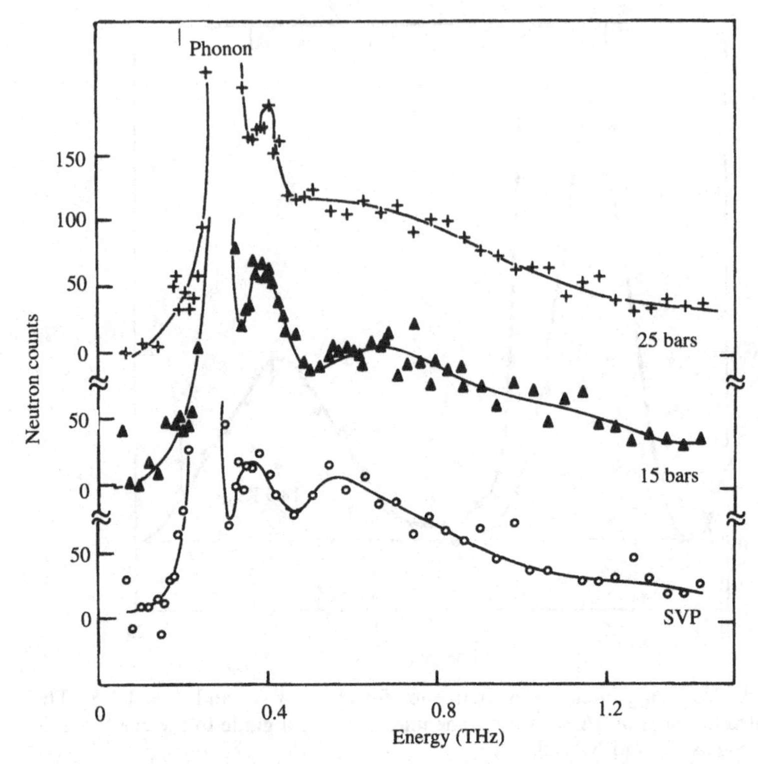

Fig. 10.5. The pressure dependence of the scattering intensity vs. frequency at $T = 1.27$ K, $Q = 1.5$ Å^{-1}. This should be compared with similar data at 1.13 Å^{-1} shown in Fig. 7.9 [Source: Stirling, 1985].

(1972), whose data are shown in Figs. 10.4 and 8.1 for $T = 1.2$ K. As do Stirling and Glyde (1990), they find the multiparticle peak to be quite symmetric for $Q = 0.3$ Å^{-1}, although at $Q = 0.8$ Å^{-1}, it has a high-energy tail extending to $\omega \sim 70$ K. This high-energy tail seems characteristic of all data at high Q (see further remarks at the end of Section 8.1).

The low-temperature $S(\mathbf{Q}, \omega)$ data always show a broad distribution centred at $\sim$ 20–25 K for $0.3 \lesssim Q \lesssim 2.5$ Å^{-1} (see Fig. 1.6). There seems little doubt that such a broad, high-energy component can arise from processes involving the creation of two quasiparticles. High-resolution neutron studies (especially in the maxon region) show detailed structure at frequencies close to that expected for the creation of two rotons (2Δ), a roton and a maxon ($\Delta + \Delta_M$) and two maxons ($2\Delta_M$). The roton and maxon regions make the dominant contribution to the two-quasiparticle

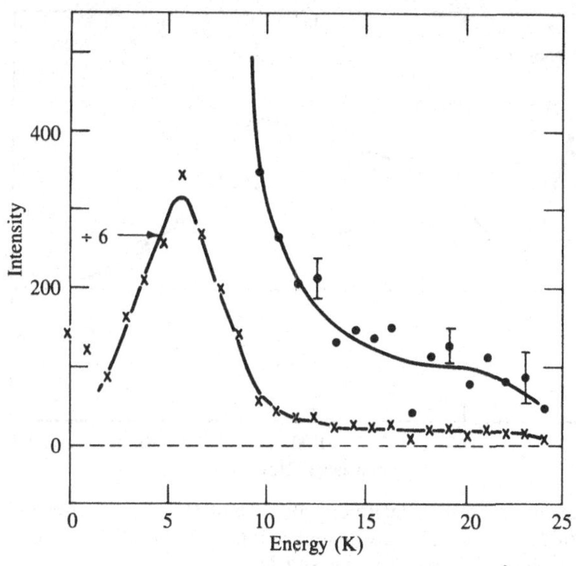

Fig. 10.6. Scattering intensity vs. energy loss for $Q = 0.3$ Å^{-1}, at a temperature 2.3 K just above T_λ. An expanded view of the high-energy region shows no evidence for the resonance which is visible in the low-temperature data in Fig. 10.4 [Source: Svensson, Martel, Sears and Woods, 1976].

scattering because of the high density of states at Q_M and Q_R. Such fine-scale structure is strikingly evident in recent ILL data of Stirling and coworkers, as shown in Fig. 10.5 as well as in Figs. 7.7, 7.9, and 7.10.

The preceding analysis gives a natural explanation of the origin of the high-frequency multiparticle component in terms of the two-quasiparticle spectrum. This spectrum is coupled (or mixed) into the density fluctuation spectrum *only* because of the effect of the Bose order parameter. As a counterexample, $S(\mathbf{Q}, \omega)$ in normal liquid ^{3}He exhibits a well defined zero sound peak for $Q \lesssim 1$ Å^{-1} but there is *no* evidence for any high-energy multiparticle structure of the kind that arises in superfluid ^{4}He in this low-momentum region.

Apart from our conclusion that the multiparticle component at high frequencies *should* disappear with the condensate fraction (i.e., it should be absent in normal liquid ^{4}He), at the present little is known about its precise temperature dependence. It deserves more study in its own right, especially in the region of low Q where it appears to separate out clearly as a distinct symmetric peak, without any high-frequency tail (see Figs. 7.1 and 10.4). Svensson (1989) has argued that the $Q = 0.3$ Å^{-1} data of

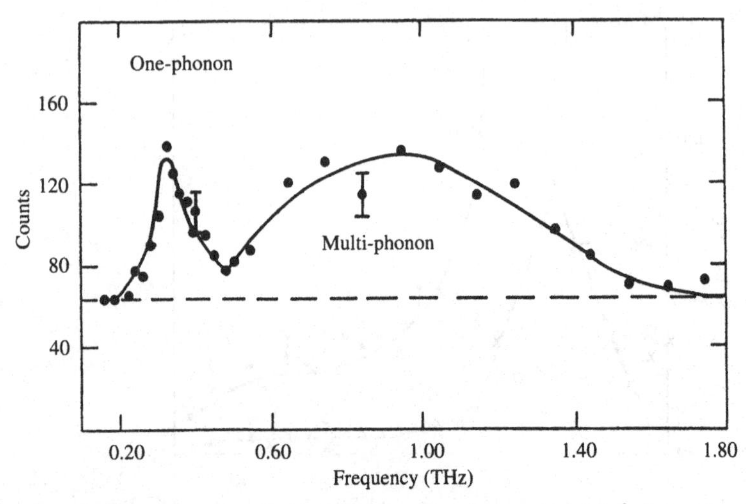

Fig. 10.7. Scattering neutron intensity distribution at $Q = 2.9$ Å^{-1} for $T = 1.1$ K, and a pressure of 24.3 atm. This can be compared with the SVP data in Fig. 7.20 [Source: Smith, Cowley, Woods, Stirling and Martel, 1977].

Svensson, Martel, Sears and Woods (1976) at 1.2 and 2.3 K (shown in Figs. 10.4 and 10.6, respectively) are consistent with the disappearance of this peak above T_λ, in agreement with the preceding remarks. Further high-resolution studies at low Q would be very useful, at a series of temperatures.

A careful study was carried out by Smith, Cowley, Woods, Stirling and Martel (1977) in the region $2.9 \leq Q \leq 3.3$ Å^{-1}, at both low and high pressures, looking for evidence of a two-roton bound state. Some of their data are shown in Figs. 10.7 and 10.8. The curves shown in Fig. 10.8 are based on the Ruvalds–Zawadowski (1970) expression for $G_1(\mathbf{Q}, \omega)$ in conjunction with (10.31),

$$S(\mathbf{Q}, \omega) \propto n_0 \operatorname{Im}\left[\frac{1}{\omega - \omega_Q - g_3^2 G_2(Q, \omega)}\right],$$

where $G_2(\mathbf{Q}, \omega)$ is given by (10.28). The hybridization coupling strength $g_3 \equiv \sqrt{n_0} g_4$, $\rho_0(Q) = \pi A/Q$ and the energy cutoff D in (10.28) were treated as fit parameters. The overall agreement is found to be quite reasonable, but this must be viewed in the context of the remarks we made following (10.31).

Many calculations (at $T = 0$) of the scattering due to the creation of two quasiparticles have been carried out (see, for example, Jackson, 1973,

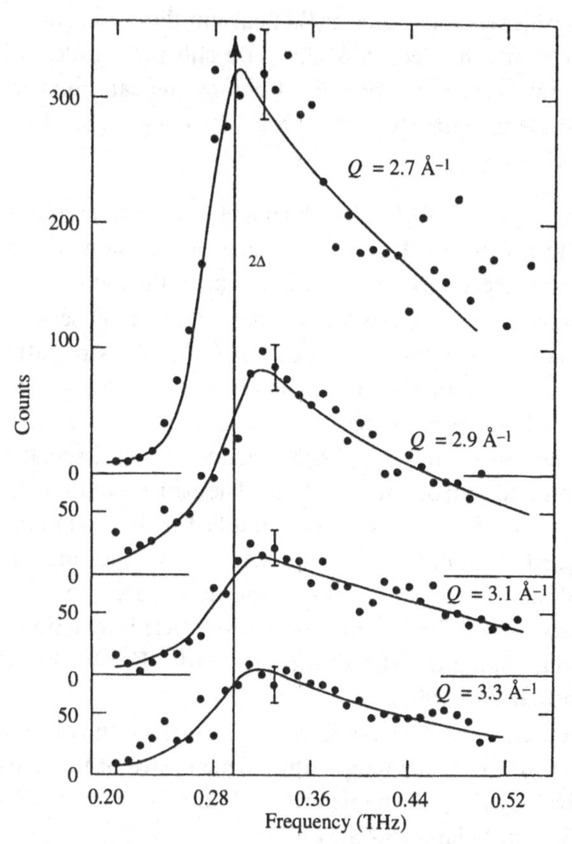

Fig. 10.8. Scattered neutron distributions (same temperature and pressure as in Fig. 10.7) for momentum transfers in the range $2.7 \leq Q \leq 3.3$ Å^{-1}. The solid lines are fits based on the RZ expressions in (10.27) and (10.28), convoluted with instrumental resolution [Source: Smith *et al.*, 1977].

1974; Götze and Lücke, 1976) using semi-phenomenological approaches. These studies involve the evaluation of the equivalent of the expression in the third line of (3.45). The most detailed study using the correlated-basis-function approach is by Manousakis and Pandharipande (1986), who evaluated the contributions to $S(\mathbf{Q}, \omega)$ due to the creation of one and two quasiparticles. Over a wide range of wavevectors, their calculations exhibited structure from roton–phonon, roton–maxon, maxon–maxon and roton–phonon pair spectrum. However, as we noted in Section 9.1, MP ignore the possibility of bound states. In the context of this chapter, (10.10) is approximated by $G_2 = 2F_0$.

In concluding this section, it is clear that current theoretical discussions of the pair spectrum exhibited by $S(\mathbf{Q},\omega)$ are still very crude, although promising. Several extensions are needed before one can hope to make a serious comparison with the data. We list a few areas for future theoretical work:

(a) We need a calculation of $S(\mathbf{Q},\omega)$ which includes the pair contribution from both terms in (3.47) or its analogue. Expressions like (10.31) cannot be expected to give a quantitative fit to the data.

(b) In using expressions like (10.28), we have limited ourselves to the structure related to exciting two rotons, $\omega \sim 2\Delta$. It is straightforward to evaluate $\rho_2^0(\mathbf{Q},\omega)$ in (10.14) to obtain a result which is valid at all Q in the whole spectral region $2\Delta < \omega < 2\Delta_M$. This allows one to include the spectrum (and possible bound states) associated with maxon–roton and maxon–maxon pairs. The pair-excitation structure exhibited by $\text{Im}\, F_0(\mathbf{Q},\omega)$ is strongly modified in $\text{Im}\, G_2(\mathbf{Q},\omega)$, which in turn is further modified by hybridization with the single-particle excitation described by $\text{Im}\, G_1(\mathbf{Q},\omega)$. The results are very dependent on the values chosen for the effective two-particle interaction g_4 and the condensate-induced hybridization strength. For further details, see Juge and Griffin (1993).

(c) At finite temperatures $T \gtrsim 1.2$ K, one must include the quasiparticle width $E_Q - i\Gamma_Q$, where the temperature dependence of the half-width $\Gamma_Q(T)$ is given by (7.2). A simple way of doing this is to use (Ruvalds, Zawadowski and Solana (RZS), 1972)

$$F_0(\mathbf{Q},\omega) = 2\int d\omega' \frac{\rho_2^0(\mathbf{Q},\omega')}{\omega - \omega' + i2\Gamma_Q} \tag{10.32}$$

in (10.13) with ρ_2^0 as defined by (10.14). The resulting generalization of (10.27) and (10.28) is given by Eq. (4.16) of ZRS for $\omega \sim 2\Delta$. In addition, one can take into account the changing condensate-induced hybridization by allowing the parameter $n_0(T)$ in (10.27) to decrease as the temperature increases.

(d) In this and the preceding section, we assume that the input phonon–maxon–roton spectrum E_Q (see (10.5)) is a single quasiparticle branch. A major extension would be to incorporate the Glyde–Griffin scenario developed in Section 7.2. This generalization is necessary if one hopes to understand the high-frequency maxon region in any detail. In this region, the particle–hole and two-particle spectrum can be expected to overlap in energy and hybridize.

10.3 Raman scattering from superfluid ^{4}He

Second-order inelastic light scattering was first suggested as a useful probe of superfluid ^{4}He by Halley (1969). Up to a constant which is of no interest here, the Raman-scattering rate or extinction coefficient for an isotropic liquid is given by ($\hat{\varepsilon}_i, \hat{\varepsilon}_f$ are the polarization vectors of the incident and scattered light)

$$h(\omega) = I_s(\omega)(\hat{\varepsilon}_i \cdot \hat{\varepsilon}_f)^2 + I_d(\omega)\left[\frac{3}{4} + \frac{1}{4}(\hat{\varepsilon}_i \cdot \hat{\varepsilon}_f)^2\right] , \tag{10.33}$$

where $\hbar\omega = \Omega_i - \Omega_f$ is the energy transferred from the photons to the liquid and

$$I_\ell(\omega) \equiv \left(\frac{4}{25}\right)^{\ell/2} \int dq\, 4\pi q^2 \int dq'\, 4\pi q'^2 t_\ell(q) t_\ell(q') S_2^\ell(q, q'; \omega) . \tag{10.34}$$

Here $t_\ell(q)$ involves a Fourier transform over the atomic polarizability tensor $\alpha_{ij}(\mathbf{r} - \mathbf{r}')$. The functions

$$S_2^\ell(q, q'; \omega) \equiv \frac{1}{2}(2\ell + 1) \int d(\cos\theta_{qq'}) P_\ell(\cos\theta_{qq'}) S_2(\mathbf{q}, \mathbf{q}', \omega) \tag{10.35}$$

are the $\ell = 0, 2$ (s, d) projections of the four-point density correlation function

$$\begin{aligned} S_2(\mathbf{q}, \mathbf{q}'; \omega) \equiv \\ \int d\mathbf{r}_1 \int d\mathbf{r}_2 \int d\mathbf{r}_3 \int d\mathbf{r}_4 \int d(t - t') e^{i\omega(t-t')} e^{-i\mathbf{q}\cdot(\mathbf{r}_1 - \mathbf{r}_2)} e^{-i\mathbf{q}'\cdot(\mathbf{r}_3 - \mathbf{r}_4)} \\ \times \langle \delta\hat{\rho}(\mathbf{r}_1, t) \delta\hat{\rho}(\mathbf{r}_2, t) \delta\hat{\rho}(\mathbf{r}_3, t') \delta\hat{\rho}(\mathbf{r}_4, t') \rangle . \end{aligned} \tag{10.36}$$

It has already been assumed that the important values of q and q' in the integration (10.34) are much larger than the wavevectors of the incident and scattered light. Apart from this approximation, the above results for the second-order scattering may be viewed as exact. In superfluid ^{4}He, the s-wave ($\ell = 0$) scattering contribution $I_s(\omega)$ is much weaker than the d-wave ($\ell = 2$) contribution $I_d(\omega)$. We restrict our attention to the d-wave Raman-scattering contribution.

For the derivation of (10.33)–(10.36), we refer to Section 4.2 of the review article by Stephen (1976) as well as Stephen (1969), Iwamoto (1970), Nakajima (1971) and Fetter (1972). The starting point of our discussion is that the measured Raman-scattering rate is given rigorously in terms of the basic correlation function $S_2(\mathbf{q}, \mathbf{q}', \omega)$ defined in (10.36). This Fourier transform is easily worked out:

$$S_2(\mathbf{q}, \mathbf{q}'; \omega) = \int_{-\infty}^{\infty} dt\, e^{i\omega t} \langle \delta\rho(\mathbf{q}, t) \delta\rho(-\mathbf{q}, t) \delta\rho(\mathbf{q}', 0) \delta\rho(-\mathbf{q}', 0) \rangle . \tag{10.37}$$

We recall (Sections 2.1 and 6.2) that inelastic neutron scattering and Brillouin light scattering are direct probes of the two-point density correlation function $\langle\delta\rho(\mathbf{q},t)\delta\rho(-\mathbf{q},0)\rangle$, although over different wavevector and time domains. In contrast, we see that the Raman-scattering experiments are more indirect, for two reasons. First of all, $S_2(\mathbf{q},\mathbf{q}',\omega)$ involves a thermal average over four density operators (or eight quantum field operators). Secondly, the Raman-scattering rate $h(\omega)$ in (10.33) involves integrations of $S_2(\mathbf{q},\mathbf{q}';\omega)$ over $\mathbf{q}$ and $\mathbf{q}'$, as given by (10.34) and (10.35). This means that, in general, it is difficult to unravel the structure of $S_2(\mathbf{q},\mathbf{q}';\omega)$ from the Raman-scattering data. Fortunately, in superfluid ^{4}He, the existence of well defined density fluctuations with a roton minimum means that the dominant contribution comes from wavevectors close to the roton wavevector Q_R (because of the high density of states there). This allows one to extract useful information from $h(\omega)$.

In the literature, the simplest approximation to (10.37) treats the density fluctuations as the basic variables and uses a simple decoupling (for $\omega \neq 0$):

$$\begin{aligned} S_2(\mathbf{q},\mathbf{q}';\omega) &= \int dt\, e^{i\omega t}\, \left[\langle\rho(\mathbf{q},t)\rho(\mathbf{q}',0)\rangle\langle\rho(-\mathbf{q},t)\rho(-\mathbf{q}',0)\rangle\right. \\ &\quad \left. + \langle\rho(\mathbf{q},t)\rho(-\mathbf{q}',0)\rangle\langle\rho(-\mathbf{q},t)\rho(\mathbf{q}',0)\rangle\right] \\ &= \delta_{\mathbf{q},\mathbf{q}'} 2\int_{-\infty}^{\infty}\frac{d\omega'}{2\pi}S(\mathbf{q},\omega')S(-\mathbf{q},\omega-\omega')\ . \end{aligned} \tag{10.38}$$

Here we have taken into account that the dynamic structure factor $S(\mathbf{q},\omega)$ is an even function of $\mathbf{q}$. The approximation reduces $S_2(\mathbf{q},\mathbf{q}';\omega)$ to a product of two density fluctuations of wavevector $\mathbf{q}$ and $-\mathbf{q}$ which propagate in the liquid without any interaction. $S_2(\mathbf{q},\mathbf{q}';\omega)$ involves a frequency convolution over two dynamic structure factors. To the extent that this approximation is valid, the Raman-scattering intensity gives us no new information about the dynamics of the liquid, over and above what can be obtained from neutron-scattering studies (at least in principle).

In evaluating $I_d(\omega)$ based on the convolution approximation (10.38), one can use the quasiparticle ansatz (see Sections 1.1 and 2.1)

$$S(\mathbf{q},\omega) = Z(q)\delta(\omega-\omega_q) \tag{10.39}$$

in the region near the roton minimum. This gives

$$S_2(\mathbf{q},\mathbf{q}';\omega) \sim \delta_{\mathbf{q},\mathbf{q}'} Z^2(q)\delta(\omega-2\omega_q) \tag{10.40}$$

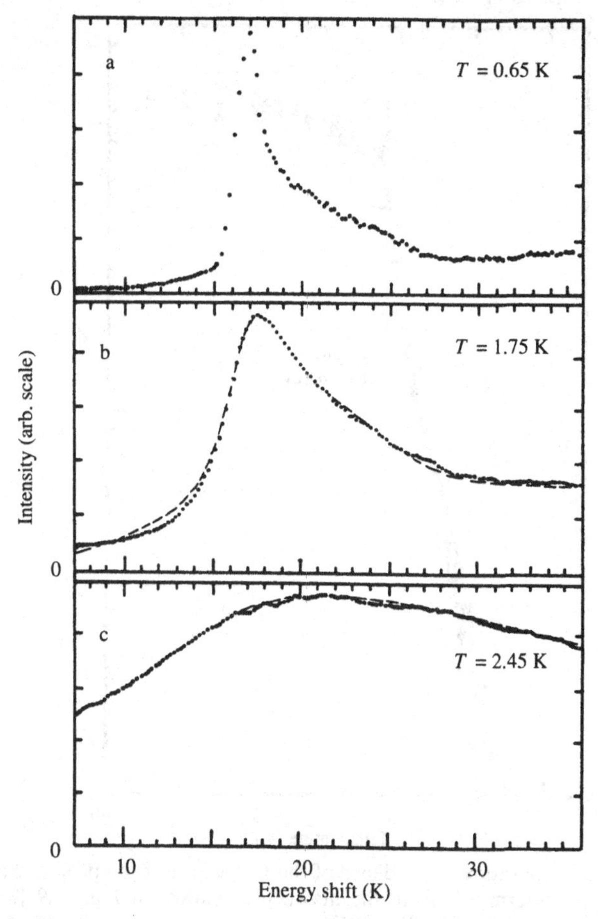

Fig. 10.9. The Raman-scattering intensity vs. energy transfer for three different temperatures, at a pressure of 5 bar. The dashed curves are fits based on a Lorentzian approximation to $S(\mathbf{Q}, \omega)$ in (10.38) [Source: Ohbayashi *et al.*, 1990].

and hence

$$I_d(\omega) \sim t_d^2(Q_R)Z^2(Q_R)\rho_2^0(\omega) \ . \tag{10.41}$$

Here the two-particle spectral density $\rho_2^0(\mathbf{Q} = 0, \omega)$ describing the excitation of two rotons with zero net momentum is given by the first term in (10.16). At temperatures above 1 K, a simple extension to include the roton finite width (due to scattering from thermally excited rotons) involves replacing the delta function in (10.39) by a Lorentzian. The convolution frequency integral (10.38) over two normalized Lorentzians

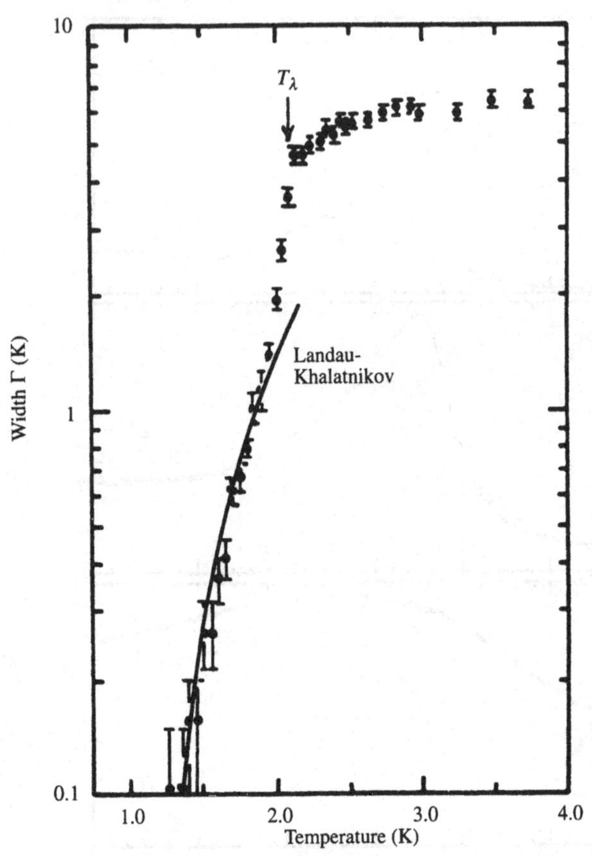

Fig. 10.10. The temperature dependence of the half-width of the peak in $S(\mathbf{Q}, \omega)$ at $Q = 2$ Å^{-1} as determined from the fits such as shown in Fig. 10.9 [Source: Ohbayashi *et al.*, 1990; Ohbayashi, 1991].

with a half-width Γ gives a normalized Lorentzian of half-width 2Γ. The resulting generalization of (10.41) for $I_d(\omega)$ can then be used to extract information about the temperature dependence of the roton half-width $\Gamma(T)$. This procedure has been used by Greytak and Yan (1971) for temperatures up to 1.8 K and more recently by Ohbayashi *et al.* (1990) to T_λ and above. As shown in Fig. 10.9, it appears to be capable of giving good fits to the Raman data as a function of the temperature.

The results so obtained for the temperature dependence of the roton energy and width (see Figs. 10.10 and 10.11) are in good agreement with those obtained directly from neutron-scattering data on $S(\mathbf{Q}, \omega)$, such as given by Stirling and Glyde (1990). However two comments should

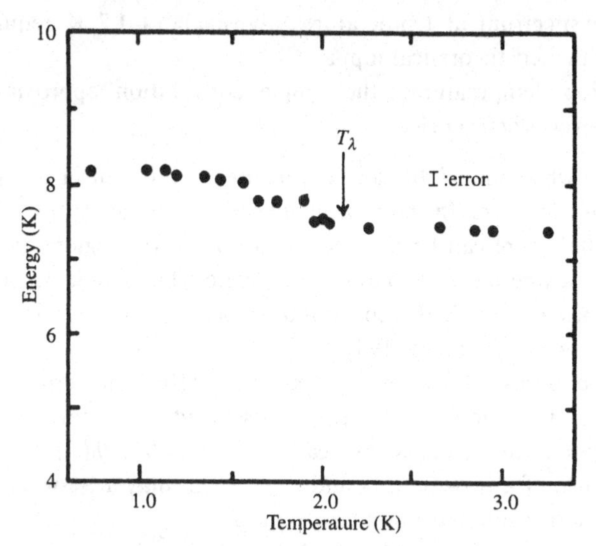

Fig. 10.11. The temperature dependence of the peak position in $S(\mathbf{Q}, \omega)$ at $Q = 2$ Å^{-1}, as determined from fits such as shown in Fig. 10.9 [Source: Ohbayashi *et al.*, 1990; Ohbayashi, 1991].

be made. First of all, results such as those shown in Figs. 10.10 and 10.11 are based on (10.38) and thus include the effect of the roton–roton interaction on the width of the resonance in $S(\mathbf{Q}, \omega)$ but not on the shape of the two-particle spectral density $\rho_2(\mathbf{Q} = 0, \omega)$ (for the case $Q \neq 0$, see Figs. 10.1 and 10.2). In many-body language, propagator renormalization effects are included but not vertex corrections (the effect of interactions between the excitations, which may lead to bound states).

The second comment concerns the interpretation of the results in Figs. 10.10 and 10.11 near and above T_λ in terms of the "roton" width and energy. In Section 7.2, we argued that a careful analysis of the $S(\mathbf{Q}, \omega)$ line shape at $Q \sim 2$ Å^{-1} in the region near T_λ (see Fig. 7.11) was consistent with the idea that the roton peak intensity was vanishing, being replaced by a broad particle–hole spectrum (associated with thermally excited rotons) which characterizes the normal phase. If one models the resulting changes in the $S(\mathbf{Q}, \omega)$ line shape in terms of a single Lorentzian with a rapidly increasing width near T_λ, one will naturally be led to results of the kind shown in Figs. 10.10 and 10.11. One cannot, however, interpret such fits as giving information about rotons near and above T_λ. As we discussed in connection with neutron-scattering data in Section 7.2, a meaningful extraction of information about the quasiparticle spectrum at temperatures above about 1.7 K requires a rather sophisticated theoretical input.

Even at low temperatures, the simple convolution approximation (10.38) has several deficiencies:

(a) It ignores what one might call the excluded-volume effect due to the ^{4}He atom hard core. In the coordinate space form given in (10.36), it is clear that there can be no contribution from the regions $|\mathbf{r}_3 - \mathbf{r}_4|$, $|\mathbf{r}_1 - \mathbf{r}_2| < a$, where a is the hard-core diameter. The simple decoupling (10.38) does not handle this constraint properly (for further discussion and references, see Halley, 1989).
(b) As we mentioned above, by approximating (10.37) in terms of two non-interacting rotons, all the effects discussed in Section 10.1 have been ignored. In particular, we cannot discuss how $h(\omega)$ will show the presence of a two-roton bound state which may arise when there is an attractive roton–roton interaction g_4.
(c) In approximations such as (10.38), the key step lies in treating the density fluctuations, rather than the field operators, as the fundamental variables.

All available calculations of $S_2(\mathbf{q},\mathbf{q}';\omega)$ which include the effect of two-roton bound states are based on identifying the density fluctuations as the elementary excitations. That is, something like

$$\hat{\rho}(\mathbf{q}) = \sqrt{Z(q)}[\hat{\alpha}^+_{\mathbf{q}} + \hat{\alpha}_{-\mathbf{q}}] \tag{10.42}$$

is used, where $\hat{\alpha}^+_{\mathbf{q}}$ is the creation operator of an excitation (see Section 9.1). In this type of approach, the correlation function $S_2(\mathbf{q},\mathbf{q}';\omega)$ involving four density operators is effectively reduced to a two-particle Green's function, such as $G_2(\mathbf{Q}=0,\omega)$ defined in (10.1). To this extent, the description of two-roton bound states can be taken over from the analysis given in Section 10.1. This approach of relating $S_2(\mathbf{q},\mathbf{q}';\omega)$ directly to $G_2(\mathbf{Q}=0,\omega)$ involves the same sort of approximation as taking the density-response function χ_{nn} to be directly proportional to the single-particle Green's function, as in (10.31). Halley and Korth (1991) have extended the analysis based on (10.42) to include backflow by calculating $S_2(\mathbf{q},\mathbf{q}';\omega)$ starting from (10.37) using the correlated-basis-function approach of Manousakis and Pandharipande (1986) discussed in Section 9.1.

What one would like to see is a calculation of $S_2(\mathbf{q},\mathbf{q}';\omega)$ based on treating it as a true four-particle Green's function (involving eight quantum field operators). In earlier chapters, we saw the importance of keeping the

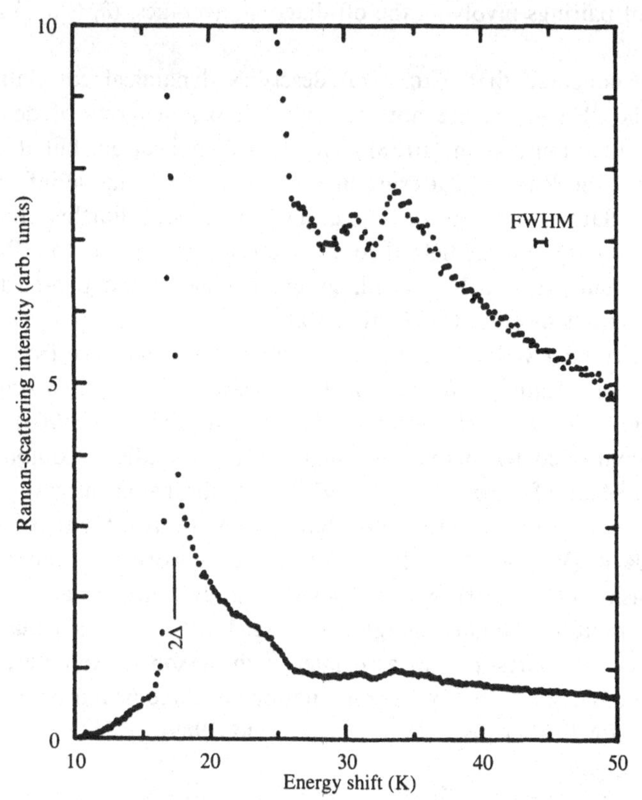

Fig. 10.12. Raman intensity vs. energy shift ω, measured at 0.65 K and SVP. The instrumental FWHM is 0.75 K. Weak structure above the two-roton peak at 2Δ is clearly evident in the expanded part of the high-energy data [Source: Ohbayashi, 1991].

distinction clear between χ_{nn} and the more fundamental single-particle Green's functions $G_{\alpha\beta}$. The analogous investigation of $S_2(\mathbf{q},\mathbf{q}';\omega)$ has not been carried out in the literature to date. Writing (10.37) in the form

$$\sum_{\substack{k_1,k_2,\\k_3,k_4}} \langle \hat{a}^+_{k_1}(t)\hat{a}_{k_1+q}(t)\hat{a}^+_{k_2}(t)\hat{a}_{k_2-q}(t)\hat{a}^+_{k_3}\hat{a}_{k_3+q'}\hat{a}^+_{k_4}\hat{a}_{k_4-q'}\rangle \ , \tag{10.43}$$

one sees that reducing this expresion to products of pairs of single-particle operators (i.e., single-particle Green's functions) results in additional terms which are not included in (10.38) even in a normal Bose fluid. Moreover, when there is a Bose condensate present, we have a whole new class of pairings involving the off-diagonal averages, $\langle \hat{a}_k^+(t)\hat{a}_{-k}^+ \rangle$ and $\langle \hat{a}_k(t)\hat{a}_{-k} \rangle$.

We thus conclude that $S_2(\mathbf{q},\mathbf{q}';\omega)$ describes dynamical correlations in superfluid ^{4}He which are not expressed simply in terms of density correlation functions, as in (10.38). This is a complication, but it also suggests that the Raman-scattering intensity may yield unique information not available from neutron-scattering experiments. Further studies of $S_2(\mathbf{q},\mathbf{q}';\omega)$ based on the field-theoretic analysis appropriate to a Bose-condensed fluid are clearly needed, as are studies of the two-particle Green's functions discussed in Section 10.1.

As we have reviewed in Section 10.1, there is good evidence from the low-temperature Raman data that, at low pressure, there is a two-roton bound state. This was first observed by Greytak and Yan (1969) and has been confirmed by increasingly high-precision studies. Combining the Raman data of Murray *et al.* (1975) with the roton energy given by neutron scattering, the two-roton binding energy is estimated to be 0.27 ± 0.04 K (Woods *et al.*, 1977). More recent work is summarized by Ohbayashi (1989, 1991), who has also found evidence for additional fine-scale structure at higher energies (see Fig. 10.12). The latter may be due to higher-order resonances associated with maxon–maxon states as well as to bound states involving combinations of three or more maxons and rotons (for further discussion, see Iwamoto, 1989).

11

Relation between excitations in liquid and solid ^{4}He

In the early 1970's, attention was drawn to the remarkable similarity between the excitation spectra exhibited by $S(\mathbf{Q}, \omega)$ in solid ^{4}He and superfluid ^{4}He at low temperatures (Werthamer, 1972; Horner, 1972a; Glyde, 1974), as shown very dramatically in the theoretical results of Figs. 11.1 and 11.2. While various suggestions have been made as to the origin of this similarity, it remains an unresolved and intriguing problem. In this brief chapter, we compare the theoretical description of excitations in a quantum solid with those of a Bose-condensed liquid. While we review the key ideas, we assume that the reader has some familiarity with an introductory account of quantum crystals. (The modern theory of excitations in quantum crystals was essentially completed in the early 1970's. For background and a more detailed discussion of solid ^{4}He than we give in this chapter, we recommend the review by Glyde, 1976.)

In both condensed phases, it is important to distinguish clearly between the elementary excitations and the density fluctuations. We argue that the phonons in solid ^{4}He are the natural analogue of the single-particle excitations in liquid ^{4}He. In Section 11.1, defining the phonons as the poles of the displacement correlation function, we briefly review theories which start with the self-consistent harmonic (SCH) approximation or something similar. In Section 11.2, we discuss the relation between the displacement–displacement and the density–density correlation functions in solid ^{4}He. This relation is based on the well known Green's function analysis of anharmonic crystals initiated by Ambegaokar, Conway and Baym (1965). Finally, in Section 11.3, we compare the expressions for $S(\mathbf{Q}, \omega)$ in solid and superfluid ^{4}He.

We do not discuss the interesting possibility of finding a Bose condensate in a quantum solid. For references, see Meisel (1992).

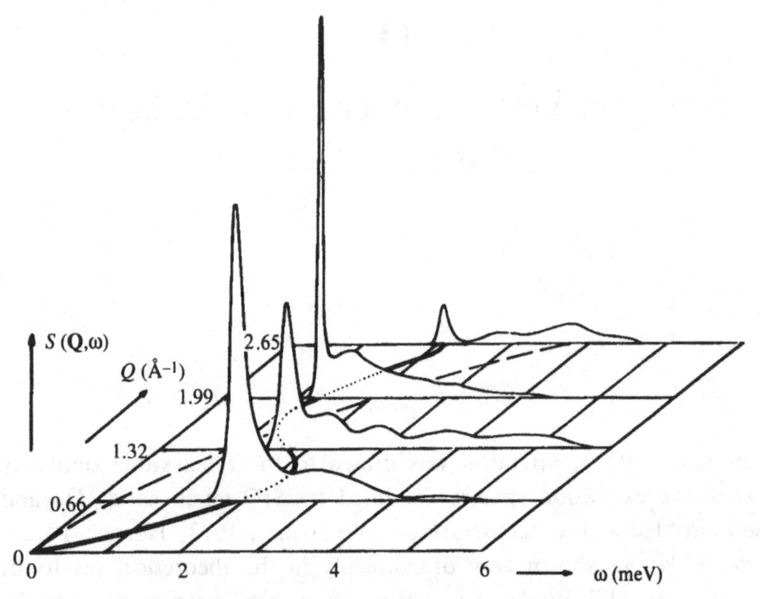

Fig. 11.1. Theoretical results for $S(\mathbf{Q},\omega)$ as a function of $\mathbf{Q}$ (along the (111) direction) and ω, in bcc solid ^{4}He. The peak intensity is seen to follow a "phonon–maxon–roton" type dispersion curve (dark line), with the shifting of spectral weight to free-particle-like behaviour (dashed line) at high Q [Source: Horner, 1974a].

11.1 Phonons as poles of the displacement correlation function

The usual Hamiltonian describing an anharmonic crystal is obtained by expanding the interatomic potential energy in powers of the atomic displacements from the (Bravais) equilibrium sites. The degrees of freedom are described by the displacement field and thus the elementary excitations (phonons) correspond to the poles of the displacement–displacement correlation function (usually called the phonon propagator). The lowest-order harmonic approximation consists of neglecting the cubic and higher-order anharmonic force constants. Expressing the atomic displacement of the l-th atom in terms of the usual phonon creation and annihilation operators, we have

$$\hat{u}_l(t) = \sum_{\mathbf{q},\lambda} \left(\frac{\hbar}{2mN\omega^0_{q\lambda}} \right)^{\frac{1}{2}} \varepsilon_{q,\lambda} e^{i\mathbf{q}\cdot\mathbf{R}_l} \hat{A}_{q\lambda}(t) \tag{11.1}$$

where $\hat{A}_{q\lambda} \equiv \hat{a}_{q\lambda} + \hat{a}^+_{-q\lambda}$ and $\varepsilon_{q,\lambda}$ is the polarization vector of the λ phonon branch. The Fourier transform of the (retarded) one-phonon Green's

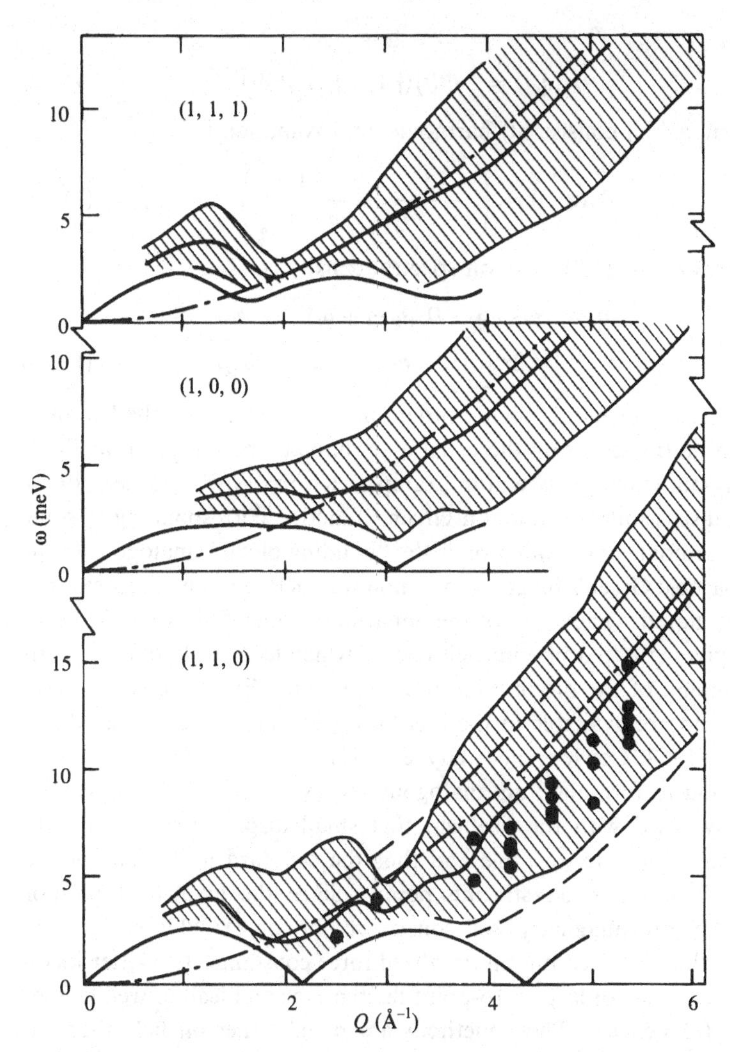

Fig. 11.2. Theoretical dispersion curves (ω vs. Q) for one-phonon and multiphonon structure in $S(\mathbf{Q},\omega)$ in bcc solid ^{4}He (in three different directions). The heavy lines and shaded region give the mean energy and width, respectively, of the multiphonon scattering. The dash–dot line gives the free-atom recoil frequency. The heavy dots represent the peaks in the $S(\mathbf{Q},\omega)$ data (Kitchens *et al.*, 1972), with a width given by the dashed lines [Source: Horner, 1974b].

function

$$D_\lambda(\mathbf{q}, t) = -i\theta(t)\langle[\hat{A}_{q\lambda}(t), \hat{A}^+_{q\lambda}(0)]\rangle \tag{11.2}$$

is given in a (self-consistent) harmonic approximation by

$$D_\lambda(\mathbf{q}, \omega) = \left(\frac{1}{\omega - \omega^0_{q\lambda}} - \frac{1}{\omega + \omega^0_{q\lambda}}\right) . \tag{11.3a}$$

The corresponding one-phonon spectral density is

$$\begin{aligned} A_\lambda(\mathbf{q}, \omega) &\equiv -2 \operatorname{Im} D_\lambda(\mathbf{q}, \omega + i\eta) \\ &= 2\pi \left[\delta(\omega - \omega^0_{q\lambda}) - \delta(\omega + \omega^0_{q\lambda})\right] . \end{aligned} \tag{11.3b}$$

The effect of including higher-order anharmonic terms in the Hamiltonian can be treated using the usual diagrammatic perturbation methods of many-body theory (see, for example, Kwok, 1967; Horner, 1967). In such discussions, the phonon Green's function corresponding to some sort of harmonic approximation is the "building block", analogous to the single-particle Green's function for a non-interacting gas in discussions of quantum liquids. The effect of the anharmonic interactions (cubic, quartic, ...) gives rise to a phonon self-energy which leads to frequency shifts and damping of the original harmonic phonons. Extensive calculations of this kind are available in the literature (see, for example, Maradudin and Fein, 1962; Cowley, 1968; Glyde, 1971).

In trying to formulate an analogous theory for solid ^{4}He, one is immediately faced with the problem that small displacements about the equilibrium sites are unstable and thus the standard harmonic approximation is not a good starting point. Since the associated phonon frequencies are imaginary, self-consistent phonon (SCP) theories have been developed, based on renormalized force constants. The latter incorporate the effects of large zero-point fluctuations and lead to well defined phonon frequencies. These methods are based either on field-theoretic resummations of higher-order effects or on variational approximations to the many-body wavefunctions. In the latter approach, the physics is clear, since a wavefunction which is Gaussian in the atomic displacements is equivalent to some effective harmonic Hamiltonian. (For a lucid summary of how this procedure is carried out, we refer to Section II of Gillis, Werthamer and Koehler, 1968.) The problem is that if the strong short-range correlations (SRC) are included by an additional Jastrow–Feenberg-type function, we do not necessarily find any simple equivalence to an SC harmonic Hamiltonian. In particular, many theories which include SRC no longer guarantee that the excitations still correspond to poles of the displacement correlation function. More or less satisfactory self-consistent phonon theories with short-range correlations were developed in the early 1970's. A Green's function approach is given by Horner (1974c).

One such theory of phonon excitations can be elegantly formulated in variational terms using the general approach of correlated basis functions (Feenberg, 1969). As we review in Section 9.1, this kind of approach has been extensively developed as a description of liquid ^{4}He at $T = 0$. Thus it is useful to discuss the analogous approach for solid ^{4}He. Here we briefly sketch the work of Koehler and Werthamer (KW, 1971), who have applied the CBF philosophy to the determination of the excited states of solid Helium in a consistent manner. Earlier work by Koehler (1967, 1968) treated the ground state in terms of a variational many-body wavefunction $|\Phi_0\rangle$ which was a product of a Jastrow function describing the short-range correlations (SRC) and a Gaussian function $|\Phi_G\rangle$ in the atomic displacements describing the long-range correlations. The KW excited states are constructed by operating on this ground state $|\Phi_0\rangle$ with polynomials in the atomic displacement operators (this is the analogue of Feynman's construction of excited states in liquid ^{4}He by operating on the ground state with the *density* operator, as discussed in Sections 1.1 and 9.1). The low-energy one- and two-phonon states are defined such that the Jastrow wavefunction remains unchanged from the ground state (any such change would require large energies).

In the context of this KW theory, Werthamer (1973) made the important observation that the SRC can be variationally constrained by the requirement that the SRC do not modify the *static* displacement correlation function, i.e.,

$$D_{ll'} \equiv \langle\Phi_0|\hat{u}_l\hat{u}_{l'}|\Phi_0\rangle = \langle\Phi_G|\hat{u}_l\hat{u}_{l'}|\Phi_G\rangle \ . \tag{11.4}$$

The optimized energies of the KW one-phonon states obtained in this manner can be proven to be identical to the SC frequencies of a harmonic crystal whose force constants are *defined* only through the coefficients of the Gaussian part $|\Phi_G\rangle$ of the ground-state wavefunction. Thus the low-lying excited states defined by KW allow for a simple but still *consistent* interpretation as phonons even when SRC are included (see also Horner, 1971). A key feature is that the minimization of the ground-state energy with respect to the Gaussian coefficients in $|\Phi_G\rangle$ automatically leads to the diagonalization of the Hamiltonian in the one-phonon-state subspace.

Several authors have emphasized that the poles of the displacement correlation function are not necessarily exhausted by the phonon frequencies given by SCP theory. A useful comparison can be made between SCH theories of quantum crystals and Hartree–Fock theories of quantum liquids. The Hartree–Fock approximation gives the best renormalized single-particle states but completely ignores collective modes. The SCH approximation, in contrast, gives the best collective modes but ignores single-particle excitations. Most SCP theories implicitly *assume* from the beginning that the elementary excitations can be classified in terms of some equivalent harmonic lattice. This assumption is built into the form used for the excited-state wavefunctions. In this connection, we note that the precise relation between collective vs. single-particle theories of excitations in solid ^{4}He has never really been resolved (Fredkin and Werthamer, 1965; Gillis and Werthamer, 1968; Werthamer, 1969). Some aspects of this relation are touched on by the numerical work of Horner (1972b), whose results for the spectral density $A_\lambda(\mathbf{Q},\omega)$ show the clear transition from phonon-like excitations at low Q and ω to more single-particle-like excitations at high Q and ω. The same transition is shown by the results for $S(\mathbf{Q},\omega)$ given in Figs. 11.1 and 11.2.

One should also recall that when anharmonic corrections are included in such theories, the phonon propagator spectral density is not usually strongly peaked as in (11.3*b*). As Q increases, the spectral density spreads out over a larger frequency region and develops a high-energy tail, as shown by the example in Fig. 11.3. Such self-energy effects inevitably arise when one corrects the SCH spectral density to include anharmonic interactions. Thus, the peak positions in Fig. 11.3 are quite different from the line centres (denoted by the arrows) defined as the normalized first frequency moment of $A_\lambda(\mathbf{Q},\omega)$. One is faced with how to define an appropriate mean or average phonon energy for given Q,λ (Horner, 1972b). Indeed, one is led to question the usefulness of any such simplified description in calculating the thermodynamic properties of a quantum (or highly anharmonic) crystal in place of the full displacement-field spectral density $A_\lambda(\mathbf{Q},\omega)$.

Finally, we note that for small Q, the phonons in a solid can be also usefully classified as either hydrodynamic or collisionless, following the terminology of Section 6.2. Typically phonons studied by ultrasonics or Brillouin light scattering are in the low-energy, hydrodynamic domain and are often referred to as first sound. Solid ^{4}He also exhibits second sound, which may be viewed as an oscillation in the local number density of phonons. As Kwok and Martin (1966) discuss, both first sound and second sound appear as poles of the displacement correlation function of an anharmonic crystal in the long-wavelength limit (see also Kwok,

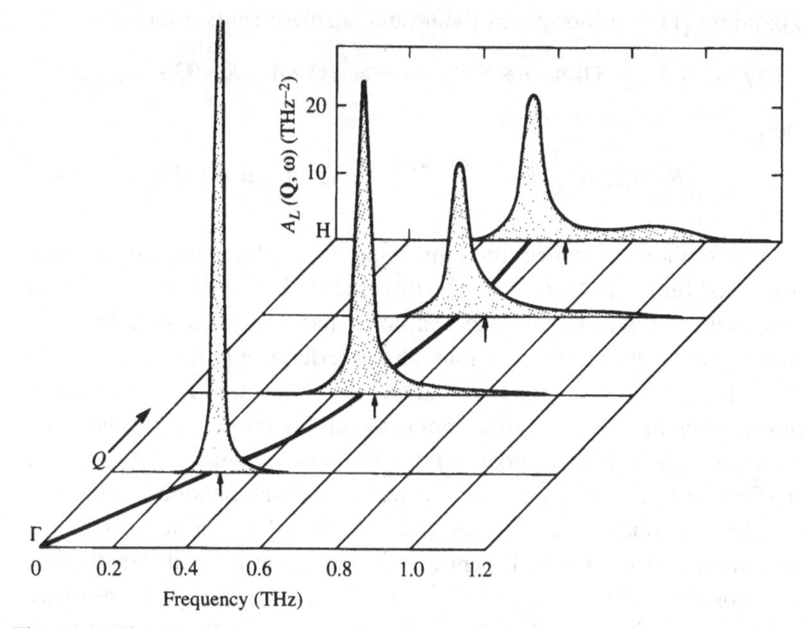

Fig. 11.3. The one-phonon spectral density $A(\mathbf{Q},\omega)$ as a function of $\mathbf{Q}$ and ω, for a longitudinal mode in bcc solid ^{4}He along the (1,0,0) direction. The phonon self-energy due to a single "bubble" has been computed self-consistently. In comparing these results with Fig. 3 of Horner (1972b), see ref. 20 in McMahan and Beck (1973) [Source: McMahan and Beck, 1973].

1967). In contrast, the low-Q phonons in solids which are excited by neutron scattering are in the high-energy collisionless region. For this reason, such excitations are often referred to as zero sound phonons, in analogy to the collective modes in quantum liquids.

11.2 Phonons vs. density fluctuations in solid ^{4}He

Since the middle 1960's, it has been realized that, in general, the dynamic structure factor $S(\mathbf{Q},\omega)$ of anharmonic crystals can be a complicated function of the underlying phonon excitations. The key simplifying feature of a crystal is that the dynamics involve displacements of the atoms with respect to a Bravais lattice, $\mathbf{r}_l = \mathbf{R}_l + \mathbf{u}_l$. Substituting (2.5) into the structure factor (2.7) gives

$$S(\mathbf{Q},t) = \frac{1}{N}\sum_{l,l'}\langle e^{-i\mathbf{Q}\cdot\mathbf{r}_l(t)} e^{i\mathbf{Q}\cdot\mathbf{r}_{l'}(0)}\rangle . \tag{11.5}$$

Expanding (11.5) in powers in the atomic displacements, we obtain

$$S(\mathbf{Q},\omega) = S_{\mathrm{Bragg}}(\mathbf{Q})\delta(\omega) + S_1(\mathbf{Q},\omega) + S_{\mathrm{int}}(\mathbf{Q},\omega) + S_{\mathrm{mp}}(\mathbf{Q},\omega) \ , \quad (11.6)$$

where

$$S_1(\mathbf{Q},t) = \frac{1}{N}d^2(Q)\sum_{l,l'} e^{i\mathbf{Q}\cdot(\mathbf{R}_l-\mathbf{R}_{l'})}\mathbf{Q}\cdot\langle\mathbf{u}_l(t)\mathbf{u}_{l'}(0)\rangle\cdot\mathbf{Q} \quad (11.7)$$

is the "one-phonon" contribution involving the displacement–displacement correlation function $D_{ll'}(t) \equiv \langle u_l(t)u_{l'}(0)\rangle$. In (11.7), $d^2(Q) = \exp[-2W(Q)]$ is the Debye–Waller factor. The additional terms in (11.6) describe two-phonon (and higher) contributions plus interference terms. The higher-order phonon contributions which arise in (11.6) complicate the situation considerably in a highly anharmonic quantum crystal like solid ^{4}He. These give rise to both a broad multiphonon continuum in $S(\mathbf{Q},\omega)$, and interference terms which contribute within the one-phonon region and considerably modify the contribution of $S_1(\mathbf{Q},\omega)$ to the total dynamic structure factor at large Q. It is now recognized that in solid ^{4}He, there is an important difference between the density fluctuation spectrum (peak positions in $S(\mathbf{Q},\omega)$) and the underlying phonon excitations (peaks in $A_\lambda(\mathbf{Q},\omega)$ or $S_1(\mathbf{Q},\omega)$).

The many-body theory of the density-response function $\chi_{nn}(\mathbf{Q},\omega)$ in solid ^{4}He can be formulated in terms of $D_\lambda(\mathbf{Q},\omega)$ following the analysis of Ambegaokar, Conway and Baym (ACB, 1965). The final result can be written schematically in the form

$$\chi_{nn}(\mathbf{Q},\omega) = d^2(Q)R_\lambda(\mathbf{Q},\omega)D_\lambda(\mathbf{Q},\omega)R_\lambda(\mathbf{Q},\omega) + \chi_{\mathrm{mp}}(\mathbf{Q},\omega) \ . \quad (11.8)$$

The density fluctuation spectrum is thus seen to include one-phonon as well as multiphonon contributions (involving two or more phonons). In addition, there are interference contributions described by the vertex functions $R_\lambda(\mathbf{Q},\omega)$ in (11.8), by which higher-order phonon processes modify the contribution of the single-phonon scattering. Thus the S_1 and S_{int} terms in (11.6) can be combined into a *single* contribution $S_P(\mathbf{q},\omega)$ which may be called the physical "one-phonon" contribution to $S(\mathbf{Q},\omega)$. This consists of all contributions to S_{int} which contain a single-phonon propagator as an intermediate state. The effect of the interference terms on the one-phonon peak was first clearly exhibited in alkali halide crystals (Cowley and Woods, 1969).

To the extent that it can be limited to $S_1(\mathbf{Q},\omega)$ in (11.6), the density fluctuation spectrum is identical to that of the displacement fluctuations (given by SCP theory, as discussed in Section 11.1). However, in a

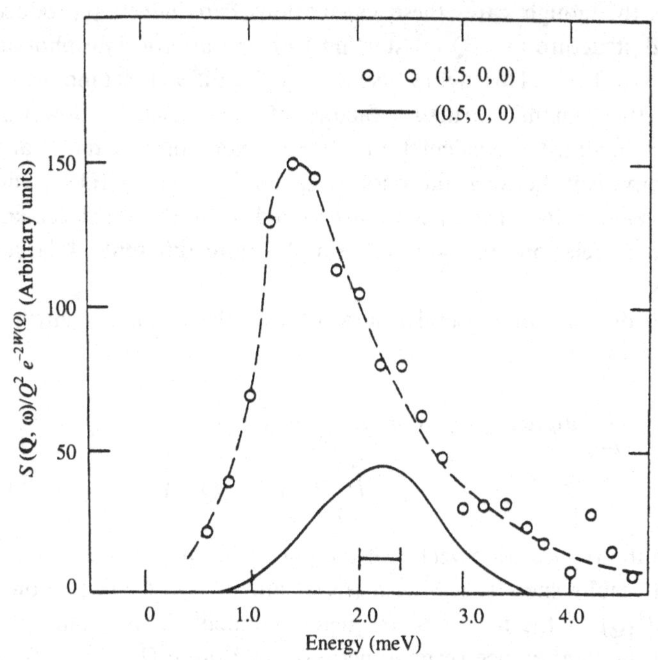

Fig. 11.4. Scaled neutron-scattering intensity vs. ω from longitudinal modes L[100] at two values of Q differing by a reciprocal lattice vector of the bcc solid. The instrumental resolution is shown by the bar. See caption of Fig. 11.5 [Source: Osgood, Kitchens, Shirane and Minkiewicz, 1972].

quantum crystal like solid ^{4}He, one cannot restrict oneself to $S_1(\mathbf{Q}, \omega)$. This is shown dramatically in Fig. 11.4 by experimental data for $S(\mathbf{Q}, \omega)$, suitably normalized, at two values of $\mathbf{Q}$, $\mathbf{Q}'$ which differ only by a reciprocal lattice vector $\boldsymbol{\tau}$ of the Bravais bcc lattice. Since we normalize the results by dividing by $Q^2 d^2(Q)$, $S_1(\mathbf{Q}, \omega)$ should be identical to $S_1(\mathbf{Q}', \omega)$, where $\mathbf{Q}' = \mathbf{Q} + \tau$. This equivalence follows immediately from the translational invariance of the displacement correlation function i.e., $D_\lambda(\mathbf{Q}, \omega) = D_\lambda(\mathbf{Q} + \tau, \omega)$. In contrast, the data in Fig. 11.4 show that $S(\mathbf{Q}, \omega)/Q^2 d^2(Q)$ is quite different at $\mathbf{Q}$ and $\mathbf{Q}'$.

Because of the features outlined above, it is difficult to extract the phonon frequencies (or, more generally, the phonon spectral density) directly from the measured values of $S(\mathbf{Q}, \omega)$ at larger values of Q. Horner (1972a, 1974a), Glyde and Goldman (1976) and others have carried out detailed calculations of $S(\mathbf{Q}, \omega)$ using (11.8), limiting themselves to one- and two-phonon processes plus their coupling. As shown in

Fig. 11.5, with enough care, these calculations can indeed reproduce the observed structure in $S(\mathbf{Q},\omega)$ starting from an appropriate phonon propagator $D_\lambda(\mathbf{Q},\omega)$. This figure also shows the different components making up the dynamic structure factor $S(\mathbf{Q},\omega)$. Clearly, however, without such microscopic calculations, one cannot hope to make any detailed comparison between the phonon spectral density $A_\lambda(\mathbf{Q},\omega)$ and $S(\mathbf{Q},\omega)$ in solid ^{4}He. The spectra associated with the displacement and density correlation functions are simply quite different at larger wavevectors.

ACB have derived some useful f-sum rules for the different contributions to $S(\mathbf{Q},\omega)$. These are

$$\begin{aligned}\int_{-\infty}^{\infty} d\omega\ \omega S_{1\lambda}(\mathbf{Q},\omega) &= \frac{(\mathbf{Q}\cdot\varepsilon_{q\lambda})^2}{2m} d^2(Q) \\ &= \int_{-\infty}^{\infty} d\omega\ \omega S_{P\lambda}(\mathbf{Q},\omega)\ , \end{aligned} \qquad (11.9)$$

where $\varepsilon_{q\lambda}$ is the polarization vector of the phonon branch being studied. These exact results show that *both* $S_1(\mathbf{Q},\omega)$ and $S_P(\mathbf{Q},\omega)$ take up only a fraction $d^2(Q)$ of the total first frequency moment. This means that the effect of the interference vertex functions $R_\lambda(\mathbf{Q},\omega)$ in (11.8) is only to rearrange the spectral weight associated with the single-phonon propagator, not to change its total contribution to the first frequency moment f-sum rule.

11.3 Relation between $S(\mathbf{Q},\omega)$ in superfluid and solid ^{4}He

Clearly one can draw an analogy between the ACB result in (11.8) for solid ^{4}He with (3.47) and the dielectric formalism expression (5.24) in the case of superfluid ^{4}He. This similarity has been discussed by Wong and Gould (1974) as well as Glyde (1984). The single-particle Green's function $G_{\alpha\beta}(\mathbf{Q},\omega)$ in Bose-condensed liquid ^{4}He plays the same role as the one-phonon (or displacement-field) correlation function $D_\lambda(\mathbf{Q},\omega)$ in solid ^{4}He. Wong (1979) has pointed out that the Debye–Waller factor $d^2(Q)$ in (11.8) plays a role analogous to the Bose-condensate order parameter $|\langle\hat{\psi}\rangle|^2 = n_0$ which enters into the Bose vertex function $\bar{\Lambda}_\alpha$ in (5.24). Despite the formal similarity of the structure of (11.8) and (5.24), however, there is a fundamental difference between an anharmonic crystal and a Bose-condensed fluid since, in the former, the phonon self-energy does *not* have a part analogous to the term in (5.75). In crystals, one has a relation similar to (5.24) but no equivalent of the inverse relation

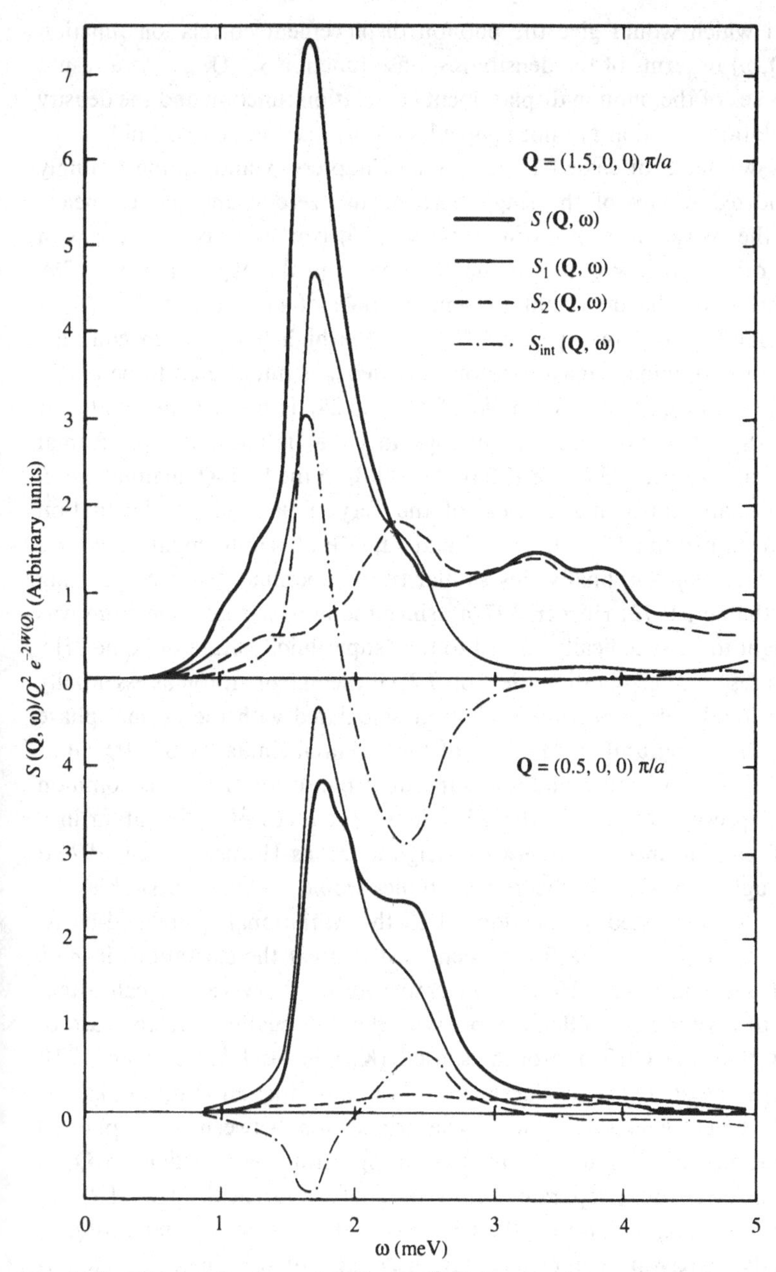

Fig. 11.5. Calculated values of $S(\mathbf{Q},\omega)/Q^2d^2(Q)$ vs. ω for two equivalent points in Q-space (see also Fig. 11.4). $S(\mathbf{Q},\omega)$ is decomposed into the bare one-phonon component S_1, the two-phonon component S_2 of the multiparticle contribution, and the interference component S_{int}. Note that the scaled S_1 component is identical for the two values of Q [Source: Horner, 1972a].

(5.31) which would give the phonon displacement correlation function $D_\lambda(\mathbf{Q},\omega)$ in terms of the density-response function $\chi_{nn}(\mathbf{Q},\omega)$. As a result, the poles of the phonon displacement correlation function and the density correlation function are not coupled as in a Bose-condensed fluid.

As we have discussed at length in Chapters 5 and 7, the strongly hybridized nature of the single-particle and zero sound modes means that the two contributions in (5.24) or (7.6) can strongly interfere with each other. At low Q it is better to start from the expression in (5.76). In this case, the dominant low-energy pole of *both* $G_{\alpha\beta}$ and χ_{nn} is a zero sound mode in superfluid ^{4}He. In the high-Q region, in contrast, where the dominant maxon–roton pole in χ_{nn} is interpreted to be an SP excitation originating as a pole of $G_{\alpha\beta}$, (5.24) is more appropriate. It is in this SP region that the analogy to (11.8) is relevant. The formal similarity between (3.47) or (5.24) and (11.8) in the high-Q, high-ω region is especially intriguing in view of the very similar density fluctuation spectrum exhibited by both solid and superfluid ^{4}He (compare Fig. 11.2 with Fig. 1.6). Originally this similarity to superfluid ^{4}He was puzzling (Werthamer, 1972; Horner, 1972a), since the maxon–roton spectrum was thought to be specifically related to the "superfluid" nature of liquid ^{4}He. Our present interpretation (Section 7.2) of the maxon–roton as essentially a renormalized atomic-like excitation associated with the normal phase allows for a natural explanation of the essential similarity of $S(\mathbf{Q},\omega)$ at large Q and ω in the solid and superfluid phases of ^{4}He. This common origin presumably lies in the vibrational dynamics of a ^{4}He atom in a small cage formed by its nearest neighbours, as Horner (1972a, 1974a) has suggested. Further theoretical studies would be highly desirable.

As we discussed in Section 11.1, the variational correlated-basis-function (CBF) approach has been used to treat the excitations in *both* solid and superfluid ^{4}He at low temperatures. However we call attention to a very basic difference between these discussions. In the case of solid ^{4}He, the CBF approach is used (Koehler and Werthamer, 1971, 1972) to evaluate the displacement correlation function $D_\lambda(\mathbf{Q},\omega)$, i.e. the displacement-field excitations. The connection between these phonon excitations and the density fluctuation spectrum described by $S(\mathbf{Q},\omega)$ is then given by a separate ACB-type analysis (as in Section 11.2). In superfluid ^{4}He, in contrast, the CBF method is used to compute $S(\mathbf{Q},\omega)$ directly, bypassing completely any discussion of the single-particle dynamics described by $G_{\alpha\beta}(\mathbf{Q},\omega)$.

In summarizing this whole chapter, it would seem that the word "phonon" should never be used in quantum solids without an appropriate adjective stating what "kind" of phonon one is dealing with. Much of the earlier literature is confusing because the distinction between various kinds of phonons was either not realized or insufficiently emphasized. The same comment is equally valid for "phonons" in quantum liquids, as we have seen in Chapters 5–7. More generally, both in Bose-condensed liquids and in quantum (or highly anharmonic) crystals, there is a subtle relation between the elementary excitations and the density fluctuations (the latter being measured in neutron scattering).

12
The new picture: some unsolved problems

In this book, we have developed the theory of the excitation spectrum of superfluid ^{4}He in which the Bose condensate plays the central role. In Chapter 5, we showed how a Bose broken symmetry inevitably leads to a mixing of the single-particle and density fluctuations. Combining the general results of the dielectric formalism with recent high-resolution neutron-scattering data over a wide range of wavevectors, energies and temperatures, we were led in Chapter 7 to a new interpretation of the well known phonon–maxon–roton dispersion curve. In Section 12.1, we briefly recapitulate this new scenario and discuss how it developed from preceding theoretical work. We also review earlier studies which had independently suggested that rotons were in fact atomic-like single-particle excitations, quite different from the long-wavelength phonons. In addition, we address the question of what Feynman's work says about the nature of rotons.

The most important topic which has not been covered in this book is superfluid ^{3}He–^{4}He mixtures. The appropriate dielectric formulation has been developed by Talbot and Griffin (1984c), as we briefly summarize in Section 12.2. Much work remains to be done in using these formal results in a detailed analysis of experimental $S(\mathbf{Q}, \omega)$ data, even at the level of Section 7.2 in the case of pure ^{4}He.

Finally, in Section 12.3, we list some specific topics where further theoretical and experimental work would be useful. This list, which brings together suggestions scattered throughout the book, also acts as a convenient summary of our major themes.

12.1 Comments on the development of the new picture

We recall that in the presence of a Bose broken symmetry, the single-particle excitations, the particle–hole excitations and the two-particle

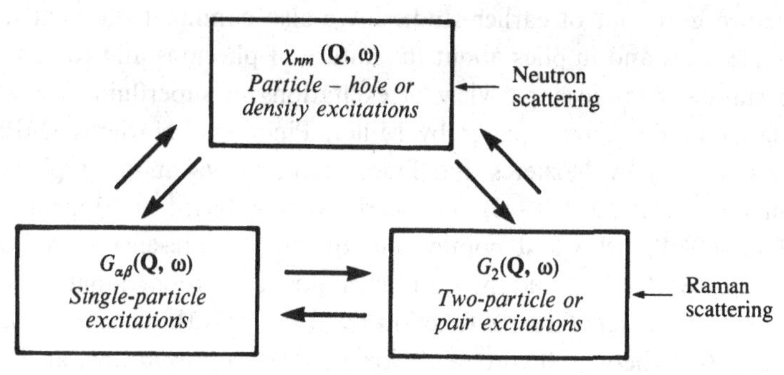

Fig. 12.1. A block diagram showing how the condensate couples the various kinds of excitations in a Bose-condensed fluid.

excitations are all coupled into each other (Fig. 12.1). This sharing of excitations is the key dynamical consequence of a Bose condensate, as was first emphasized by Gavoret and Nozières (1964) and Hohenberg and Martin (1965). The formal relations between the single-particle Green's function $G_{\alpha\beta}(\mathbf{Q}, \omega)$, the particle–hole density-response function $\chi_{nn}(\mathbf{Q}, \omega)$ and the two-particle Green's functions $G_2(\mathbf{Q}, \omega)$ have been discussed at length in Chapters 5 and 10. The precise implications of these coupled equations of motion, however, require the solution of very complex equations for self-energies, vertex functions, polarization parts etc. This has only been successfully carried out in the limit of small Q and ω at $T = 0$ (as reviewed in Section 6.3). Even in the absence of rigorous calculations, however, the general structure of the coupled correlation functions made evident using the dielectric formalism (Chapter 5) does allow one to develop various scenarios concerning the excitations in a Bose fluid like superfluid ^{4}He. A quote from p. 23 of Bogoliubov (1947) applies equally well here: "All we can require from a molecular theory of superfluidity, at least at the first stage of investigation, is to be able to account for the qualitative picture of this phemonenon being based on a certain simplified scheme."

In the interpretation of phonons, maxons and rotons developed in Chapter 7, the phonon is a zero sound (ZS) mode while the maxon–roton is a single-particle (SP) excitation which has weight in $S(\mathbf{Q}, \omega)$ only because of the Bose broken symmetry. Physically, these two modes are interpreted as quite different excitation branches which are hybridized through the condensate to produce the observed phonon–maxon–roton dispersion curve (Glyde and Griffin, 1990). In this section, we review how this picture grew out of earlier studies. We also comment on what this picture assumes and implies about the nature of phonons and rotons.

The standard microscopic view of excitations in superfluid ^{4}He was formulated in the classic paper by Miller, Pines and Nozières (MPN, 1962) as well as by Nozières and Pines in their 1964 monograph (see Section 7.3 of NP, 1964, 1990). This early work is based on Hugenholtz and Pines (1959), who first pointed out that in the presence of a Bose condensate, $S(\mathbf{Q}, \omega)$ would have a term directly proportional to the Beliaev single-particle Green's function $G_{\alpha\beta}(\mathbf{Q}, \omega)$. Working to all orders in perturbation theory, Gavoret and Nozières (1964) proved that, at small Q and ω and at $T = 0$, both of these correlation functions exhibit the same phonon pole. (This behaviour already shows up in a dilute, weakly interacting Bose gas.) While explicit calculations were limited to the low Q, ω region at low T, it was assumed that in superfluid ^{4}He:

(a) The density fluctuation spectrum described by $S(\mathbf{Q}, \omega)$ is directly proportional to $G_{\alpha\beta}(\mathbf{Q}, \omega)$ at all values of Q.
(b) Both functions exhibit a single excitation branch at high Q as well as low Q. The only difference between a gas and a liquid is that in a Bose liquid, the dispersion relation has a roton minimum at $Q \simeq 2$ Å^{-1}.
(c) This common excitation branch exhibited by both $G_{\alpha\beta}(\mathbf{Q}, \omega)$ and $S(\mathbf{Q}, \omega)$ is crucially dependent on the existence of a Bose condensate.
(d) The situation above T_λ is completely different.

This set of assumptions has had wide currency and still has strong proponents in the field-theoretic literature (see, in particular, Nepomnyashchy, 1992). It was first challenged by Pines (1966), who argued that in superfluid ^{4}He the phonon part of the quasiparticle spectrum in $S(\mathbf{Q}, \omega)$ was best described as a zero sound mode which exists both below and above T_λ without much change. On the other hand, Pines did not discuss what this implied about the corresponding single-particle Green's functions $G_{\alpha\beta}(\mathbf{Q}, \omega)$.

Szépfalusy and Kondor (1974) and Griffin and Cheung (1973) first used the dielectric formalism of Ma and Woo (1967) to discuss the inter-related structure of *both* $G_{\alpha\beta}(\mathbf{Q}, \omega)$ and $S(\mathbf{Q}, \omega)$, for all $\mathbf{Q}, \omega$ and T. A key feature of the dielectric formalism is how it draws attention to the possibility that a single-particle excitation and the zero sound can be hybridized by the Bose condensate $n_0(T)$, allowing both renormalized excitations to appear in $S(\mathbf{Q}, \omega)$ *below* T_λ. Precisely how this hybridization is realized in superfluid ^{4}He, however, is not at all obvious (see the review by Griffin, 1991).

The single-particle (SP) scenario described in the preceding paragraph was extended to finite temperatures by Griffin (1979a) as well as Griffin and Talbot (1981), in an attempt to explain the finite-temperature neutron-scattering data in the region $0.8 \lesssim Q \lesssim 2$ Å^{-1} (Woods and Svensson, 1978). However, subsequent studies (Griffin, 1987, 1989) for small Q emphasized that, in fact, the phonon region is better described by the zero sound (ZS) limit (as suggested by Pines, 1966). Glyde and Svensson (1987) and Svensson (1989) suggested that these contradictory results could be understood if $S(\mathbf{Q}, \omega)$ described quite *different* excitation branches in the low- and high-Q regions. This was a very revolutionary concept since, starting with Landau (1947), the guiding assumption has always been that the phonon–maxon–roton quasiparticle dispersion relation described a *single* excitation branch in which there was no fundamental distinction between a phonon and a roton. The completely different temperature dependence of the intensity of the sharp peak observed in $S(\mathbf{Q}, \omega)$ at low Q (Woods, 1965b) and high Q (Woods and Svensson, 1978) was interpreted by Stirling and Glyde (1990) in terms of the dielectric formalism, based on the idea that the low-Q phonon was a collective density fluctuation while the high-Q maxon–roton was an atomic-like excitation. This physical distinction between phonons and rotons goes back to Feynman (1954), MPN (1962) and Chester (1963). None of this early work, however, considered the specific role of the condensate $n_0(T)$ in hybridizing the two excitations. Glyde and Griffin (1990) first illustrated how the coupling induced by the condensate could explain the continuous phonon–maxon–roton dispersion relation which is observed.

As discussed in Section 6.3, the low-energy phonon part of the spectrum is somewhat subtle in the superfluid phase. Above T_λ, such phonon resonances in $S(\mathbf{Q}, \omega)$ are zero sound density fluctuations, which also occur in normal liquid ^{3}He as well as in classical liquids (see Section 7.1). Below T_λ, this zero sound mode still exists, although it is now associated with the effective fields produced by the condensate and non-condensate atoms, i.e., the two components of $\bar{\chi}_{nn}$ as given by (5.11*b*). If the frequency spectrum associated with $\bar{G}_{\alpha\beta}$ and $\bar{\chi}^R_{nn}$ occurs at relatively low frequencies, the theory naturally leads to a zero sound phonon resonance in both χ_{nn} and $G_{\alpha\beta}$ which has a velocity which is temperature-independent (see Sections 6.3 and 7.2).

It is perhaps not surprising that a zero sound density fluctuation dominates the spectrum of χ_{nn} at low Q , both below and above T_λ. Because of the Bose broken symmetry which sets in below T_λ, however, this phonon mode also has finite weight in the single-particle Green's function $G_{\alpha\beta}(\mathbf{Q},\omega)$ and in fact dominates the *low*-energy elementary excitation spectrum at low temperatures. In a liquid, Bose statistics do *not* play a crucial role in the existence of this zero sound mode, but a condensate is crucial to ensure that this mode has finite weight as an elementary excitation in the superfluid phase. Thus we can say that the unique feature of superfluid ^{4}He is not so much that phonons exist as that the condensate allows these modes to play the role of elementary excitations *below* T_λ. In contrast, the zero sound phonons in normal liquid ^{4}He or liquid ^{3}He do not play the role of elementary excitations.

In the high-energy, short-wavelength region, we have argued that the maxon–roton peak in $S(\mathbf{Q},\omega)$ has its origin as a pole in the single-particle spectrum described by $G_{\alpha\beta}(\mathbf{Q},\omega)$. This excitation probably exists above T_λ without much change in the dispersion relation. These single-particle states have high energy (>8 K) and exist only at relatively large wavevectors ($\gtrsim$ 0.8 Å^{-1}), and thus it seems physically reasonable that they would not be modified very much when the liquid goes from the normal to the superfluid phase (or for that matter, to the solid phase discussed in Chapter 11).

The idea that a roton is physically quite different from a phonon has a long history. We recall that Feynman (1954) originally argued that the roton corresponds to a high-energy excited state involving the motion of a single atom in the potential well of its neighbors. Miller, Pines and Nozières (1962) also viewed the roton as an atomic-like excitation modified by backflow. Chester gave several suggestive arguments that indicated that phonons and rotons were physically quite distinct excitations, concluding that (to quote from p. 65 of Chester, 1963):

> From $Q = 0$ up to $Q \simeq 1$ Å^{-1} the motions in the fluid consist of longitudinal density waves .... On the other hand, for $Q \simeq 2$ Å^{-1} we find that there is essentially no local density fluctuation and we have a single particle propagating fast through the liquid. The increase in effective mass of this atom comes from a large scale hydrodynamic backflow – with little or no change in the local fluid density. The region between $Q = 1$ Å^{-1} and $Q = 2$ Å^{-1} must be interpreted as a rather complicated region in which we pass from a high frequency phonon mode at 1 Å^{-1} to a single particle propagation at 2 Å^{-1}. This is clearly a possible transition in a liquid with an open structure and with low potential barriers.

Our present analysis gives a reasonable scenario which allows for realization of this kind of picture and extends these ideas to higher temperatures, to T_λ and above.

One of the most interesting implications of this new picture is that it suggests one should be able to understand the properties of *normal* liquid ^{4}He just above T_λ in terms of a gas of roton-like excitations. Indeed, models based on excitations with a roton energy gap were used with partial success in the early literature (see, for example, Henshaw and Woods, 1961). Within the Glyde–Griffin picture, the "roton liquid" theory developed by Bedell, Pines and Fomin (1982) might also be appropriate *above* T_λ. We recall that this phenomenological theory attempts to describe the thermodynamic properties of liquid ^{4}He in the region just below T_λ in terms of an interacting gas of rotons, using a self-consistent field approach modelled after that used in the Landau Fermi liquid theory description of quasiparticle interactions in normal liquid ^{3}He. If the roton excitations are indeed intrinsic poles of the single-particle Green's functions (with an energy largely unaffected by the appearance of a condensate), the roton liquid theory formulation of Bedell, Pines and Fomin might be as relevant in the region immediately above T_λ as it is below. (Of course, singularities in the specific heat and other thermodynamic quantities in the vicinity of T_λ, specifically associated with the critical behaviour at a second-order phase transition, are not described by such a theory.)

The relation of our scenario to the well known analysis of the excited states in a Bose liquid by Feynman (1954) is not completely clear. We recall that Feynman's analysis showed how Bose statistics was the key in ensuring that there were no *other* low-energy, long-wavelength excitations, apart from a phonon-like density fluctuation which by itself is *not* dependent on the nature of the quantum statistics involved. This work, however, made no reference to any role of a Bose condensate. The following remarks may be useful in understanding Feynman's work.

In Section 1.1, we briefly reviewed the ideas behind the variational calculation of Feynman (1954). While Feynman (1953b, 1954) was motivated by a detailed microscopic examination of the various possible excited states in a Bose liquid, the variational theory he was led to is now viewed as a generic "single-mode approximation" for the density-response function. That is, if one starts with the assumption that $S(\mathbf{Q}, \omega)$ is dominated at all Q by a *single* undamped excitation of frequency ω_Q, the dispersion relation is given by the Feynman formula (1.6). This result is thus equally valid for normal liquid ^{4}He and liquid ^{3}He (for further discussion, see p. 915 of Mahan, 1990). The "dip" that such a disper-

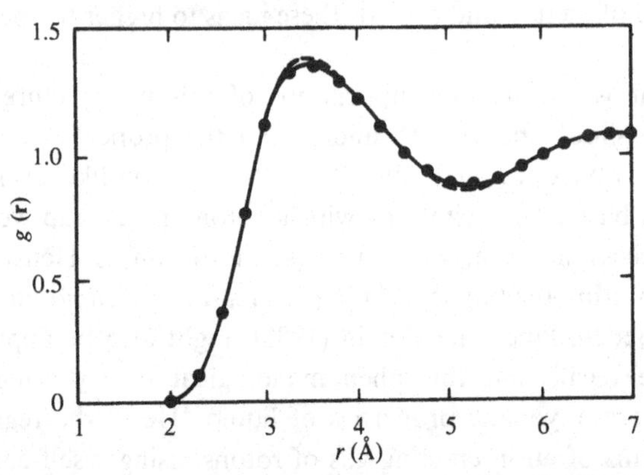

Fig. 12.2. The pair distribution function g(**r**) as a function of distance at 2.0 K, SVP. The solid line represents high-precision neutron data (Svensson *et al.*, 1980) and the circles are path-integral Monte Carlo results. At this temperature, the calculated condensate fraction is 8%. When Bose symmetrization is not done, the only change in g(**r**) is shown by the dashed line [Source: Ceperley and Pollock, 1986].

sion relation exhibits occurs at the wavevector $Q \sim 2$ Å^{-1} where the static structure factor $S(\mathbf{Q})$ has its first maximum. Such a "roton dip" is thus a consequence of the short-range spatial correlations between atoms and has little to do with the quantum statistics the atoms obey. This fact is further emphasized by the path integral Monte Carlo simulations of Ceperley and Pollock (1986). As shown in Fig. 12.2, the *short-range* behaviour of the pair distribution function $g(\mathbf{r})$ in superfluid ^{4}He is essentially unaffected when one does not carry out the Bose symmetrization. However, we recall the arguments given in Section 7.2 (see Fig. 7.17) against the assumption that the peak in $S(\mathbf{Q}, \omega)$ in normal liquids is associated with a single continuous excitation branch valid at both small and large Q.

On the other hand, Bose statistics do play a crucial (but indirect) role in understanding why there are no *other* low-lying states in liquid ^{4}He besides the one generated by (1.4) (or the more modern versions such as (9.10)), at least at small wavevectors. By way of contrast, in normal liquid ^{3}He, at small wavevectors there are, in addition to the zero sound phonon mode, many other low-energy modes corresponding to the single-particle Fermi quasiparticle excitations. It is Bose symmetrization which forces the excitations involving the motion of a single atom to have finite energy in the Feynman analysis. In turn, it is the Bose condensate that allows the density fluctuation spectrum (phonons) to have a significant overlap on the field fluctuation spectrum described by the single-particle Green's function $G(\mathbf{Q},\omega)$, justifying the implicit assumption of Feynman and Landau.

Needless to say, by itself, our interpretation of the roton as an intrinsic single-particle excitation of normal liquid ^{4}He does not shed any light on why it has the precise energy spectrum hypothesized in Fig. 7.23. In this connection, Feynman's explicit discussion of the motion of a single atom in liquid ^{4}He is of great interest and deserves careful study. Very recently, Stringari (1992) has suggested a promising way of studying this question in a quantitative manner (see remarks at end of Section 9.1).

12.2 Dielectric formalism for superfluid ^{3}He–^{4}He mixtures

In this book, the dynamics of superfluid ^{4}He has been discussed starting from the microscopic theory of Bose-condensed liquids. The various scenarios we have developed concerning the nature of the excitations have the common feature that the Bose condensate plays a pivotal role. In order to test these ideas, we have seen that neutron-scattering experiments done at varying temperatures and pressures have been of special importance. Indeed the recent picture put forward by Glyde and Griffin (1990) was developed in an attempt to understand the temperature-dependent line shapes as the liquid goes from the superfluid to normal phase (see Section 7.2).

In developing these ideas, the study of superfluid ^{3}He–^{4}He mixtures is particularly promising since the ^{3}He concentration gives a new parameter which can be varied, in addition to the pressure and temperature. In particular, recent high-momentum-transfer neutron-scattering data (see Chapter 4) indicate that the condensate fraction in a superfluid ^{3}He–^{4}He solution of 12% is considerably increased (Wang, Sosnick and Sokol, 1992). This suggests that hybridization effects induced by the ^{4}He condensate may be much stronger in mixtures than in pure superfluid ^{4}He.

Talbot (1983) and Talbot and Griffin (1984c) have worked out the formal extension of the dielectric formalism given in Section 5.1 to a fully interacting system of Fermions and Bosons, with due allowance for a Bose condensate. This formalism can be applied to superfluid ^{3}He–^{4}He mixtures. Just as in the case of pure liquid ^{4}He, the Bose broken symmetry leads to a coupling of the various correlation functions which is made most manifest by working in terms of irreducible, proper diagrams. The formal analysis is more complicated than that given in

Section 5.1 by the necessity to include a whole new class of response functions involving the ^{3}He atoms.

There is already a considerable body of literature on the elementary excitations and collective modes in superfluid ^{3}He–^{4}He mixtures. A useful introduction to the literature is given in the review by Glyde and Svensson (1987). Fåk *et al.* (1990) give a detailed analysis of recent neutron scattering data over a wide range of temperatures (0.07–1.5 K), with extensive references. The phenomenological quasiparticle theory of dilute ^{3}He–^{4}He mixtures is discussed in detail by Baym and Pethick (1978), Ruvalds (1978) as well as in Chapter 24 of Khalatnikov (1965).

Much of the work on mixtures to date has been concerned with low ^{3}He concentrations at low temperatures, with an emphasis on using the system as a way of studying a dilute Fermi gas of quasiparticles. In contrast, our major interest is to see how the phonon–maxon–roton excitation associated with the ^{4}He atoms is modified by the ^{3}He atoms. With this end in mind, it would be useful to have more neutron-scattering data at high ^{3}He concentrations, over a wide range of temperature (including above the superfluid transition).

Besides the dielectric formalism we review in this section, the variational and phenomenological approaches reviewed in Chapter 9 have all been extended to superfluid ^{3}He–^{4}He mixtures at $T = 0$. In connection with neutron-scattering data, the polarization potential approach of Section 9.2 is discussed by Hsu, Pines and Aldrich (1985) while the memory function formalism has been used by Lücke and Szprynger (1982) and Szprynger and Lücke (1985). As in the case of pure superfluid ^{4}He, these approaches can be brought into contact with the dielectric formalism by suggesting useful parameterizations of the "regular" functions in the latter theory.

The coherent (spin-independent) neutron-scattering differential cross-section is proportional to (see Section 2.1)

$$|b_4|^2(1-x_3)S_{44}(\mathbf{Q},\omega) + |b_3|^2 x_3 S_{33}(\mathbf{Q},\omega) + 2\,\mathrm{Re}(b_3 b_4^*)\sqrt{x_3(1-x_3)}S_{34}(\mathbf{Q},\omega)\ , \tag{12.1}$$

where b_3 and b_4 are the coherent neutron-scattering lengths for ^{3}He and ^{4}He atoms, respectively, and x_3 is the molar concentration of ^{3}He (see Fåk *et al.*, 1990, for further details). The dynamic structure factors are related to various number density response functions in the usual way:

$$S_{ab}(\mathbf{Q},\omega) = -\frac{1}{\pi\sqrt{n_a n_b}}[N(\omega)+1]\ \mathrm{Im}\ \chi_{ab}(\mathbf{Q},\omega)\ , \tag{12.2}$$

where $\chi_{ab}(\mathbf{Q},\omega)$ can be related to the Bose Matsubara Fourier components

$$\chi_{ab}(\mathbf{Q},i\omega_n) = \frac{1}{\Omega}\int_0^\beta d\tau\, e^{i\omega_n\tau}\langle\hat{\rho}_a(\mathbf{Q},\tau)\hat{\rho}_b(-\mathbf{Q})\rangle \qquad (12.3)$$

(see (3.41) and (5.4) for analogous definitions for pure ^{4}He). Here a, b can stand for ^{3}He or ^{4}He and $\hat{\rho}_a(\mathbf{Q})$ is the Fourier component of the number density operator of ^{3}He ($a = 3$) and ^{4}He ($a = 4$), defined as in (2.5) and (3.12). In the interpretation of neutron-scattering data on dilute mixtures, the cross-term S_{34} in (12.1) is often neglected, in which case the resonances in the total scattering intensity can be identified with renormalized ^{4}He and ^{3}He excitations associated with S_{44} and S_{33}, respectively.

Since ^{3}He and ^{4}He atoms have the same electronic polarizability, the cross-section for Brillouin light scattering (Stephen, 1976) can be expressed in terms of the dynamic structure factor associated with the total number density $n = n_3 + n_4$,

$$S_{nn}(\mathbf{Q},\omega) \equiv \sum_{a,b} S_{ab}(\mathbf{Q},\omega)\ . \qquad (12.4)$$

As we discuss in Section 6.2, Brillouin light scattering probes the hydrodynamic domain. In superfluid ^{3}He–^{4}He mixtures, this region is especially interesting since the intensity of second sound in $S_{nn}(\mathbf{Q},\omega)$ is greatly enhanced by the presence of the ^{3}He concentration fluctuations (Khalatnikov, 1965). The hydrodynamic spectrum of superfluid ^{3}He–^{4}He mixtures has been thoroughly mapped out by light scattering (see, for example, Stephen, 1976; Rockwell, Benjamin and Greytak, 1975). It would be interesting to use the two-fluid equations for mixtures (Khalatnikov, 1965) to derive the single-particle Green's functions for the ^{4}He atoms in the asymptotic region of low Q and ω (the analogous calculation for pure ^{4}He is sketched in Section 6.2).

We now briefly summarize the results of Talbot and Griffin (1984c). These formally rigorous expressions are valid for all $\mathbf{Q},\omega$ and T and are very relevant to inelastic neutron-scattering studies. In terms of irreducible contributions $\bar{\chi}_{ab}$, one finds

$$\left.\begin{aligned} \chi_{44} &= \frac{\bar{\chi}_{44} - A}{\epsilon}\ , \quad \chi_{33} = \frac{\bar{\chi}_{33} - A}{\epsilon}\ , \\ \chi_{34} &= \chi_{43} = \frac{\bar{\chi}_{34} + A}{\epsilon}\ , \end{aligned}\right\} \qquad (12.5)$$

where we define

$$\left.\begin{aligned} \epsilon(\mathbf{Q},\omega) &\equiv 1 - V(Q)[\bar{\chi}_{44} + 2\bar{\chi}_{34} + \bar{\chi}_{33}] \ , \\ A(\mathbf{Q},\omega) &\equiv \bar{\chi}_{44} V(Q) \bar{\chi}_{33} - \bar{\chi}_{34} V(Q) \bar{\chi}_{43} \ . \end{aligned}\right\} \tag{12.6}$$

We make use of the fact that the bare interatomic potentials between the Helium atoms are identical, $V_{ab} = V$, for $a, b = 3, 4$. As expected, all four response functions χ_{ab} in (12.5) share the same poles, which are given by the zeros of the dielectric function $\epsilon(\mathbf{Q},\omega)$ (Bartley *et al.*, 1973). We also note that

$$\chi_{nn} = \sum_{a,b} \chi_{ab} = \frac{\bar{\chi}_{nn}}{1 - V\bar{\chi}_{nn}} \ , \tag{12.7}$$

where $\bar{\chi}_{nn} \equiv \sum_{a,b} \bar{\chi}_{ab}$. The simplest MFA approximation (see Section 3.4) corresponds to using the free-gas response functions for $\bar{\chi}_{ab}$, i.e., $\bar{\chi}_{44} = \chi^0_{44}$, $\bar{\chi}_{33} = \chi^0_{33}$ and $\bar{\chi}_{34} = \chi^0_{34} = 0$.

The crucial step in the dielectric formalism for Bose-condensed systems is to split irreducible contributions into proper (regular) and improper (condensate) parts, as discussed in Section 5.1. We first define ($a = 3$ or 4) the mixed density-field correlation functions

$$\chi_{a\mu}(\mathbf{Q}, i\omega_n) = -\frac{1}{\sqrt{\Omega}} \int_0^\beta d\tau \, e^{i\omega_n\tau} \langle \hat{\rho}_a(\mathbf{Q},\tau) \hat{b}^+_{\mathbf{Q}\mu} \rangle \ , \tag{12.8}$$

the generalization of (5.4). One finds that the irreducible parts are given by (see (5.11*a*))

$$\chi_{a\nu}(\mathbf{Q}, i\omega_n) = \bar{\Lambda}_{a\mu}(\mathbf{Q}, i\omega_n) \bar{G}^{(4)}_{\mu\nu}(\mathbf{Q}, i\omega_n) \ , \tag{12.9}$$

where $\bar{G}^{(4)}_{\mu\nu}$ is the matrix ^{4}He single-particle Green's function which only contains proper self-energies $\bar{\Sigma}^{(4)}_{\mu\nu}$ and the summation convention is used for repeated Greek subscripts. Thus in mixtures, we have two Bose symmetry-breaking vertex functions $\bar{\Lambda}_{4\mu}$ and $\bar{\Lambda}_{3\mu}$, which play a key role in that they determine the hybridization of the poles of $G^{(4)}_{\mu\nu}$ and χ_{ab}. The irreducible contribution to the response function separates into two parts,

$$\bar{\chi}_{ab} = \bar{\chi}^R_{ab} + \bar{\chi}^c_{ab} \ , \tag{12.10}$$

with (see (5.11*b*))

$$\bar{\chi}^c_{ab} = \bar{\Lambda}_{\mu a} \bar{G}^{(4)}_{\mu\nu} \bar{\Lambda}_{\nu b} \ . \tag{12.11}$$

The improper ^{4}He self-energy is given by

$$\bar{\Sigma}^{(4)c}_{\mu\nu} = (\bar{\Lambda}_{\mu 4} + \bar{\Lambda}_{\mu 3})\frac{V(Q)}{\epsilon^R}(\bar{\Lambda}_{\nu 4} + \bar{\Lambda}_{\nu 3}) \ , \tag{12.12}$$

where (see (5.10))

$$\epsilon^R \equiv 1 - V(Q)\bar{\chi}^R_{nn} \ . \tag{12.13}$$

Similarly, one can derive the equivalent of (3.47) or (5.24) for the response functions χ_{ab} in mixtures,

$$\chi_{ab} = \Lambda_{\mu a} G^{(4)}_{\mu\nu} \Lambda_{\nu b} + \chi^R_{ab} \ , \tag{12.14}$$

where $\Lambda_{\mu a} = \bar{\Lambda}_{\mu a}/\epsilon^R$ and ϵ^R is defined in (12.13). This is the key result of the dielectric formalism for mixtures since it shows how the spectrum of the ^{4}He single-particle Green's function $G^{(4)}_{\mu\nu}$ is coupled into the number-density-response functions χ_{ab}. The zeros of $\epsilon(\mathbf{Q}, \omega)$ in (12.6) determine the poles of both χ_{ab} and $G^{(4)}_{\mu\nu}$. In contrast, the spectrum of the ^{3}He single-particle Green's function $G^{(3)}(\mathbf{Q}, \omega)$ is not determined by the zeros of $\epsilon(\mathbf{Q}, \omega)$. This difference can already be seen by the fact that $G^{(4)}$ and χ_{ab} involve Bose Matsubara frequencies while $G^{(3)}$ involve Fermi Matsubara frequencies.

Talbot and Griffin (1984c) have derived the equivalent of the Ward identities given in Section 5.1 and used them to obtain various exact results in the zero-frequency limit. As one example, one can show that the vertex function $\bar{\Lambda}_{3\mu}(\mathbf{Q}, \omega = 0) = 0$. At the present time, this formalism is mainly useful in showing how the Bose broken symmetry leads to a certain inevitable coupling between the poles of χ_{ab} and $G^{(4)}_{\mu\nu}$ in mixtures. The equivalent of the analysis given in Section 7.2 for pure superfluid ^{4}He requires further high-resolution neutron-scattering data as a function of temperature, especially in the region below and above the superfluid transition. This seems like one of the most promising areas of research in the near future.

Most phenomenological studies of the zeros of $\epsilon(\mathbf{Q}, \omega)$ defined in (12.6) approximate $\bar{\chi}_{33}$ as the Lindhard function of a non-interacting gas of Fermi quasiparticles and take $\bar{\chi}_{44}$ in a single-mode approximation appropriate to low temperatures (see, for example, Bartley *et al.*, 1973; Hsu, Pines and Aldrich, 1985; Szprynger and Lücke, 1985). One immediately sees that the resulting response functions χ_{ab} will exhibit complicated hybridization effects between the Fermi quasiparticle p–h spectrum and the Boson pole of $\bar{\chi}_{44}$. In the polarization approach, for example, $\bar{\chi}_{44}$ is modelled by an expression similar to (9.28). In neutron-scattering studies, we need to know the ^{3}He Lindhard function at relatively large values of Q (say 1–1.5 Å^{-1}).

The original Landau–Pomeranchuk (LP) spectrum for a single ^{3}He atom in bulk liquid ^{4}He is $Q^2/2m_3^*$, with an effective mass $m_3^* \simeq 2.4m_3$. With this LP quadratic spectrum, the quasiparticle particle–hole spectrum is a band described by $Q^2/2m_3^* \pm Qv_F$. For a dilute solution, this is a very narrow band centred on the LP ^{3}He quasiparticle branch (at $x_3 = 0.06$, we have $k_F \simeq 0.3$ Å^{-1}). At such low concentrations, there is no collective zero sound mode associated with the ^{3}He atoms (in contrast to pure liquid ^{3}He, shown in Figs. 7.5, 7.6 and 7.18). Thus if we use the LP single-particle spectrum, the only ^{3}He density fluctuation branch in dilute mixtures is a narrow band which would cross the maxon–roton ^{4}He excitation at around $Q \simeq 1.7$ Å^{-1} and $\omega \simeq 10$ K. This would be expected to produce strong hybridization effects between these two branches in this cross-over region and a strong-level repulsion of the p–h branch (Bartley *et al.*, 1973). This cross-over region has been extensively discussed in the literature (Hsu *et al.*, 1985; Szprynger and Lücke, 1985) with the conclusion that it is crucial to include the lower density of ^{4}He atoms in mixtures and also that the LP single-particle spectrum must be modified at the large values of Q probed in the neutron-scattering experiments.

In a formal sense, these hybridization effects are analogous to the coupling between the single-particle maxon–roton spectrum and the ^{4}He p–h branch which we considered in the case of pure ^{4}He (see Section 7.2). A natural extension of the model calculations given at the end of Section 7.2 (see (7.23)) would be to add in the contribution χ_{33}^0 to describe the p–h spectrum of the ^{3}He atoms. The neutron-scattering line shapes from superfluid mixtures are clearly a rich area for future studies, especially at higher temperatures and concentrations where the collective dynamics of both ^{3}He and ^{4}He are important. The dielectric formalism should give a rigorous basis for understanding and describing the various hybridization effects which occur.

12.3 Suggestions for future research

In this concluding section, we pull together a few suggestions as to where further work would be especially useful. Many of these suggestions have already been made in the earlier chapters. Our selection of topics for further research is influenced by what we believe is the most interesting feature of superfluid ^{4}He and ^{3}He–^{4}He mixtures, namely the unique dynamical structure of various correlation functions which arises because of the Bose condensate.

Theoretical

(1) Develop parameterized forms for $S(\mathbf{Q}, \omega)$, as a function of temperature, which are consistent with the general structure induced by the Bose condensate. The preliminary analysis given at the end of Section 7.2 must be extended to include the two-excitation spectra before one can successfully deal with the very interesting maxon region $Q \sim 1$ Å^{-1}.

(2) Work out the finite-temperature version of the Zawadowski, Ruvalds and Solana (1972) analysis of the condensate-induced hybridization between the one-roton and two-roton branches (Section 10.2).

(3) The polarization potential approach (see Section 9.2) seems to give a useful way of parameterizing the general structure of correlation functions as predicted by the dielectric formalism. The PP approach should be extended to finite temperatures (> 1 K) by including the thermally excited particle–hole excitations. The high-energy multiparticle excitations have been discussed (at $T = 0$) by Hess and Pines (1988).

(4) Formulate the Glyde–Griffin scenario at $T = 0$ in terms of a variational many-body wavefunction (see Section 9.1). The recent work of Stringari (1992) seems promising in this connection.

(5) Make use of the first frequency-moment sum rules specific to Bose-condensed fluids (see Section 8.1 and 8.3). Extension of these to third frequency-moment sum rules would be of interest.

(6) Work out the energy of single-particle excitations in a Bose liquid from first principles (Feynman, 1953b; Stringari, 1992).

(7) The interaction $V(\mathbf{Q})$ which appears in the final formulas of the dielectric formalism is assumed to be renormalized to some appropriate t-matrix when one includes multiple scattering. It would be useful to carry out this procedure more explicitly, summing up all contributions to regular quantities which involve two isolated propagator lines (see Section 5.4).

(8) The one-loop approximation to the dielectric formalism (see Section 5.3) should be worked out in detail for superfluid ^{3}He–^{4}He mixtures discussed in Section 12.2.

Experimental

(1) Measure the temperature dependence of the $S(\mathbf{Q},\omega)$ line shape at high energies in the cross-over region around $Q \sim 2.4$ Å^{-1} (see Figs. 7.19 and 7.20). This should be done at a series of temperatures above and below T_λ.

(2) Measure the temperature dependence and dispersion of the high-energy peak which shows up in the low-Q neutron data (see Figs. 7.1 and 10.4). Is this a two-roton bound state (as usually assumed) or is it a remnant of a single-particle excitation, as suggested in Figs. 7.22 and 7.23?

(3) High-resolution experiments on the temperature dependence of the $S(\mathbf{Q},\omega)$ line shape at a series of different pressures would be very useful, especially in the intermediate maxon region. As shown very dramatically by Fig. 7.8, at a pressure of 20 bar, the normal distribution is peaked at a much higher energy than the maxon peak. As a result, $S(\mathbf{Q},\omega)$ under high pressure exhibits the Glyde–Griffin scenario in a clearer fashion than the SVP data (where the superfluid and normal distributions are peaked at very similar energies).

(4) Measure the pressure and temperature dependence of the neutron-scattering line shapes in ^{3}He–^{4}He mixtures in the vicinity of the tricritical point (i.e., at high concentrations and temperatures).

(5) Develop new experimental probes of the existence and properties of rotons above the superfluid transition temperature. In the Glyde–Griffin picture, rotons (more generally, maxon–rotons) are intrinsic single-particle excitations of normal liquid ^{4}He. The rotons develop finite weight in $S(\mathbf{Q},\omega)$ only as a result of the Bose broken symmetry. Thus neutron scattering is not a useful probe of rotons above T_λ. Quantum evaporation studies would seem one way of studying single-particle excitations above and below T_λ (see Caroli *et al.*, 1976; Maris, 1992).

References

Abrikosov, A.A., Gor'kov, L.P. and Dzyaloshinskii, I.E. (1963), *Methods of Quantum Field Theory in Statistical Physics* (Prentice-Hall, Englewood Cliffs, N.J.).
Ahlers, G. (1978), in *Quantum Liquids*, ed. by J. Ruvalds and T. Regge (North-Holland, New York), p. 1.
Aldrich III, C.H., Pethick, C.G. and Pines, D. (1976), *Phys. Rev. Lett.* **37**, 845.
Aldrich III, C.H. and Pines, D. (1976), *Journ. Low Temp. Phys.* **25**, 673.
Aldrich III, C.H. and Pines, D. (1978), *Journ. Low Temp. Phys.* **32**, 689.
Allen, J.F. and Misener, A.D. (1938), *Nature* **141**, 75.
Ambegaokar, V., Conway, J. and Baym, G. (1965), in *Lattice Dynamics*, ed. by R.F. Wallis (Pergamon, New York), p. 261.
Andersen, K.H., Stirling, W.G., Scherm, R., Stunault, A., Fåk, B., Dianoux, A.J. and Godfrin, H. (1991), to be published.
Anderson, P.W. (1966), *Rev. Mod. Phys.* **38**, 298.
Anderson, P.W. (1984), *Basic Notions of Condensed Matter Physics* (Benjamin/Cummings, Menlo Park, Calif.).
Bartley, D.L., Robinson, J.E. and Wong, V.K. (1973), *Journ. Low Temp. Phys.* **12**, 71.
Bartley, D.L. and Wong, V.K. (1975), *Phys. Rev.* **B12**, 3775.
Baym, G. (1969), in *Mathematical Methods in Solid State and Superfluid Theory*, ed. by R.C. Clark and G.H. Derrick (Oliver and Boyd, Edinburgh), p. 121.
Baym, G., and Kadanoff, L.P. (1961), *Phys. Rev.* **124**, 287.
Baym, G. and Pethick, C.J., (1978), in *The Physics of Liquid and Solid Helium*, ed. by K.H. Bennemann and J.B. Ketterson (Wiley, New York), Vol. II.
Bedell, K., Pines, D. and Fomin, I. (1982), *Journ. Low Temp. Phys.* **48**, 417.
Bedell, K., Pines, D. and Zawadowski, A. (1984), *Phys. Rev.* **B29**, 102.
Beliaev, S.T. (1958a), *Sov. Phys. – JETP* **7**, 289.
Beliaev, S.T. (1958b), *Sov. Phys. – JETP* **7**, 299.
Bendt, P.J., Cowan, R.D. and Yarnell, J.H. (1959), *Phys. Rev.* **113**, 1386.
Bijl, A. (1940), *Physica* **7**, 869.
Bogoliubov, N.N. (1947), *J. Phys. U.S.S.R.* **11**, 23.
Bogoliubov, N.N. (1963, 1970), *Lectures on Quantum Statistics* (Gordon and Breach, New York), Vol. 2.
Boon, J.P. and Yip, S. (1980), *Molecular Hydrodynamics* (McGraw-Hill, New York).
Brown, G.V. and Coopersmith, M.H. (1969), *Phys. Rev.* **178**, 327.

Brueckner, K.A. and Sawada, K. (1957), *Phys. Rev.* **106**, 1117.
Buyers, W.J.L., Sears, V.F., Lonngi, P.A. and Lonngi, D.A. (1975), *Phys. Rev.* **A11**, 697.
Campbell, C.E. (1978), in *Progress in Liquid Physics*, ed. by C.A. Croxton (Wiley, New York), p. 213.
Campbell, C.E. and Clements, B.E. (1989), in *Elementary Excitations in Quantum Fluids*, ed. by K. Ohbayashi and M. Watabe (Springer-Verlag, Berlin), p. 8.
Caroli, C., Roulet, B. and Saint-James, D. (1976), *Phys. Rev.* **B13**, 3875.
Carraro, C. and Koonin, S.E. (1990), *Phys. Rev. Lett.* **65**, 2792.
Carraro, C. and Rinat, A.S. (1992), *Phys. Rev.* **B45**, 2945.
Ceperley, D.M. and Pollock, E.L. (1986), *Phys. Rev. Lett.* **56**, 351.
Ceperley, D.M. and Pollock, E.L. (1987), *Can. J. Phys.* **65**, 1416.
Chester, G.V. (1963), in *Liquid Helium*, ed. by G. Careri (Academic Press, New York), p. 51.
Chester, G.V. (1969), in *Lectures in Theoretical Physics* (Vol. XI-B): *Quantum Fluids and Nuclear Matter*, ed. by K.T. Mahanthappa and W.E. Britten (Gordon and Breach, New York), p. 253.
Chester, G.V. (1975), in *The Helium Liquids*, ed. by J.G.M. Armitage and I.E. Farquhar (Academic Press, London), p. 1.
Cheung, T.H. (1971), Ph.D. Thesis, University of Toronto, unpublished.
Cheung, T.H. and Griffin, A. (1970), *Can. Journ. Phys.* **48**, 2135.
Cheung, T.H. and Griffin, A. (1971a), *Phys. Lett.* **A35**, 141.
Cheung, T.H. and Griffin, A. (1971b), *Phys. Rev.* **A4**, 237.
Clark, J.W. and Ristig, M.L. (1989), in *Momentum Distributions*, ed. by R.N. Silver and P.E. Sokol (Plenum, New York), p. 39.
Cohen, M. and Feynman, R.P. (1957), *Phys. Rev.* **107**, 13.
Copley, J.R.D. and Lovesey, S.W. (1975), *Rep. Prog. Phys.* **38**, 461.
Cowley, R.A. (1968), *Rep. Prog. Phys.* **31**, 123.
Cowley, R.A. and Buyers, W.J.L. (1969), *J. Phys. C: Solid State Phys.* **2**, 2262.
Cowley, R.A. and Woods, A.D.B. (1968), *Phys. Rev. Lett.* **21**, 787.
Cowley, R.A. and Woods, A.D.B. (1971), *Can. J. Phys.* **49**, 177.
Cummings, F.W., Hyland, G.J. and Rowlands, G. (1970), *Phys. Condensed Matter* **12**, 90.
Dietrich, O.W., Graf, E.H., Huang, C.G. and Passell, L. (1972), *Phys. Rev.* **A5**, 1377.
Donnelly, R.J. (1991), *Quantized Vortices in Helium II* (Cambridge University Press, New York).
Dzugutov, M. and Dahlborg, U. (1989), *Phys. Rev.* **40A**, 4103.
Einstein, A. (1925), *Sitz. Berlin Preuss. Akad. Wiss.* , p. 3.
Fåk, B. and Andersen, K.H. (1991), *Phys. Lett.* **A160**, 468.
Fåk, B., Guckelsberger, K., Körfer, M., Scherm, R. and Dianoux, A.J. (1990), *Phys. Rev.* **B41**, 8732.
Family, F. (1975), *Phys. Rev. Lett.* **34**, 1374.
Feenberg, E. (1969), *Theory of Quantum Fluids* (Academic Press, New York).
Ferrell, R.A., Menyhard, N., Schmidt, H., Schwabl, F. and Szépfalusy, P. (1968), *Ann. Phys. (New York)* **47**, 565.
Fetter, A.L. (1970), *Ann. Phys. (New York)* **60**, 464.
Fetter, A.L. (1972), *Journ. Low Temp. Phys.* **6**, 487.
Fetter, A.L. and Walecka, J.D. (1971), *Quantum Theory of Many Particle Systems* (McGraw-Hill, New York).
Feynman, R.P. (1953a), *Phys. Rev.* **91**, 1291.

Feynman, R.P. (1953b), *Phys. Rev.* **91**, 1301.
Feynman, R.P. (1954), *Phys. Rev.* **94**, 262.
Feynman, R.P. and Cohen, M. (1956), *Phys. Rev.* **102**, 1189.
Forster, D. (1975), *Hydrodynamic Fluctuations, Broken Symmetry and Correlation Functions* (W.A. Benjamin, Reading, Mass.).
Fredkin, D.R. and Werthamer, N.R. (1965), *Phys. Rev.* **138A**, 1528.
Fukushima, K. and Iseki, F. (1988), *Phys. Rev.* **B38**, 4448.
Gavoret, J. (1963), Thèse de Doctorat ès Sciences Physiques, Paris, *Annales de Physique* **8**, 441.
Gavoret, J. and Nozières, P. (1964), *Ann. Phys. (New York)* **28**, 349.
Gay, C. and Griffin, A. (1985), *Journ. Low Temp. Phys.* **58**, 479.
Gersch, H.A. and Rodriguez, L.J. (1973), *Phys. Rev.* **A8**, 905.
Gillis, N.S. and Werthamer, N.R. (1968), *Phys. Rev.* **167**, 607.
Gillis, N.S., Werthamer, N.R. and Koehler, T.R. (1968), *Phys. Rev.* **165**, 951.
Ginzburg, V.L. (1943), *J. Exp. Theor. Phys. (U.S.S.R.)* **13**, 243.
Giorgini, S., Pitaevskii, L. and Stringari, S. (1992), *Phys. Rev.* **B46**, 6374.
Giorgini, S. and Stringari, S. (1990), *Physica* **B 165–166**, 511.
Girardeau, M. and Arnowitt, R. (1959), *Phys. Rev.* **113**, 755.
Glyde, H.R. (1971), *Can. Journ. Phys.* **49**, 761.
Glyde, H.R. (1974), *Can. Journ. Phys.* **52**, 2281.
Glyde, H.R. (1976), in *Rare Gas Solids*, ed. by M.L. Klein and J.A. Venables (Academic Press, New York), vol. 1, p. 382.
Glyde, H.R. (1984), in *Condensed Matter Research Using Neutrons*, ed. by S.W. Lovesey and R. Scherm (Plenum, New York) p. 95.
Glyde, H.R. (1992a), *Phys. Rev.* **B45**, 7321.
Glyde, H.R. (1992b), to be published.
Glyde, H.R. and Goldman, V.V. (1976), *Journ. Low Temp. Phys.* **25**, 601.
Glyde, H.R. and Griffin, A. (1990), *Phys. Rev. Lett.* **65**, 1454.
Glyde, H.R. and Svensson, E.C. (1987), in *Methods of Experimental Physics*, ed. by D.L. Price and K. Sköld (Academic Press, New York), vol. 23, ch. 13 .
Götze, W. and Lücke, M. (1976), *Phys. Rev.* **B13**, 3825.
Greytak, T.J. (1978), in *Quantum Liquids*, ed. by J. Ruvalds and T. Regge (North-Holland, Amsterdam), p. 121.
Greytak, T.J. and Kleppner, D. (1984), in *New Trends in Atomic Physics*, ed. by G. Grynberg and R. Stora (North-Holland, New York), Vol. II, p. 1127.
Greytak, T.J. and Yan, J. (1969), *Phys. Rev. Lett.* **22**, 987.
Greytak, T.J. and Yan, J. (1971), *Proc. of the 12th International Conference on Low Temperature Physics*, Kyoto, 1970, ed. by E. Kanda (Keigaku Publ. Comp., Tokyo), p. 89.
Griffin, A. (1972), *Phys. Rev.* **A6**, 512.
Griffin, A. (1979a), *Phys. Rev.* **B19**, 5946.
Griffin, A. (1979b), *Phys. Lett.* **71A**, 237.
Griffin, A. (1980), *Phys. Rev.* **B22**, 5193.
Griffin, A. (1981), *Journ. Low Temp. Phys.* **44**, 441.
Griffin, A. (1984), *Phys. Rev.* **B30**, 5057.
Griffin, A. (1985), *Phys. Rev.* **B32**, 3289.
Griffin, A. (1987), *Can. Journ. Phys.* **65**, 1368.
Griffin, A. (1988), *Physica* **C156**, 12.
Griffin, A. (1989), in *Elementary Excitations in Quantum Fluids*, ed. by K. Ohbayashi and M. Watabe (Springer-Verlag, Berlin), p. 23.

Griffin, A. (1991), in *Excitations in 2D and 3D Quantum Fluids*, ed. by A.F.G. Wyatt and H.J. Lauter (Plenum, New York), p. 15.
Griffin, A. and Cheung, T.H. (1973), *Phys. Rev.* **A7**, 2086.
Griffin, A. and Payne, S. (1986), *Journ. Low Temp. Phys.* **64**, 155.
Griffin, A. and Svensson, E.C. (1990), *Physica* **B165–166**, 487.
Griffin, A. and Talbot, E. (1981), *Phys. Rev.* **B24**, 5075.
Griffin, A., Wu, W.C. and N. Lambert (1992), to be published.
Halley, J.W. (1969), *Phys. Rev.* **181**, 338.
Halley, J.W. (1989), in *Elementary Excitations in Quantum Fluids*, ed. by K. Ohbayashi and M. Watabe (Springer-Verlag, Berlin), p. 106.
Halley, J.W. and Korth, M.S. (1991), in *Excitations in 2D and 3D Quantum Fluids*, ed. by A.F.G. Wyatt and H.J. Lauter (Plenum, New York), p. 91.
Hansen, J.P. and McDonald, I.R. (1986), *Theory of Simple Liquids*, Second Ed. (Academic Press, London).
Harling, O.K. (1970), *Phys. Rev. Lett.* **24**, 1046.
Harling, O.K. (1971), *Phys. Rev.* **A3**, 1073.
Henshaw, D.G. and Woods, A.D.B. (1961), *Phys. Rev.* **121**, 1266.
Herwig, K.W., Sokol, P.E., Snow, W.M. and Blasdell, R.C. (1991), *Phys. Rev.* **B44**, 308.
Herwig, K.W., Sokol, P.E., Sosnick, T.R., Snow, W.M. and Blasdell, R.C. (1990), *Phys. Rev.* **B41**, 103.
Hess, D.W. and Pines, D. (1988), *Journ. Low Temp. Phys.* **72**, 247.
Hilton, P.A., Cowley, R.A., Scherm, R. and Stirling, W.G. (1980), *J. Phys. C: Solid State Phys.* **13**, L295.
Hohenberg, P.C. (1967), *Phys. Rev.* **158**, 383.
Hohenberg, P.C. (1973), *Journ. Low Temp. Phys.* **11**, 745.
Hohenberg, P.C. and Martin, P.C. (1964), *Phys. Rev. Lett.* **12**, 69.
Hohenberg, P.C. and Martin, P.C. (1965), *Ann. Phys. (New York)* **34**, 291.
Hohenberg, P.C. and Platzman, P.M. (1966), *Phys. Rev.* **152**, 198.
Horner, H. (1967), *Z. Physik* **205**, 72.
Horner, H. (1971), *Z. Physik* **242**, 432.
Horner, H. (1972a), *Phys. Rev. Lett.* **29**, 556.
Horner, H. (1972b), *Journ. Low Temp. Phys.* **8**, 511.
Horner, H. (1974a), in *Proceedings of the 13th International Conference on Low Temperature Physics*, ed. by K.D.T. Timmerhaus, W.J. O'Sullivan and E.F. Hammel (Plenum Press, New York), vol. 1, p. 3.
Horner, H. (1974b), in *Proceedings of the 13th International Conference on Low Temperature Physics*, ed. by K.D.T. Timmerhaus, W.J. O'Sullivan and E.F. Hammel (Plenum Press, New York), vol. 1, p. 125.
Horner, H. (1974c), in *Dynamical Properties of Solids*, ed. by G.K. Horton and A.A. Maradudin (North-Holland, Amsterdam), vol. I, p. 451.
Hsu, W., Pines, D. and Aldrich III, C.H. (1985), *Phys. Rev.* **B32**, 7179.
Huang, K. (1964), in *Studies in Statistical Mechanics*, ed. by J. De Boer and G.E. Uhlenbeck (North-Holland, Amsterdam), vol. II.
Huang, K. (1987), *Statistical Mechanics*, Second Edition (John Wiley, New York).
Huang, K. and Klein, A. (1964), *Ann. Phys. (New York)* **30**, 203.
Hugenholtz, N. and Pines, D. (1959), *Phys. Rev.* **116**, 489.
Inkson, J.C. (1984), *Many-Body Theory of Solids* (Plenum, New York).
Iwamoto, F. (1970), *Prog. Theoret. Phys.* (Kyoto) **44**, 1135.

Iwamoto, F. (1989), in *Elementary Excitations in Quantum Fluids*, ed. by K. Ohbayashi and M. Watabe (Springer-Verlag, Berlin), p. 117.
Jackson, H.W. (1969), *Phys. Rev.* **185**, 186.
Jackson, H.W. (1973), *Phys. Rev.* **A8**, 1529.
Jackson, H.W. (1974), *Phys. Rev.* **A10**, 278.
Juge, K.J. and Griffin, A. (1993), to be published.
Kadanoff, L.P. and Baym, G. (1962), *Quantum Statistical Mechanics* (Benjamin, New York).
Kadanoff, L.P. and Martin, P.C. (1963), *Ann. Phys. (New York)* **24**, 419.
Kalos, M.H., Lee, M.A., Whitlock, P.A. and Chester, G.V. (1981), *Phys. Rev.* **B24**, 115.
Kapitza, P. (1938), *Nature* **141**, 74.
Khalatnikov, I.M. (1965), *An Introduction to the Theory of Superfluidity* (Benjamin, New York).
Kirkpatrick, T.R. (1984), *Phys. Rev.* **B30**, 1266.
Kirkpatrick, T.R. and Dorfman, J.R. (1985), *Journ. Low Temp. Phys.* **58**, 399, 301; **59**, 1.
Kitchens, T.A., Shirane, G., Minkiewicz, V.J. and Osgood, E.B. (1972), *Phys. Rev. Lett.* **29**, 552.
Koehler, T.R. (1967), *Phys. Rev. Lett.* **18**, 654.
Koehler, T.R. (1968), *Phys. Rev.* **17**, 942.
Koehler, T.R. and Werthamer, N.R. (1971), *Phys. Rev.* **A3**, 2074.
Koehler, T.R. and Werthamer, N.R. (1972), *Phys. Rev.* **A5**, 2230.
Kondor, I. and Szépfalusy, P. (1968), *Acta Phys. Hung.* **24**, 81.
Kwok, P.C. (1967), in *Solid State Physics*, ed. by F. Seitz, D. Turnbull and H. Ehrenreich (Academic Press, New York), vol. 20, p. 306.
Kwok, P.C. and Martin, P.C. (1966), *Phys. Rev.* **142**, 495.
Landau, L.D. (1941), *J. Phys. U.S.S.R.* **5**, 71.
Landau, L.D. (1947), *J. Phys. U.S.S.R.* **11**, 91.
Landau, L.D. (1949), *Phys. Rev.* **75**, 884.
Landau, L.D. and Khalatnikov, I.M. (1949), *Zh. Eksp. Teor. Fiz.* **19**, 637.
Landau, L.D. and Lifshitz, E.M. (1959), *Statistical Physics* (Pergamon, Oxford).
Lee, T.D., Huang, K. and Yang, C.N. (1957), *Phys. Rev.* **106**, 1135.
Leggett, A.J. (1975), *Rev. Mod. Phys.* **47**, 331.
Lifshitz, E.M. and Pitaevskii, L.P. (1980), *Statistical Physics*, Part 2, (Pergamon, Oxford).
London, F. (1938a), *Nature* **141**, 643.
London, F. (1938b), *Phys. Rev.* **54**, 947.
London, F. (1950), *Superfluids*, Vol. 1: Macroscopic Theory of Superconductivity (Wiley, New York).
London, F. (1954), *Superfluids*, Vol. 2: Macroscopic Theory of Superfluid Helium (Wiley, New York).
Lücke, M. (1980), in *Modern Trends in the Theory of Condensed Matter*, Proc. of the 16th Karpacz Winter School of Theoretical Physics, ed. by A. Pekalski and J. Przystawa (Springer-Verlag, Berlin), p. 74.
Lücke, M. and Szprynger, A. (1982), *Phys. Rev.* **B26**, 1374.
Ma, S.K. (1971), *Journ. Math. Phys.* **12**, 2157.
Ma, S.K., Gould, H. and Wong, V.K. (1971), *Phys. Rev.* **A3**, 1453.
Ma, S.K. and Woo, C.W. (1967), *Phys. Rev.* **159**, 165.
Mahan, G. (1990), *Many-Particle Physics*, Second Edition (Plenum, New York).

Manousakis, E. (1989), in *Momentum Distributions*, ed. by R.N. Silver and P.E. Sokol (Plenum, New York) p. 81.
Manousakis, E. and Pandharipande, V.R. (1984), *Phys. Rev.* **B30**, 5064.
Manousakis, E. and Pandharipande, V.R. (1986), *Phys. Rev.* **B33**, 150.
Manousakis, E., Pandharipande, V.R. and Usmani, Q.N. (1985), *Phys. Rev.* **B33**, 7022.
Maradudin, A.A. and Fein, A.E. (1962), *Phys. Rev.* **128**, 2589.
Maris, H.J. (1977), *Rev. Mod. Phys.* **49**, 341.
Maris, H.J. (1992), *Journ. Low Temp. Phys.* **87**, 773.
Maris, H.J. and Massey, W.E. (1970), *Phys. Rev. Lett.* **25**, 220.
Marshall, W. and Lovesey, S.W. (1971), *Theory of Thermal Neutron Scattering* (Oxford University Press, New York).
Martel, P., Svensson, E.C., Woods, A.D.B., Sears, V.F. and Cowley, R.A. (1976), *Journ. Low Temp. Phys.* **32**, 285.
Martin, P.C. (1965), in *Low Temperature Physics – LT9*, ed. by J.G. Daunt, D.O. Edwards, F.J. Milford and M. Yaqub (Plenum, New York), p. 9.
Martin, P.C. (1968), in *Many-Body Physics*, ed. by C. De Witt and R. Balian (Gordon and Breach, New York), p. 37.
McMahan, A.K. and Beck, H. (1973), *Phys. Rev.* **A8**, 3247.
McMillan, W.L. (1965), *Phys. Rev.* **138**, A442.
Meisel, M.W. (1992), *Physica* **B178**, 121.
Mezei, F. and Stirling, W.G. (1983), in *75th Jubilee Conference on Helium-4*, ed. by J.G.M. Armitage (World Scientific, Singapore), p. 111.
Miller, A., Pines, D. and Nozières, P. (1962), *Phys. Rev.* **127**, 1452.
Mineev, V.P. (1980), *Sov. Phys. – JETP Lett.* **32**, 489.
Minkiewicz, V.J., Kitchens, T.A., Shirane, G. and Osgood, E.B. (1973), *Phys. Rev.* **A8**, 1513.
Mook, H.A. (1974), *Phys. Rev. Lett.* **32**, 1167.
Mook, H.A. (1983), *Phys. Rev. Lett.* **51**, 1454.
Mook, H.A., Scherm, R. and Wilkinson, M.K. (1972), *Phys. Rev.* **A6**, 2268.
Murray, R.L., Woerner, R.L. and Greytak, T.J. (1975), *J. Phys. C: Sol. State Phys.* **C8**, L90.
Nakajima, S. (1971), *Prog. Theoret. Phys. (Kyoto)* **45**, 353.
Nelson, D.R. (1983), in *Phase Transitions and Critical Phenomena*, ed. by C. Domb and J.L. Lebowitz (Academic Press, London), Vol. 7, p. 1.
Nepomnyashchii, Y.A. (1983), *Sov. Phys. – JETP* **58**, 722.
Nepomnyashchy, Y.A. (1992), *Phys. Rev.* **B46**, 6611.
Nepomnyashchii, Y.A. and Nepomnyashchii, A.A. (1974), *Sov. Phys. – JETP* **38**, 134.
Nepomnyashchii, Y.A. and Nepomnyashchii, A.A. (1978), *Sov. Phys. – JETP* **48**, 493.
Nozières, P. (1964), *Theory of Interacting Fermi Systems* (Benjamin, New York).
Nozières, P. (1966), in *Quantum Fluids*, ed. by D.F. Brewer (North-Holland, Amsterdam), p. 1.
Nozières, P. (1983), *Quantum Liquids and Solids*, lecture notes, Collège de France.
Nozières, P. and Pines, D. (1964, 1990), *Theory of Quantum Liquids*, Vol. II: Superfluid Bose Liquids (Addison-Wesley, Redwood City, Calif.).
O'Connor, J.T., Palin, C.J. and Vinen, W.F. (1975), *J. Phys. C: Solid State Phys.* **8**, 101.
Ohbayashi, K. (1989), in *Elementary Excitations in Quantum Fluids*, ed. by K. Ohbayashi and M. Watabe (Springer-Verlag, Berlin), p. 32.

Ohbayashi, K. (1991), in *Excitations in 2D and 3D Quantum Fluids*, ed. by A.F.G. Wyatt and H.J.L. Lauter (Plenum, New York), p. 77.
Ohbayashi, K., Udagawa, K., Yamashita, H., Watabe, M. and Ogita, N. (1990), *Physica* **B165–166**, 485.
Osgood, E.B., Kitchens, T.A., Shirane, G. and Minkiewicz, V.J. (1972), *Phys. Rev.* **A5**, 1537.
Palevsky, H., Otnes, K., Larsson, K.E., Pauli, R. and Stedman, R. (1957), *Phys. Rev.* **108**, 1346.
Parry, W.E. and ter Haar, D. (1962), *Ann. Phys. (New York)* **19**, 496.
Payne, S.H. and Griffin, A. (1985), *Phys. Rev.* **B32**, 7199.
Penrose, O. (1951), *Phil. Mag.* **42**, 1373.
Penrose, O. (1958), in *Proceedings of the International Conference on Low Temperature Physics*, ed. by J.R. Dillinger (University of Wisconsin Press, Madison), p. 117.
Penrose, O. and Onsager, L. (1956), *Phys. Rev.* **104**, 576.
Pike, E.R. (1972), *Journ. de Physique* **33**, C1, 25.
Pike, E.R., Vaughan, J.M. and Vinen, W.F. (1970), *J. Phys. C: Sol. State Phys.* **3**, L40.
Pines, D. (1963), in *Liquid Helium*, ed. by G. Careri (Academic Press, New York), p. 147.
Pines, D. (1965), in *Low Temperature Physics – LT9*, ed. by J.G. Daunt, D.O. Edwards, F.J. Milford and M. Yaqub (Plenum, New York), p. 61.
Pines, D. (1966), in *Quantum Fluids*, ed. by D.F. Brewer (North-Holland, Amsterdam), p. 257.
Pines, D. (1985), in *Highlights of Condensed Matter Theory*, ed. by F. Bassani, F. Fumi and M.P. Tosi (North-Holland, New York), p. 580.
Pines, D. (1987), *Can. Journ. Phys.* **65**, 1357.
Pines, D. (1989), *Physics Today* **42**, 61.
Pines, D. and Nozières, P. (1966), *Theory of Quantum Liquids*, Vol. I: Normal Fermi Liquids (Addison-Wesley, Redwood City, Calif.).
Pitaevskii, L.P. (1959), *Sov. Phys. – JETP* **9**, 830.
Pitaevskii, L.P. (1987), *Sov. Phys. – JETP Lett.* **45**, 185.
Pitaevskii, L.P. and Fomin, I.A. (1974), *Sov. Phys. – JETP* **38**, 1257.
Pitaevskii, L.P. and Stringari, S. (1991), *Journ. Low Temp. Phys.* **85**, 377.
Placzek, G. (1952), *Phys. Rev.* **86**, 337.
Platzman, P. and Tzoar, N. (1965), *Phys. Rev.* **139**, A410.
Pollock, E.L. and Ceperley, D.M. (1987), *Phys. Rev.* **B36**, 8343.
Popov, V.N. (1965), *Sov. Phys. – JETP* **20**, 1185.
Popov, V.N. (1983), *Functional Integrals in Quantum Field Theory and Statistical Physics* (Reidel, Dordrecht).
Popov, V.N. (1987), *Functional Integrals and Collective Excitations* (Cambridge University Press).
Popov, V.N. and Serendniakov, A.V. (1979), *Sov. Phys. – JETP* **50**, 193.
Puff, R.D. (1965), *Phys. Rev.* **137**, 17406.
Puff, R.D. and Tenn, J.S. (1970), *Phys. Rev.* **A1**, 125.
Rahman, A., Singwi, K.S. and Sjölander, A. (1962), *Phys. Rev.* **126**, 986.
Reatto, L. and Chester, G.V. (1967), *Phys. Rev.* **155**, 88.
Rickayzen, G. (1980), *Green's Functions and Condensed Matter* (Academic Press, New York).
Rinat, A.S. (1989), *Phys. Rev.* **B41**, 4247.

Rockwell, D.A., Benjamin, R.F. and Greytak, T.J. (1975), *Journ. Low Temp. Phys.* **18**, 389.
Ruvalds, J. (1978), in *Quantum Liquids*, ed. by J. Ruvalds and T. Regge (North-Holland, New York), p. 263.
Ruvalds, J. and Zawadowski, A. (1970), *Phys. Rev. Lett.* **25**, 233.
Scherm, R., Dianoux, A.J., Fåk, B., Guckelsberger, K., Körfer, M. and Stirling, W.G. (1989), *Physica* **B156 & 157**, 311.
Scherm, R., Guckelsberger, K., Fåk, B., Sköld, K., Dianoux, A.J., Godfrin, H. and Stirling, W.G. (1987), *Phys. Rev. Lett.* **59**, 217.
Schrieffer, J.R. (1964), *Theory of Superconductivity* (W.A. Benjamin, Reading, Mass.).
Sears, V.F. (1969), *Phys. Rev.* **185**, 200.
Sears, V.F. (1981), *Can J. Phys.* **59**, 555.
Sears, V.F. (1985), *Can J. Phys.* **63**, 68.
Sears, V.F., Svensson, E.C., Martel, P. and Woods, A.D.B. (1982), *Phys. Rev. Lett.* **49**, 279.
Senger, G., Ristig, M.L., Campbell, C.E. and Clark, J.W. (1992), *Ann. Phys. (N.Y.)* **218**, 160.
Silver, R.N. (1988), *Phys. Rev.* **B37**, 3794.
Silver, R.N. (1989), *Phys. Rev.* **B38**, 2283.
Silver, R.N. and Sokol, P.E., eds. (1989), *Momentum Distributions* (Plenum, New York).
Sköld, K. and Pelizzari, C.A. (1980), *Phil. Trans. of the Roy. Soc. of London* **B290**, 605.
Sköld, K., Pelizzari, C.A., Kleb, R. and Ostrowski, G.E. (1976), *Phys. Rev. Letters* **37**, 842.
Smith, A.J., Cowley, R.A., Woods, A.D.B., Stirling, W.G. and Martel, P. (1977), *J. Phys. C: Solid State Phys.* **10**, 543.
Snow, W.M. (1990), Ph.D. thesis, Harvard University, unpublished.
Snow, W.M., Wang, Y. and Sokol, P.E. (1992), *Europhys. Lett.* **19**, 403.
Sokol, P.E. (1987), *Can. J. Phys.* **65**, 1393.
Sokol, P.E. and Snow, W.M. (1991), in *Excitations in 2D and 3D Quantum Fluids*, ed. by A.F.G. Wyatt and H.J. Lauter (Plenum, New York), p. 47.
Sokol, P.E., Sosnick, T.R. and Snow, W.M. (1989), in *Momentum Distributions*, ed. by R.N. Silver and P.E. Sokol (Plenum, New York), p. 139.
Sosnick, T.R., Snow, W.M., Silver, R.N. and Sokol, P.E. (1991), *Phys. Rev.* **B43**, 216.
Sosnick, T.R., Snow, W.M. and Sokol, P.E. (1990), *Phys. Rev.* **B41**, 11185.
Stephen, M.J. (1969), *Phys. Rev.* **187**, 279.
Stephen, M.J. (1976), in *The Physics of Liquid and Solid Helium*, ed. by K.H. Benneman and J.B. Ketterson (Wiley, New York) Vol. 1, p. 307.
Stirling, W.G. (1983), in *75th Jubilee Conference on Helium-4*, ed. J.G.M. Armitage (World Scientific, Singapore), p. 109.
Stirling, W.G. (1985), in *Proc. 2nd International Conference on Phonon Physics*, ed. by J. Kollar, N. Kroó, N. Menyhard, and T. Siklos (World Scientific, Singapore), p. 829.
Stirling, W.G. (1991), in *Excitations in 2D and 3D Quantum Fluids*, ed. by A.F.G. Wyatt and H.J. Lauter (Plenum, New York), p. 25.
Stirling, W.G. and Glyde, H.R. (1990), *Phys. Rev.* **B41**, 4224.
Stirling, W.G., Talbot, E.F., Tanatar, B. and Glyde, H.R. (1988), *Journ. Low Temp. Phys.* **73**, 33.

Straley, J.P. (1972), *Phys. Rev.* **A5**, 338.
Stringari, S. (1991), *Journ. Low Temp. Phys.* **84**, 279.
Stringari, S. (1992), *Phys. Rev.* **B46**, 2974.
Svensson, E.C. (1989), in *Elementary Excitations in Quantum Fluids*, ed. by K. Ohbayashi and M. Watabe (Springer-Verlag, Heidelberg), p. 59.
Svensson, E.C. (1991), in *Excitations in 2D and 3D Quantum Fluids*, ed. by A.F.G. Wyatt and H.J. Lauter (Plenum, New York), p. 59.
Svensson, E.C., Martel, P., Sears, V.F. and Woods, A.D.B. (1976), *Can. J. Phys.* **54**, 2178.
Svensson, E.C., Martel, P. and Woods, A.D.B. (1975), *Phys. Lett.* **55A**, 151.
Svensson, E.C., Sears, V.F., Woods, A.D.B. and Martel, P. (1980), *Phys. Rev.* **B21**, 3638.
Svensson, E.C. and Tennant, D.C. (1987), *Japan Journ. Applied Phys.* **26**, (Supp. 26-3), 31.
Szépfalusy, P. (1965), *Acta Phys. Hungarica* **19**, 109.
Szépfalusy, P. and Kondor, I. (1974), *Ann. Phys. (New York)* **82**, 1.
Szprynger, A. and Lücke, M. (1985), *Phys. Rev.* **B32**, 4442.
Talbot, E.F. (1983), Ph.D. Thesis, University of Toronto, unpublished.
Talbot, E., Glyde, H.R., Stirling, W.G. and Svensson, E.C. (1988), *Phys. Rev.* **B38**, 11229.
Talbot, E. and Griffin, A. (1983), *Ann. Phys. (New York)* **151**, 71.
Talbot, E. and Griffin, A. (1984a), *Phys. Rev.* **B29**, 2531.
Talbot, E. and Griffin, A. (1984b), *Phys. Rev.* **B29**, 3952.
Talbot, E. and Griffin, A. (1984c), *Journ. Low Temp. Phys.* **56**, 141.
Tanatar, B., Talbot, E.F. and Glyde, H.R. (1987), *Phys. Rev.* **B36**, 8376.
Tarvin, J.A., Vidal, F. and Greytak, T.J. (1977), *Phys. Rev.* **B15**, 4193.
Tisza, L. (1938), *Nature* **141**, 913.
Tserkovnikov, Yu. A. (1965), *Sov. Phys. – Doklady* **9**, 1095.
van Hove, L. (1954), *Phys. Rev.* **95**, 249.
Vinen, W.F. (1969), in *Superconductivity*, ed. by R.D. Parks (M. Dekker, New York) Vol. II, p. 1167.
Vinen, W.F. (1971), *J. Phys. C: Solid State Physics* **4**, L287.
Vollhardt, O. and Wölfle, P. (1990), *The Superfluid Phases of Helium 3* (Taylor and Francis, London).
Wagner, H. (1966), *Z. Physik* **195**, 273.
Wang, Y., Sosnick, T.R. and Sokol, P.E. (1992), to be published.
Weichman, P.B. (1988), *Phys. Rev.* **B38**, 8739.
Werthamer, N.R. (1969), *Amer. Journ. Phys.* **37**, 373.
Werthamer, N.R. (1972), *Phys. Rev. Lett.* **28**, 1102.
Werthamer, N.R. (1973), *Phys. Rev.* **A7**, 254.
West, G.B. (1975), *Phys. Rep.* **18C**, 263.
Whitlock, P.A. and Panoff, R.M. (1987), *Can. J. Phys.* **65**, 1409.
Wilks, J. (1967), *The Properties of Liquid and Solid Helium* (Clarendon, Oxford).
Wong, V.K. (1977), *Phys. Lett.* **61A**, 455.
Wong, V.K. (1979), *Phys. Lett.* **73A**, 398.
Wong, V.K. and Gould, H. (1974), *Ann. Phys. (New York)* **83**, 252.
Wong, V.K. and Gould, H. (1976), *Phys. Rev.* **B14**, 3961.
Woo, C.W. (1976), in *The Physics of Liquid and Solid Helium*, ed. by J.B. Ketterson and K.H. Bennemann (Wiley, New York) Vol. I, p. 349.
Woods, A.D.B. (1965a), in *Inelastic Scattering of Neutrons* (IAEA, Vienna), p. 191.

Woods, A.D.B. (1965b), *Phys. Rev. Lett.* **14**, 355.
Woods, A.D.B. and Cowley, R.A. (1973), *Rep. Prog. Phys.* **36**, 1135.
Woods, A.D.B., Hilton, P.A., Scherm, R. and Stirling, W.G. (1977), *J. Phys. C: Solid State Physics* **10**, L45.
Woods, A.D.B and Sears, V.F. (1977), *Phys. Rev. Lett.* **39**, 415.
Woods, A.D.B. and Svensson, E.C. (1978), *Phys. Rev. Lett.* **41**, 974.
Woods, A.D.B., Svensson, E.C. and Martel, P. (1972), in *Neutron Inelastic Scattering* (IAEA, Vienna), p. 359.
Woods, A.D.B., Svensson, E.C. and Martel, P. (1975), in *Low Temperature Physics – LT14*, ed. by M. Krusius and M. Vuorio (North-Holland, Amsterdam), Vol. 1, p. 187.
Woods, A.D.B., Svensson, E.C. and Martel, P. (1976), *Phys. Lett.* **57A**, 439.
Woods, A.D.B., Svensson, E.C. and Martel, P. (1978), *Can. J. Phys.* **56**, 302.
Yang, C.N. (1962), *Rev. Mod. Phys.* **34**, 694.
Yarnell, J.L., Arnold, G.P., Bendt, P.J. and Kerr, E.C. (1959), *Phys. Rev.* **113**, 1379.
Yoshida, F. and Takeno, S. (1987), *Prog. Theor. Phys. (Kyoto)* **77**, 864.
Yoshida, F. and Takeno, S. (1989), *Phys. Reports* **C173**, 301.
Zawadowski, A. (1978), in *Quantum Liquids*, ed. by J. Ruvalds and T. Regge (North-Holland, New York), p. 293.
Zawadowski, A., Ruvalds, J. and Solana, J. (1972), *Phys. Rev.* **A5**, 399.

Author index

Subject index